NELSON

FIT FOR LIFE

HEALTH AND PHYSICAL EDUCATION FOR THE AUSTRALIAN CURRICULUM

LEVELS 7+8

2ND ED

ROB MALPELI
AMANDA TELFORD
CLAIRE STONEHOUSE
LEE ANTON-HEM
DEAN DUDLEY
SAM WATKINS
EMMÉ WILD
DAVID BAKKER

Nelson Fit For Life Health and Physical Education for The Australian Curriculum
Levels 7 and 8
2nd Edition
Rob Malpeli
Amanda Telford
Claire Stonehouse
Lee Anton-Hem
Dean Dudley
Sam Watkins
Emmé Wild
David Bakker

Product manager: Sarah Craig/Cathy Beswick-Davison
Content developer: Rachael Pictor
Project editor: Sutha Surenddar
Editor: Anne Mulvaney/MPS Limited
Proofreader: MPS Limited
Production controller: Karen Young/Sutha Surenddar
Permissions/Photo researcher: Liz McShane
Cover designer: Mariana Maccarini
Text designer: Mariana Maccarini
Project designer: Mariana Maccarini
Typeset by: MPS Limited
Cover: Shutterstock.com/Nicetoseeya
MPS Limited

Any URLs contained in this publication were checked for currency during the production process. Note, however, that the publisher cannot vouch for the ongoing currency of URLs.

Acknowledgements
We respectfully refer to Aboriginal and Torres Strait Islander Peoples as First Nations Peoples throughout the book.

ACKNOWLEDGEMENT OF COUNTRY
Nelson acknowledges the Traditional Owners and Custodians of the lands of all First Nations Peoples of Australia. We pay respect to their Elders past and present. We recognise the continuing connection of First Nations Peoples to the land, air and waters, and thank them for protecting these lands, waters and ecosystems since time immemorial.

WARNING:
First Nations Peoples are advised that this book and associated learning materials may contain images, videos or voices of deceased persons.

For product information and technology assistance,
in Australia call 1300 790 853;
in New Zealand call 0800 449 725

For permission to use material from this text or product, please email aust.permissions@cengage.com

National Library of Australia Cataloguing-in-Publication Data
A catalogue record for this book is available from the National Library of Australia.

Cengage Learning Australia
Level 5, 80 Dorcas Street
Southbank VIC 3006 Australia

Cengage Learning New Zealand
Unit 4B Rosedale Office Park
331 Rosedale Road, Albany, North Shore 0632, NZ

For learning solutions, visit cengage.com.au

Printed in China by 1010 Printing International Limited.
3 4 5 6 7 26 25 24

CONTENTS

ABOUT THIS TITLE

This book has been written for the Australian Curriculum for Health and Physical Education Years 7 & 8. Each chapter explores the focus areas, with a range of activities and investigations to guide your learning.

Investigations

Investigations scaffold research topics and experiments, and allow you to conduct your own investigation into a topic. Practise your data analysis skills and how to present results with these.

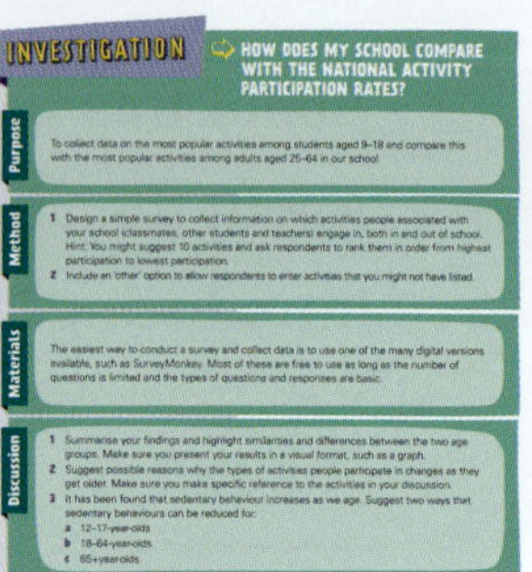

INVESTIGATION
HOW DOES MY SCHOOL COMPARE WITH THE NATIONAL ACTIVITY PARTICIPATION RATES?

Purpose
To collect data on the most popular activities among students aged 9–18 and compare this with the most popular activities among adults aged 25–64 in our school.

Method
1 Design a simple survey to collect information on which activities people associated with your school (classmates, other students and teachers) engage in, both in and out of school. Hint: You might suggest 10 activities and ask respondents to rank them in order from highest participation to lowest participation.
2 Include an 'other' option to allow respondents to enter activities that you might not have listed.

Materials
The easiest way to conduct a survey and collect data is to use one of the many digital versions available, such as SurveyMonkey. Most of these are free to use as long as the number of questions is limited and the types of questions and responses are basic.

Discussion
1 Summarise your findings and highlight similarities and differences between the two age groups. Make sure you present your results in a visual format, such as a graph.
2 Suggest possible reasons why the types of activities people participate in changes as they get older. Make sure you make specific reference to the activities in your discussion.
3 It has been found that sedentary behaviour increases as we age. Suggest two ways that sedentary behaviours can be reduced for:
a 12–17-year-olds
b 18–64-year-olds
c 65+-year-olds

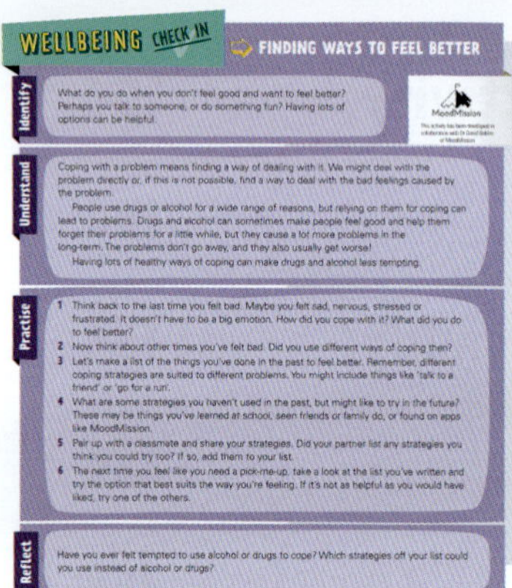

WELLBEING CHECK IN
FINDING WAYS TO FEEL BETTER

Identify
What do you do when you don't feel good and want to feel better? Perhaps you talk to someone, or do something fun? Having lots of options can be helpful.

Understand
Coping with a problem means finding a way of dealing with it. We might deal with the problem directly or, if this is not possible, find a way to deal with the bad feelings caused by the problem.
People use drugs or alcohol for a wide range of reasons, but relying on them for coping can lead to problems. Drugs and alcohol can sometimes make people feel good and help them forget their problems for a little while, but they cause a lot more problems in the long-term. The problems don't go away, and they also usually get worse!
Having lots of healthy ways of coping can make drugs and alcohol less tempting.

Practise
1 Think back to the last time you felt bad. Maybe you felt sad, nervous, stressed or frustrated. It doesn't have to be a big emotion. How did you cope with it? What did you do to feel better?
2 Now think about other times you've felt bad. Did you use different ways of coping then?
3 Let's make a list of the things you've done in the past to feel better. Remember, different coping strategies are suited to different problems. You might include things like 'talk to a friend' or 'go for a run'.
4 What are some strategies you haven't used in the past, but might like to try in the future? These may be things you've learned at school, seen friends or family do, or found on apps like MoodMission.
5 Pair up with a classmate and share your strategies. Did your partner list any strategies you think you could try too? If so, add them to your list.
6 The next time you feel like you need a pick-me-up, take a look at the list you've written and try the option that best suits the way you're feeling. If it's not as helpful as you would have liked, try one of the others.

Reflect
Have you ever felt tempted to use alcohol or drugs to cope? Which strategies off your list could you use instead of alcohol or drugs?

Wellbeing Check-in

In collaboration with Dr David Bakker of MoodMission, Wellbeing check-in activities introduce you to a specific strategy, based on scientific research, that is designed to improve your wellbeing. Practise and use these to discover which strategies work best for you!

Case studies

New and updated case studies further your learning by showing you how Health and PE knowledge can be applied in the real world. Learn how you can apply your Health and PE knowledge in the real world.

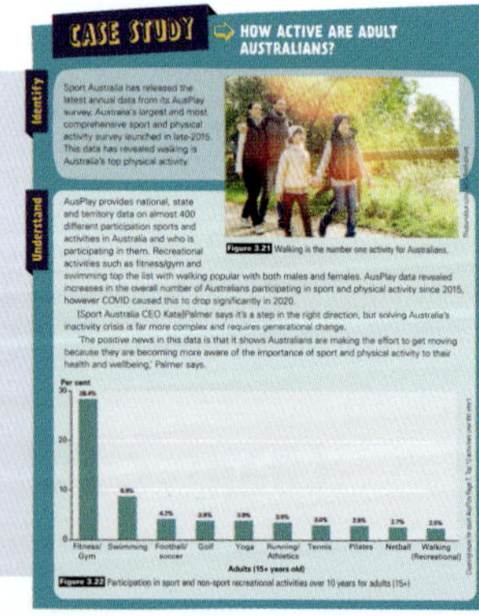

CASE STUDY
HOW ACTIVE ARE ADULT AUSTRALIANS?

Identify
Sport Australia has released the latest annual data from its AusPlay survey, Australia's largest and most comprehensive sport and physical activity survey launched in late-2015. This data has revealed walking is Australia's top physical activity.

Figure 3.21 Walking is the number one activity for Australians.

Understand
AusPlay provides national, state and territory data on almost 400 different participation sports and activities in Australia and who is participating in them. Recreational activities such as fitness/gym and swimming top the list with walking popular with both males and females. AusPlay data revealed increases in the overall number of Australians participating in sport and physical activity since 2015, however COVID caused this to drop significantly in 2020.
[Sport Australia CEO Kate] Palmer says it's a step in the right direction, but solving Australia's inactivity crisis is far more complex and requires generational change.
'The positive news in this data is that it shows Australians are making the effort to get moving because they are becoming more aware of the importance of sport and physical activity to their health and wellbeing,' Palmer says.

Figure 3.22 Participation in sport and non-sport recreational activities over 10 years for adults (15+)

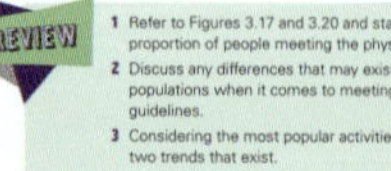

REVIEW
1 Refer to Figures 3.17 and 3.20 and state the general trends observed regarding the proportion of people meeting the physical activity and sedentary guidelines with age.
2 Discuss any differences that may exist between Indigenous and non-Indigenous populations when it comes to meeting the physical activity and sedentary behaviour guidelines.
3 Considering the most popular activities in Australia (Table 3.8), summarise at least two trends that exist.

REFLECT
1 Muscle strengthening activities are recommended for various age groups in order to experience health benefits. But not everyone can access a gym to work out. Propose four different exercises people your age might engage in to achieve muscle strengthening without necessarily going to a gym.
2 Create a list of three activities that contribute to your sedentary behaviour – remember, sleep is not considered sedentary behaviour! How could you modify these three activities to reduce the amount of time they contribute to your sedentary behaviour? Suggest one modification for each activity.
3 You have been approached by the local council to design an 'ideal' outdoor space for your community. The design needs to be aimed at increasing physical activity levels among all age groups within the community. You must consider access to the space, safety and aesthetics (the way it looks), as well as the types of facilities you will include.
Present your design as an annotated poster or multimedia presentation.

EXTEND
1 Many new housing estates require the developers to include a proportion of land that is dedicated to both passive and active leisure. Research a recent development and provide specific examples of each of these options.
2 Devise a policy or an environmental idea for your school that could be implemented to increase the amount of physical activity students are involved in. You could score 'bonus points' by also including a couple of strategies to reduce sedentary behaviours within schools.

Review, reflect and extend

New review, reflect and extend activities end each section of a chapter. These provide differentiated summary questions to review your knowledge and explore further on each section. Cognitive verbs are highlighted throughout to help familarise you with these task words. Definitions for each are provided online in Nelson MindTap.

GET HANDS ON!

You can use this textbook alongside your print Workbook or the digital versions online in Nelson MindTap – look for the Workbook icon WB to complete the activities.

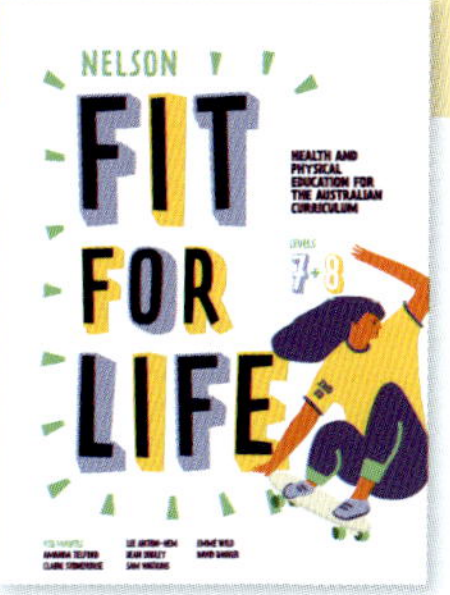

Face to face

These popular activities from our last edition are here to stay. Designed to complete in pairs, or a group, these provide opportunities for debate and collaboration and improve your ability to work in a team.

Up and moving

Health and Physical Education is a practical subject and these up and moving activities are designed to get you on your feet and moving! Explore different movement skills and practise games and sports with these short activities.

Fast facts

Grow your knowledge base with these weird or interesting facts! These short bites of information will help you stay awake with their exciting revelations!

A CUSTOMISABLE DIGITAL SOLUTION

Fit for Life for the Australian Curriculum 2e is designed so that it can be customised to suit any Health and PE classroom. The variety of digital resources will help you ensure the success of all learners. Everything is hosted in your online learning space, Nelson MindTap, with LMS integration options.

Videos

A range of videos are available in Nelson MindTap to assist with your learning. From animations detailing important or difficult concepts, and guided meditations, to interviews and case studies to analyse, the videos and inquiry questions will enhance your learning in a visual and engaging way.

Nelson MindTap StudyHub

Accompanying the print book is access to the online eText in Nelson MindTap. Directly linking to weblinks, videos, worksheets and quizzes throughout, you can easily access all the information you need immediately. In addition, you can highlight or bookmark important content or take notes using the StudyHub.

Quizzes and Gradebook

Nelson MindTap provides both pre- and post-chapter quizzes so you can track your own learning throughout the course.

Teachers can also use the Gradebook in Nelson MindTap to get an understanding of their students' knowledge allowing them to better manage and track class progress.

For teachers

Using Nelson MindTap, gain access to all of the student digital resources, access to the eText as well as access to additional teacher resources: explore the project-based learning outlines, the curriculum grids and assessment guide. Nelson MindTap's course customisation tools also offer teachers complete flexibility with adding links, resequencing, or hiding any content based on your school's approach and students' needs.

ABOUT THE AUTHORS

Rob Malpeli is regarded as a pre-eminent Physical Education leader and educator who has had over 40 published books in multiple Australian States as well as New Zealand. He has been a leading light in senior Physical Education and Years 7–10 Health Education for over 30 years who is recognised as developing contemporary and engaging resources for both teachers and students alike. Rob further supports teachers by conducting professional learning events throughout Australia and regularly runs student seminar/webinars that make learning fun, relevant and linked to real-world settings and examples.

Amanda Telford is Professor of Educational Leadership at Australian Catholic University. Over the past 30 years Amanda has worked within both the secondary and tertiary sectors in educational leadership roles. Amanda has been an advisor for state and federal governments within health and physical education for children and youth. She has a strong background in learning and teaching and known for her vision to provide leadership to the teaching profession for both pre- and in-service teachers in Australia. Amanda has co-authored over 45 books and over 100 teaching and learning publications used throughout Australia and New Zealand.

Claire Stonehouse lectures at Deakin University in Health Education, Student Wellbeing and Sexuality Education. She is currently studying to gain her PhD. Claire has worked in many sectors of the community and has experience writing curriculum and educating children, young people and professionals across the board. Her areas of interest include parents as sexuality educators of their children, capacity building in sexuality education, respectful relationships education in schools and elite sports and helping young people open up conversations about mental health.

Lee Anton-Hem is a passionate educator with over 30 years of experience specialising in health, physical education and well-being. Lee currently lectures at universities and teaches in schools. When not teaching, she manages a health and physical education consultancy. Lee has written many educational resources and co-authored award-winning books. In 2008, Lee was awarded an Australian Learning and Teaching Council Citation for: Outstanding Contribution to Student Learning and in 2007 was the recipient of the RMIT University Teaching Award – Early Career Academic.

Dr Dean Dudley is an Associate Professor (Health and Physical Education) in the Macquarie School of Education at Macquarie University. He is also an Honorary Associate Professor in the School of Human Movement and Nutrition Sciences at the University of Queensland.

Dr Dudley is a 2012 Churchill Fellow and was an Expert Consultant for the United Nations Educational, Scientific, and Cultural Organization's Quality Physical Education Guidelines for Policy Makers (2015) and the Kazan Action Plan ratified at MINEPS VI in 2017. In 2018, Dr Dudley was appointed as an Independent Specialist in Health and Physical Education by UNESCO's International Bureau of Education. Internationally, he is recognised for his work in physical education, learning assessment, pedagogy, and physical literacy.

Sam Watkins has been a teacher of Physical, Health and Outdoor Education in both Victoria and Western Australia since 2003. He has worked in both private and public schools, teaching Years 7–12, and in various school leadership roles. Sam is currently Assistant Principal at Manor Lakes P-12 College.

Emmé Wild has taught Health and Physical Education for more than 20 years in England and Australia. She has actively been involved in curriculum writing in her capacity as Head of Department and educator, both at a school level and with the School Curriculum and Standards Authority of Western Australia (SCSA). Emmé was a board member for ACHPER WA and during her time in Western Australia, also chaired the 7–10 Health & Physical Education Course Advisory Committee for SCSA. After 18 years teaching and leading departments in Perth, she has since relocated to Far North Queensland. Emmé continues to teach Health and Physical Education with passion and purpose.

THANK YOU!

Nelson Cengage and our author team would like to thank the co-authors of the previous edition for their original contributions: Rachael Whittle, Kim Vandervelde, and Jonathan Fender. The publisher and authors would also like to thank Dr Mark Lock (Ngiyampaa) for First Nations sensitivity reading.

INTRODUCING MOODMISSION

Who is MoodMission?

Dr David Bakker is the Founding Director of MoodMission. He is a Clinical Psychologist working in private practice in Hobart, and an Adjunct Research Associate at Monash University. He has experience in rural mental health outreach, youth mental health, and disability support. David is particularly interested in how technology can be used to improve mental health. With many mental health apps focusing on low-tech interventions like diaries, David is working to utilise technology to enhance mental health. He has been involved in creating MoodPrism, a mood tracking app, as well as MoodMission.

Dr David Bakker

What does MoodMission do?

MoodMission is an app available for both iPhone and Android, that helps deal with stress, low mood and anxiety. It uses evidence-based strategies to overcome feelings of depression or anxiety by helping you find new ways of coping. MoodMission currently focuses on reducing feelings of depression, anxiety and stress, and is looking at expanding into positive psychology and maintaining good wellbeing as well as improving low mood.

MoodMission's vision is to be the best provider of easily accessible, evidence-based mental health and wellbeing support by:

- showing people new and better ways of dealing with mood and anxiety problems
- educating and enlightening people about their own psychology and the importance of practising self-care
- linking people in with professional or clinical supports.

MoodMission has conducted several peer review studies that support the use of MoodMission to improving mental wellbeing.

HOW TO USE MOODMISSION

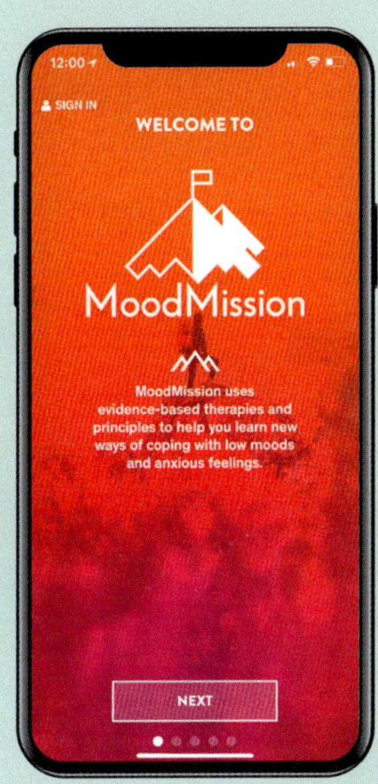

1. You can purchase the MoodMission app from the App Store or Google Play store.
2. On your first use of MoodMission, you will need to fill in a couple of surveys to get an idea of your initial mood.
3. When you want to use MoodMission, choose how you are currently feeling. Use the scales to indicate how distressing your current feelings are. You can also describe how you're feeling in more detail.
4. Five quick and easy to do strategies will appear on screen. Click on one to learn about why this strategy works.
5. If you don't like any of these strategies, or don't have the ability to complete them currently, you can choose the 'Show me 5 more' button.
6. Click 'Accept Mission' and complete the mission.
7. After completion, you can rate how successful the strategy was. The app then learns what strategies work best for you in specific situations.

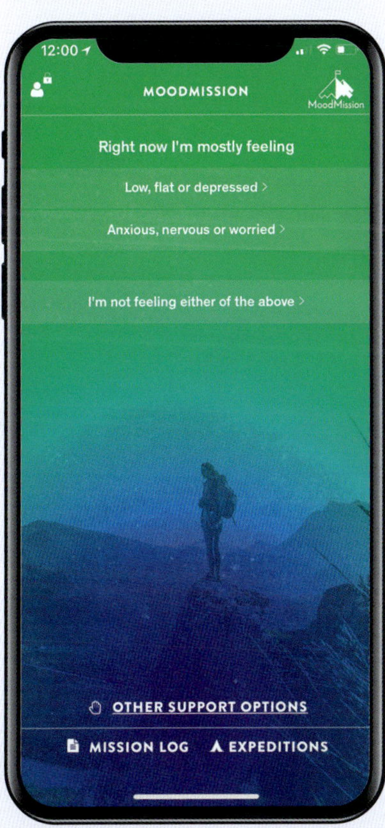

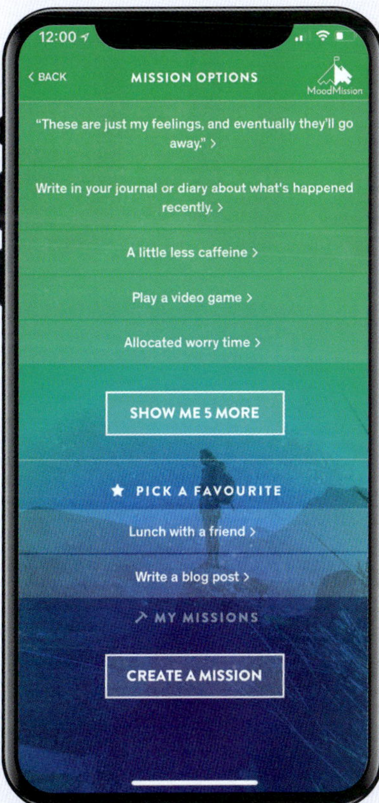

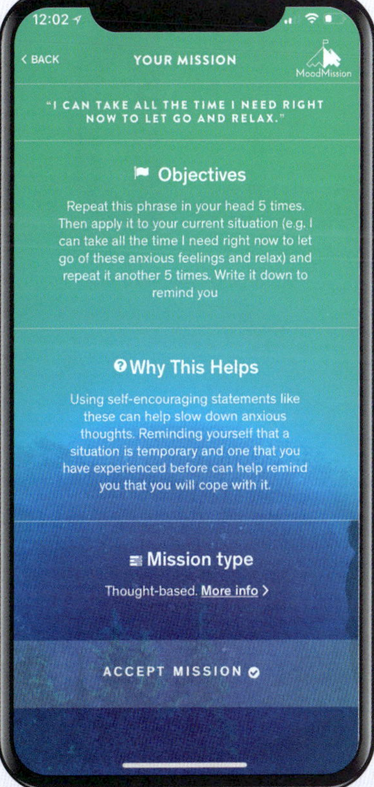

GET SMART ABOUT DRUGS

1

IN THIS CHAPTER

You will learn about drugs and alcohol – what they are and what effect they can have on you, your family and the community. Legal and illegal drugs are examined, as well as the role of drugs in sport.

Source: Shutterstock.com/Nicetoseeya

By the end of this chapter, you should be able to:

- investigate the factors that influence the use of alcohol and other drugs
- analyse how behaviours, actions and responses to drug-related situations can change depending on whether you are by yourself, with friends or with family
- explore and refine skills and strategies needed to communicate assertively and engage in relationships in respectful ways
- examine different strategies or resources to support or persuade others to seek help
- examine scenarios to highlight how emotions, dispositions and decision-making can affect a person's willingness to seek help
- explore help-seeking scenarios young people encounter and share strategies for dealing with each situation
- collaborate with peers to propose strategies they could use if they are being encouraged to use a substance such as alcohol, cannabis or inhalants
- consider different social and cultural perspectives in relation to illegal drugs, legal drugs, prescription medicines, over-the-counter products, bush medicines, alternatives to medicines and performance-enhancing drugs
- analyse the credibility, validity and relevance of health messages conveyed by different sources and apply credible information to drug-related decisions
- explore and evaluate the accessibility and reliability of health information sources.

WHAT ARE DRUGS?

Quiz
Pre-chapter

Before you start, take the pre-chapter quiz to find out how much you already know.

Worksheet 1.1

drugs chemical substances that can alter the biological functioning and structure of a living organism. Drugs can either be synthetic (human-made) or natural.

psychoactive having an effect on your mind and your senses

synthetic a manufactured substance, not natural

The human body is a complicated organism. You begin life as a single cell and develop into a highly complex, multicellular machine. To remain healthy as you grow, you must balance your physical activity with eating nutritious food and getting enough sleep.

Unfortunately, although the human body is very strong, the effects of alcohol, tobacco and other **drugs** can be damaging and long lasting, leaving the body vulnerable to illness, disease and dangerous situations.

Drugs are substances containing chemicals that can affect the way you think, feel and behave. These are known as **psychoactive** substances, because they work on the mind and your senses, and can change the way the body functions. Once a drug reaches the brain, it can change the messages the brain cells send, both to each other and to the rest of the body.

Some drugs are legal, but many are illegal. Examples of legal drugs include prescription medication, over-the-counter medicines (OTCs), caffeine, tobacco and alcohol. Illegal drugs include substances such as cannabis, ecstasy and cocaine.

There are thousands of different types of drugs. Some drugs occur naturally, coming from plants and animals, while others are **synthetic**.

Penicillin is an example of a medicine derived from a plant, in this case the *Penicillium* fungus.

FAST FACT

In the 1940s, a team led by Australian scientist Howard Florey discovered the healing properties of penicillin. Penicillin is an antibiotic that is now widely used to treat many serious diseases and infections. You may have been prescribed penicillin by your doctor when you had a bacterial infection such as tonsillitis, or a chest infection.

While many young people avoid taking drugs that may be harmful to their health and wellbeing, some adolescents experiment with a variety of drugs, including alcohol, tobacco and cannabis. These substances can become addictive, which means it is very hard to stop using them. Addiction can lead to many physical, social and mental problems.

FAST FACT

Smoking is the leading cause of preventable deaths worldwide, killing more than 5 million people annually. This is approximately one death every six seconds.

Source: https://www.tobaccoinaustralia.org.au/

9780170463096

WELLBEING CHECK IN

FINDING WAYS TO FEEL BETTER

Identify

What do you do when you don't feel good and want to feel better? Perhaps you talk to someone, or do something fun? Having lots of options can be helpful.

This activity has been developed in collaboration with Dr David Bakker of MoodMission

Understand

Coping with a problem means finding a way of dealing with it. We might deal with the problem directly or, if this is not possible, find a way to deal with the bad feelings caused by the problem.

People use drugs or alcohol for a wide range of reasons, but relying on them for coping can lead to problems. Drugs and alcohol can sometimes make people feel good and help them forget their problems for a little while, but they cause a lot more problems in the long-term. The problems don't go away, and they also usually get worse!

Having lots of healthy ways of coping can make drugs and alcohol less tempting.

Practise

1 Think back to the last time you felt bad. Maybe you felt sad, nervous, stressed or frustrated. It doesn't have to be a big emotion. How did you cope with it? What did you do to feel better?

2 Now think about other times you've felt bad. Did you use different ways of coping then?

3 Let's make a list of the things you've done in the past to feel better. Remember, different coping strategies are suited to different problems. You might include things like 'talk to a friend' or 'go for a run'.

4 What are some strategies you haven't used in the past, but might like to try in the future? These may be things you've learned at school, seen friends or family do, or found on apps like MoodMission.

5 Pair up with a classmate and share your strategies. Did your partner list any strategies you think you could try too? If so, add them to your list.

6 The next time you feel like you need a pick-me-up, take a look at the list you've written and try the option that best suits the way you're feeling. If it's not as helpful as you would have liked, try one of the others.

Reflect

Have you ever felt tempted to use alcohol or drugs to cope? Which strategies on your list could you use instead of alcohol or drugs?

1 Identify three legal drugs.
2 Identify three illegal drugs.
3 What is addiction?

1 Make a poster listing all the different types of drugs you have heard of.
   a Classify each of these drugs as legal or illegal.
   b Can you think of other names for these drugs? For example, cannabis is sometimes called weed. Discuss as a class and make sure your poster includes all the different names for each drug.

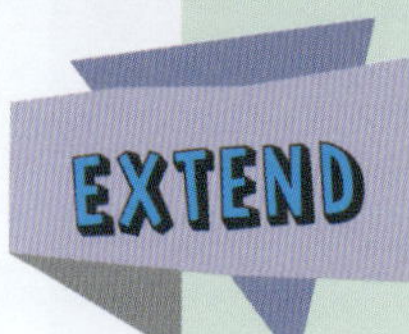

1 Everyone has personal character strengths, and these can help you deal with challenging situations in life, including situations you might find yourself in with drugs and alcohol.
   a Can you think of five character strengths you have?
   b Complete an online test, such as the VIA Character Strengths Survey for Youth For Ages 10–17, to help you identify your strengths.
   c Compare your answer in 1a, to the results of the VIA Character Strengths Survey, 1b. Were there any similarities or surprises?

Weblink
VIA Character Strengths Survey

Worksheet 1.2

HOW CAN I USE MEDICINE SAFELY?

Advances in science have led to many new medicines being manufactured. Medicines are made up of chemicals and compounds. They are used to treat a variety of illnesses and ailments and, ultimately, to improve people's lives.

PRESCRIPTION MEDICATION

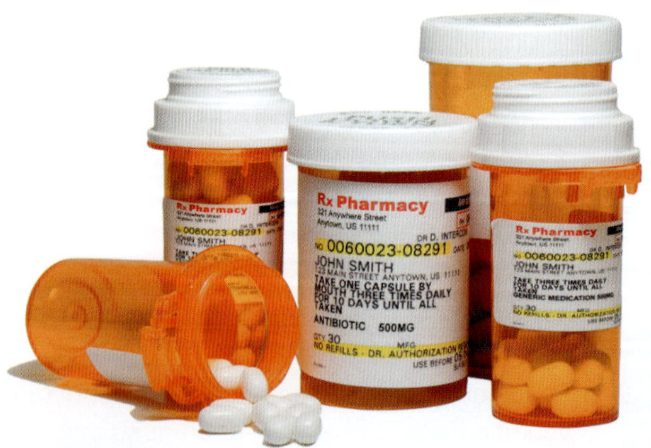

Figure 1.1 Prescription drugs

Shutterstock.com/Rob Byron

Medicines are generally prescribed by an authorised healthcare professional such as a doctor, but some can be purchased 'over the counter' from a pharmacy or supermarket. Prescription medicines are licensed and regulated by law; it is illegal to supply prescription medication without a prescription from a doctor. There are heavy penalties for people who do this, including fines and prison sentences.

Over-the-counter medicines (OTCs) are readily available and do not require a prescription from a healthcare professional. They are also known as non-prescription medicines.

Examples of OTCs include cough medicine and painkillers that don't contain paracetamol.

How medicines are administered

Drugs used to treat illnesses and infections are usually taken in the form of a tablet or pill. However, there are many different ways drugs can be administered. These are listed in Table 1.1.

Shutterstock.com/Tyler Olson

Figure 1.2 You can purchase over-the-counter medication without a prescription.

Table 1.1 Different forms of prescribed or OTC medication

Form of medicine	Example
Liquids	Cough mixture
Sprays/inhalers	For asthma or hay fever control
Drops	For ears and eyes
Patches	Skin patches to control smoking
Creams, gels or other ointments	**Steroid** creams for skin disorders
Tablets/capsules	Antibiotics
Injections	**Vaccinations**, such as the flu vaccine
Intravenous	Fluids inserted into veins by medically trained staff

steroid a human-made chemical substance that closely resembles cortisol, a naturally occurring hormone found in the human body; used to reduce inflammation

vaccinations medicines used to help the body's immune system prevent disease

opioids any legal or illegal drug made from the opium poppy, including pain-relievers and heroin

overdose the excessive use of a drug, either accidental or intentional, resulting in serious illness or death

drug abuse the harmful misuse of illegal, prescription or over-the-counter drugs that can ultimately lead to adverse health effects, addiction or dependency

FAST FACT

Prescription opioids, including heroin, opiate-based analgesics (such as codeine and oxycodone) and synthetic opioid prescriptions (such as tramadol and fentanyl), cause the majority deaths by overdose in Australia.

HEALTHY, SAFE AND ACTIVE CHOICES

When people think of **drug abuse** in Australia, they often focus on illegal street drugs such as ecstasy and cannabis. But the misuse of prescription drugs is actually a major drug issue in Australia. Prescription drug abuse occurs when medicine is obtained with a prescription but used in a manner not 'prescribed' by the healthcare professional. Using prescription drugs that have not been prescribed by a doctor can be just as dangerous as using illegal drugs. The most commonly abused prescription drugs include benzodiazepines and opiate-based drugs.

Prescription drug use

Benzodiazepines – also known as 'minor tranquillisers' – are depressants, which slow down the messages sent between the brain and the body. They are prescribed by doctors to relieve the symptoms of stress and anxiety, and also help people to sleep. Side effects can include depression, confusion, memory loss and slurred speech.

narcotic dulling the senses and promoting drowsiness

Opioids have a **narcotic** effect and are highly addictive. There are many drugs that fall into this category, including codeine, morphine, fentanyl and OxyContin. Opioids are commonly prescribed by doctors because of their effectiveness in providing pain relief. Side effects can include nausea, drowsiness, mental fog, constipation and slowed breathing.

Quiz
How can I use medicine safely?

Worksheet 1.3

There are many reasons why people choose to abuse prescription drugs. Some people believe that prescription drugs are safer than illicit street drugs. Unfortunately, this is a very common misconception. However, prescription drugs are intended to be used only by the individual the drugs were prescribed for. Second, prescription drugs are seen as being more readily available than illicit drugs. Users may see more than one doctor at a time and obtain a number of prescriptions for the same product. Finally, many people don't believe that taking prescription drugs is against the law. This is another misconception; any form of problematic prescription drug use is against the law.

CASE STUDY  PRESCRIPTION DRUG USE

Identify

People think prescription drugs are safe because they are sourced from a doctor. Consider the following scenario to determine what Fiona should do.

Understand

Fiona felt unwell and had a headache. She took some over-the-counter pain relieving tablets, but the pain didn't go away. While Fiona lay in bed, she remembered that her mother had been prescribed OxyContin, an opioid medication, for back pain. Fiona decided to take a couple of her mother's OxyContin tablets, thinking that if they had been prescribed by a doctor, they would be safe to take.

Discuss

1 State the drug classification OxyContin is in.
2 Research some of the short- and long-term effects of OxyContin.
3 If Fiona was your friend, decide what advice would you give her?

Weblink
Watch The Feed's video report about Australia's prescription drug crisis.

FAST FACT

One in ten Australians who take prescription medication are addicted.

1 Explain the difference between over-the-counter medication and prescription medication.
2 Many people believe that prescription drugs are 'safe' because they have been prescribed by a doctor. Explain why this view is incorrect.
3 Summarise the side effects of:
   a opioids
   b benzodiazepines.

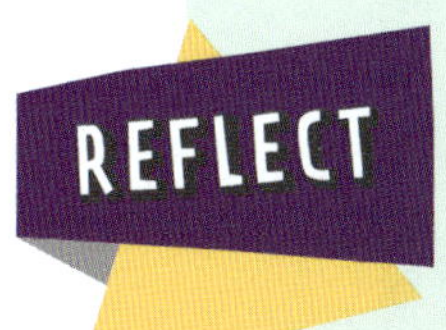

1 When you visit the doctor, explain why is it important to tell them about any medical conditions you may have, and the various medicines and supplements, both over-the-counter and prescribed, that you may currently be taking.
2 At a gathering, your friend offers you an opioid tablet that was prescribed for their father. Determine what you would do and why.

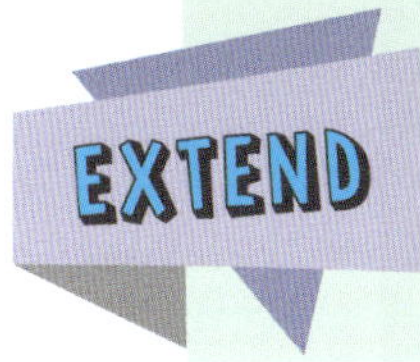

1 Your friend appears stressed about an upcoming maths test. She appears to be having trouble concentrating and has mentioned taking pills that were prescribed for her sister to help her concentrate in Year 12. Decide what advice would you give to your friend and why?

WHAT ARE TRADITIONAL AND ALTERNATIVE MEDICINES?

In addition to the medicines available from doctors and pharmacies, there is a wide range of other medicines and treatments available to the consumer, including traditional and alternative medicines.

TRADITIONAL MEDICINES

Traditional medicines are used by many cultures around the world. For thousands of years, the Chinese have used **herbal remedies** to cure a wide range of ailments and diseases, from headaches to stomach problems. Chinese herbal remedies include a tea made from the leaves of the sweet wormwood tree, which is used to treat chills and fevers, and the dried and ground roots of the ginseng plant, which have been used for more than 2000 years to help boost energy, increase endurance and reduce stress. The ancient Chinese also treated coughs and colds using the root of the liquorice plant, which was ground into a powder and drunk as a tea. The ancient Egyptians also used herbs and plants to produce traditional medicines. To cure coughs and colds, they used

herbal remedies
medicines made from plants or parts of plants

extracts from the hibiscus plant, which were placed in a hot bath so the patient could inhale the steam. The ancient Egyptians also used garlic to give them vitality.

Worksheet 1.4

Australian bush medicines

First Nations Peoples have long relied on the environment around them to provide the medicines they need for daily life. The term 'bush medicine' refers to the traditional medicinal practices that First Nations Peoples have used for thousands of years to promote healing and maintain health and wellbeing.

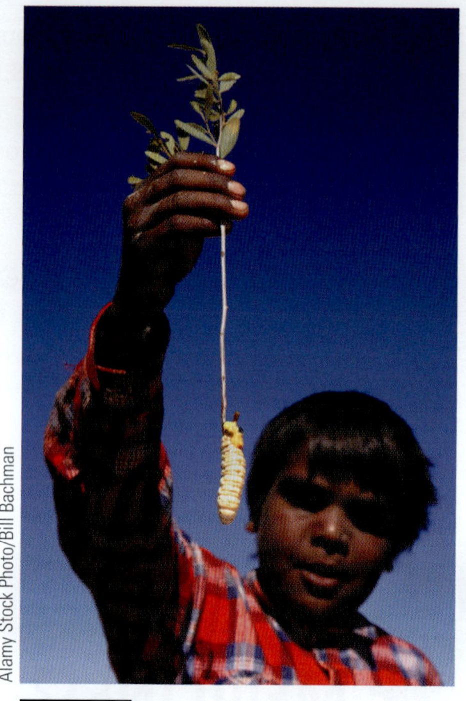

Figure 1.3 Witchetty grubs are a traditional treatment used by First Nations Peoples for burns and wounds.

Figure 1.4 A traditional Chinese medicine practitioner at work

CASE STUDY ⇨ THE AKEYULERRE HEALING CENTRE

Identify

The Akeyulerre healing centre in Alice Springs was established by Arrernte elders as a place to practise and enjoy their cultural life on a daily basis.

Figure 1.5 Arrethe leaves are ground down in a mortar and pestle to create a rub used to treat colds and flu.

Understand

Arrernte woman Teresa Alice is heavily involved in the collection and preparation of local plants into bush medicine, which plays a pivotal role as the social enterprise at the centre.

'We share these with people and send them to shops and at markets.'

The local women work with four main bush medicines:

- Apere (red river gum) – a skin cleanser
- Utnerrenge (emu bush) – used for dry and irritated skin
- Aherre-Intenhe (harlequin fuchsia bush) – used for skin sores and muscle pain
- Arrethe (rock fuchsia bush) – used as a rub to ease cold and flu symptoms

Arrethe mainly grows around the hilly and rocky areas of Central Australia.

The process of grinding up the plant provides the older women an opportunity to pass on their cultural knowledge and Arrernte language to the next generation …

To ensure the longevity of the cultural knowledge behind the bush medicines, a set of colour-coded cards have been released by the centre … 'What we hope is that we put them into the schools so they can learn about traditional medicine.'

Discuss

1 Define the role of the Akeyulerre healing centre.

2 Describe what the four main bush medicines created at the centre are used for.

3 How does the enterprise ensure cultural knowledge is passed on?

4 Research a bush medicine and create your own card that shows how it is used.

Alamy Stock Photo/blickwinkel

Alamy Stock Photo/Geraldine Buckle

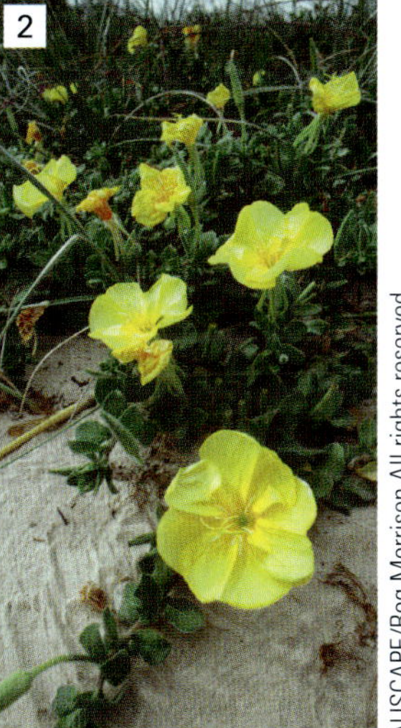

AUSCAPE/Reg Morrison All rights reserved

Shutterstock.com/Damsea

Figure 1.6 Plants used for bush medicine include (1) Emu bush, (2) snake vine, (3) goat's foot and (4) kangaroo apple

Top ten First Nations Peoples' bush medicines

Tea tree oil: crushed tea tree leaves are used to treat wounds and throat ailments. Tea tree oil has been scientifically proven to have strong antiseptic properties.

Eucalyptus oil: used to treat pains, fevers and chills.

Billygoat plum/kakadu plum: this fruit contains 50 times more vitamin C than an orange. It is the world's richest source of vitamin C.

Desert mushroom: when sucked, these mushrooms cure sore mouths and lips. Also used as a natural teething ring for babies.

Emu bush: the leaves of this bush are used to wash sores and cuts, and are occasionally gargled. The leaves have been found to have similar properties to some antibiotics.

Witchetty grub: crushed witchetty grubs are used to treat burns. The grubs are made into a paste, then applied to the wound and covered with a bandage. They are also a good source of food.

Snake vine: crushed vine is used to treat headaches and arthritis; the sap and leaves are used to treat wounds.

Sandpaper fig and stinking passion flower: the rough leaves of the sandpaper fig and the crushed fruit of the stinking passion flower are used together to relieve itching and to treat fungal skin infections.

Kangaroo apple: crushed fruit is used to treat swollen joints.

Goat's foot leaves: used to relieve the pain of stonefish and stingray stings.

Source: Adapted from 'Top 10 Aboriginal bush medicines', *Australian Geographic*, 8 February 2011

ALTERNATIVE MEDICINES IN AUSTRALIA

conventional medicine treatment of illnesses and injuries by healthcare professionals such as doctors and nurses, using drugs, radiation or surgery

holistic health an approach that considers the health of the whole body, including mental, physical, spiritual, emotional and social health

Increasing numbers of Australians use alternative methods of healing in addition to **conventional medicine**. Many alternative practices use natural and **holistic** processes that claim to have healing properties, including ayurvedic medicine, yoga, acupuncture and massage.

Ayurvedic medicine

Ayurvedic medicine refers to the traditional medicinal practices of ancient India. Ayurveda is often referred to as the 'science of life' (*ayu* means 'life' and *veda* means 'knowledge'). This 5000-year-old medicinal practice relates to the health and wellbeing of the mind and body, using a combination of diet, herbal medicine, massage, meditation, yoga and breathing exercises.

Alamy Stock Photo/robertharding

Figure 1.7 Many ayurvedic medicines are based on herbal ingredients, including bark from various trees and shrubs, plant and tree roots, seeds and herbs.

Yoga

Yoga is another practice that originated in ancient India. Yoga exercises focus on developing strength, increasing flexibility and improving posture. Yoga can also develop mental and emotional wellbeing by relaxing the mind and raising spiritual awareness. Many different practices of yoga exist, and each has a particular emphasis. Bikram Yoga is a 90-minute yoga workout

iStock.com/YinYang

Figure 1.8 Yoga is practised to strengthen the body and increase overall wellbeing.

completed in a 40°C studio with 40 per cent humidity, while yin yoga is slower paced, and postures are held for longer.

FAST FACT

1. A 2020 study found that approximately two-thirds of Australians had used alternative medicines in the past year.
2. Use was greater in women than men, with 56.2 per cent of women reporting use of alternative medicines, compared to 43.7 per cent of men.
3. Sixty per cent agreed that alternative medicines helped improve their wellbeing.

Acupuncture

Acupuncture is used to alleviate pain and nausea. It is the world's oldest form of medicine and is very popular in Australia. Acupuncture is an ancient Chinese system of healing that stimulates the mind and the body's healing response. This traditional practice involves inserting very fine needles into the skin at specific points. The needles are left in place for around 30 minutes.

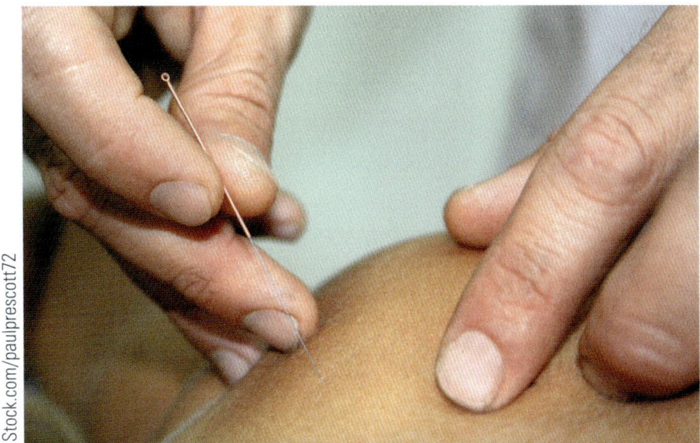

iStock.com/paulprescott72

Figure 1.9 Acupuncture originated in ancient China and is one of the best-known alternative medicine practices.

Massage

Massage therapy has been around for thousands of years and is used to treat a variety of health-related issues. Its benefits include pain relief, management of stress and anxiety, rehabilitation of sports injuries and general wellbeing.

Pilates

Pilates is a very popular workout that focuses on building core muscle strength to improve flexibility, posture and balance. Pilates was developed in the 1920s and is named after its creator, Joseph Pilates. While the moves are similar to yoga, the main emphasis of Pilates is on building the body's core and improving muscle endurance and control.

iStock.com/turk_stock_photographer

Figure 1.10 Pilates can use special machines to assist with a full body workout.

Quiz
What are traditional and alternative medicines?

1 Name and describe the use of four different First Nations Peoples' bush medicines.
2 Describe two herbs used as traditional herbal remedies.
3 Explain the difference between yoga and Pilates.

1 Bikram and yin are just two types of yoga. Research three other types of yoga. Ensure you identify the main health-related focus of each yoga type.
2 In small groups, design a garden for your school with indigenous plants that are known to have medicinal properties.
   - Investigate the plants used by First Nations Peoples that provide medicine to include in your garden.
   - Consider the layout of your garden (position of plants in the sun or shade, water access, and preferred soil type.
   - Create informative signage for your plants. Include plant information and medicinal properties.

1 Create a school-wide campaign to raise awareness of alternative medicines. Subject to time and resources available, campaign materials could include:
   - campaign message
   - slogans
   - posters/flyers
   - social media: Twitter/Facebook
   - school announcements
   - guest speakers.

Be creative with your campaign. You may wish to enlist school administrators and other teachers to support your campaign.

HOW WILL DRUG USE AFFECT MY BODY?

A psychoactive drug contains chemical substances that can adversely affect the functioning of the central nervous system and can alter a person's conscious state. This can result in confusion, as well as changes in mood, behaviour and level of consciousness.

CLASSIFICATION OF DRUGS

Drugs can be grouped within three main categories: depressants, stimulants and hallucinogens. It is important to understand that legal and illegal drugs can be found within each category, and that all drugs can lead to addiction and dependency.

Depressants

Depressants do not necessarily make people feel depressed; in fact, they are among the most commonly used drugs in the world. Depressants affect the body's central nervous system by slowing down the messages sent between the brain and the body. People who take depressants in small quantities may feel more relaxed and drowsy and have an increased heart rate. Their concentration and coordination may also be affected. When taken in larger quantities, depressants can induce panic attacks, paranoia, headaches, aggression, vomiting, comas and, ultimately, death. Some common examples of depressants include:

- alcohol
- cannabis
- opiates and opioids (heroin, morphine, codeine)
- barbiturates
- inhalants (solvents, aerosols, gases, nitrites).

Stimulants

In contrast to depressants, stimulants speed up brain activity. Stimulants can make people feel more alert and awake by increasing their heart rate, body temperature and blood pressure. There are many side effects of stimulant abuse, including suppressed appetite, anxiety and insomnia. Caffeine is an example of a natural legal stimulant; it is found in coffee, tea, energy drinks and even chocolate. Other common examples of stimulants include:

- nicotine (found in tobacco)
- cocaine
- amphetamines (ice, methamphetamine, speed)
- ecstasy
- ephedrine
- khat.

Figure 1.11 Examples of stimulants: (1) cocaine plant, (2) khat plant, (3) ecstasy pills

Shutterstock.com/William Casey

Alamy Stock Photo/Janine Wiedel Photolibrary

Figure 1.12 (1) Cannabis and (2) 'magic' mushrooms are examples of hallucinogenic drugs.

Hallucinogens

Hallucinogens, also referred to as 'psychedelics', can alter the way a person perceives reality. Hallucinogens alter the way the mind works by affecting all the senses and emotions. They can initiate hallucinations, causing people to see or hear things that don't actually exist. Hallucinogens are either made in laboratories or occur naturally in some trees, vines, seeds, fungi and leaves. Examples of hallucinogens include:

- 'magic' mushrooms
- cannabis
- MDMA (ecstasy)
- ketamine
- LSD.

Polydrug use

Polydrug use is when more than one drug is used at the same time, or during the same occasion. It can also refer to the practice of using one drug to counteract the effects of another. A polydrug user may mix legal with illegal drugs.

One common legal drug used by polydrug users is alcohol. The effects of combining multiple drugs can be very dangerous and unpredictable, especially when drugs of unknown content and purity are mixed together. Polydrug use can cause overdose, violence and aggression, unwanted sexual activity, **psychosis** and even death.

psychosis mental illness associated with the loss of contact with reality and severe changes in mood and personality

FAST FACT
Alcohol is known by a variety of names, including booze, grog, bevvie and coldie.

ALCOHOL

WB Worksheet 1.5

Drug classification: depressant

Alcohol is a popular recreational drug. Pure alcohol is so strong that its concentration in most alcoholic drinks is relatively low. There are four main types of alcoholic drink:

Shutterstock.com/Maria Fomina

Figure 1.13 There are four main types of alcoholic drinks.

1 **Wine**: made from fermented fruits (usually grapes). The alcohol content of wine is around 9–16 per cent.
2 **Beer**: prepared by brewing and fermenting water, barley, yeast and hops together. The alcohol content of beer is usually around 4–6 per cent.
3 **Spirits**: made from grains such as barley and rye and produced by fermentation and distillation. Examples include vodka, rum, whiskey and gin. Often spirits are mixed with other beverages to produce cocktails. Spirits have a high alcohol content, usually 20–40 per cent.
4 **Liqueur**: made by adding flavourings and sugar to spirits. Generally sweet, liqueurs typically contain 15–30 per cent alcohol.

Alcohol-related harm

Worksheet 1.6

FAST FACT

1. Alcohol consumption among 12–17 year olds has declined over recent years.
2. Six minutes is all it takes for the brain to start feeling the effects of alcohol.
3. In some states in Australia, it is illegal to provide alcohol to a person under 18 in a private home without their parents' approval.
4. Risky drinkers are defined as those who drink more than the recommended levels. Of risky drinkers aged 14–19, 83 per cent reported being injured due to their drinking over the past year, with 7 per cent needing to attend hospital for an alcohol-related injury.
5. Alcopops are sweet, fruit-flavoured, fizzy alcoholic drinks popular among young people. Some alcopops can contain as much as three standard drinks in one bottle!

After tobacco, alcohol is the second-largest contributor to drug-related harm in Australia. Unfortunately, the social acceptance of alcohol prevents many people from recognising the harmful long-term effects of this drug. Alcohol has been culturally and socially accepted since British colonisation of Australia in 1788. In those days, convicts were partially paid with rum!

For many young people in Australia, drinking alcohol is regarded as a sign of maturity. Many adults mark the transition from work to home with an alcoholic drink. Many people consume alcohol with food, to celebrate a special occasion, to relax or to unwind and have fun.

What is a hangover?

A hangover is the body's reaction to drinking too much alcohol. There are many unpleasant effects of a hangover.

Table 1.2 Symptoms of a hangover

Mental symptoms of a hangover	Physical symptoms of a hangover
➪ poor motivation	➪ headache
➪ poor concentration	➪ tiredness
➪ vertigo (loss of balance or dizziness)	➪ dehydration
➪ anxiety	➪ nausea
➪ irritability	➪ vomiting
➪ depression	➪ sweating

There are two main reasons why people suffer a hangover after a night of drinking. First, alcohol is a diuretic, meaning it causes the drinker to become dehydrated. Second, alcohol being broken down in the body creates toxic by-products that can affect many of the body's internal systems. The more alcohol consumed, the worse the hangover will be.

Alamy Stock Photo/Image Source

Figure 1.14 Hangovers cause both mental and physical symptoms.

What is a standard drink?

In Australia, a standard drink contains approximately 10 grams of alcohol. One standard drink will always contain the same amount of alcohol, regardless of the size of the bottle or can, or the type of alcoholic drink.

Table 1.3 These are the standard drink logos that appear on bottles and cans of beer, spirits and wine. The number on the logo identifies the number of standard drinks contained in the bottle or can – this varies depending on the size of the container and the type of alcohol involved.

Beer	Spirit	Wine
Standard Drinks 1.3	STANDARD DRINKS 1.0 APPROX	Standard Drinks 8.3

Source: Distilled Spirits Industry Council of Australia, the Australasian Associated Brewers Inc. and the Winemakers Federation of Australia

NUMBER OF STANDARD DRINKS – BEER

1.1	0.8	0.6	1.6	1.2	0.9	1.4	1	0.8
285ml Full Strength 4.8% Alc. Vol	285ml Mid Strength 3.5% Alc. Vol	285ml Low Strength 2.7% Alc. Vol	425ml Full Strength 4.8% Alc. Vol	425ml Mid Strength 3.5% Alc. Vol	425ml Low Strength 2.7% Alc. Vol	375ml Full Strength 4.8% Alc. Vol	375ml Mid Strength 3.5% Alc. Vol	375ml Low Strength 2.7% Alc. Vol

1.4	1	0.8	34	24	19
375ml Full Strength 4.8% Alc. Vol	375ml Mid Strength 3.5% Alc. Vol	375ml Low Strength 2.7% Alc. Vol	24 x 375ml Full Strength 4.8% Alc. Vol	24 x 375ml Mid Strength 3.5% Alc. Vol	24 x 375ml Low Strength 2.7% Alc. Vol

Standard drinks guide © Commonwealth of Australia | Department of Health

NUMBER OF STANDARD DRINKS – WINE

1.6	1	0.9	1.4	1	1.4	7.5
150ml Average Restaurant Serving of Red Wine 13.5% Alc. Vol	100ml Standard Serve of Red Wine 13.5% Alc. Vol	60ml Standard Serve of Port 18% Alc. Vol	150ml Average Restaurant Serving of White Wine 11.5% Alc. Vol	100ml Standard Serve of White Wine 11.5% Alc. Vol	150ml Average Restaurant Serve of Champagne 12% Alc. Vol	750ml Bottle of Champagne 12.5% Alc. Vol

8	43	21	7.5	39	19.5	28
750ml Bottle of Red Wine 13.5% Alc. Vol	4 Litres Cask Red Wine 13.5% Alc. Vol	2 Litres Cask Red Wine 13.5% Alc. Vol	750ml Bottle of hite Wine 12.5% Alc. Vol	4 Litres Cask White Wine 12.5% Alc. Vol	2 Litres Cask White Wine 12.5% Alc. Vol	2 Litres Cask of Port 17.5% Alc. Vol

Standard drinks guide © Commonwealth of Australia | Department of Health

9780170463096

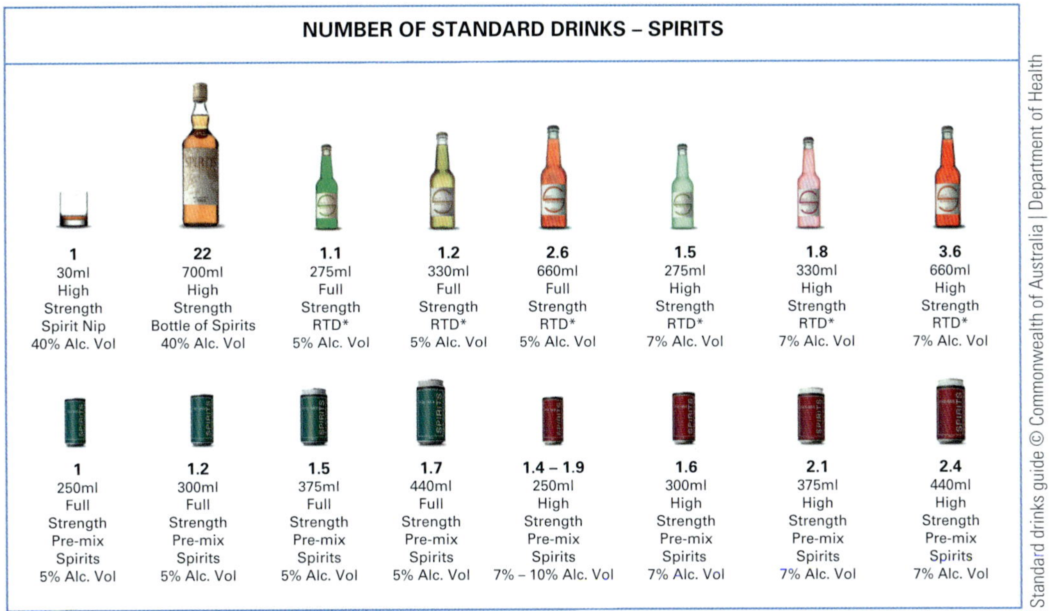

Figure 1.15 Standard servings of beer, wine and spirits, adapted from *Australian guidelines to reduce health risks from drinking alcohol*

Effects of alcohol

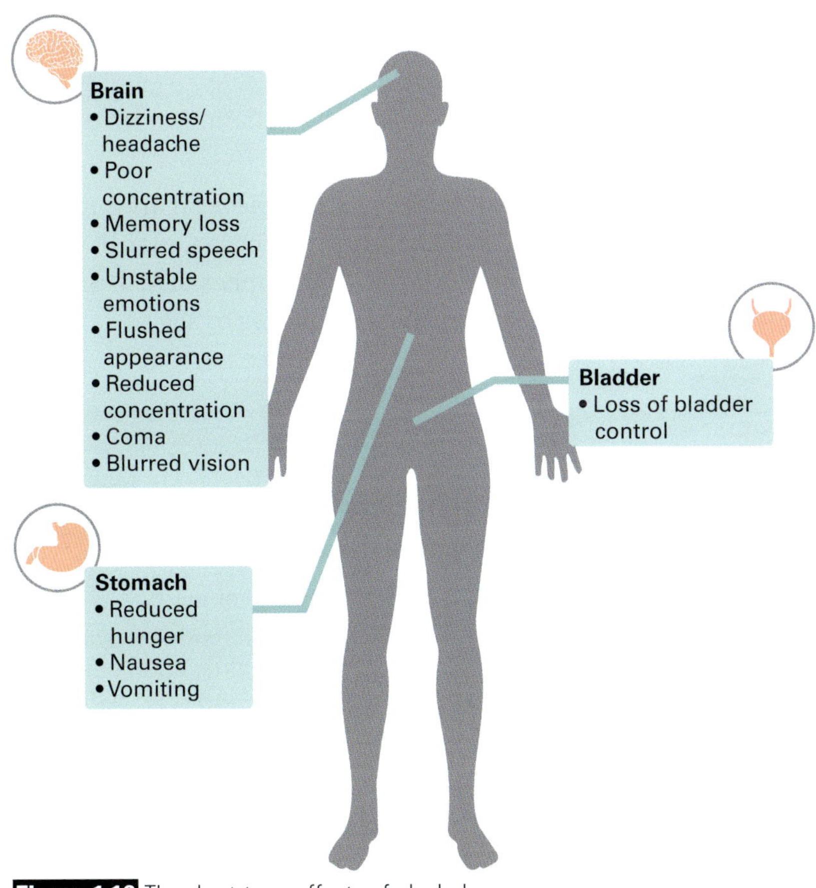

Figure 1.16 The short-term effects of alcohol

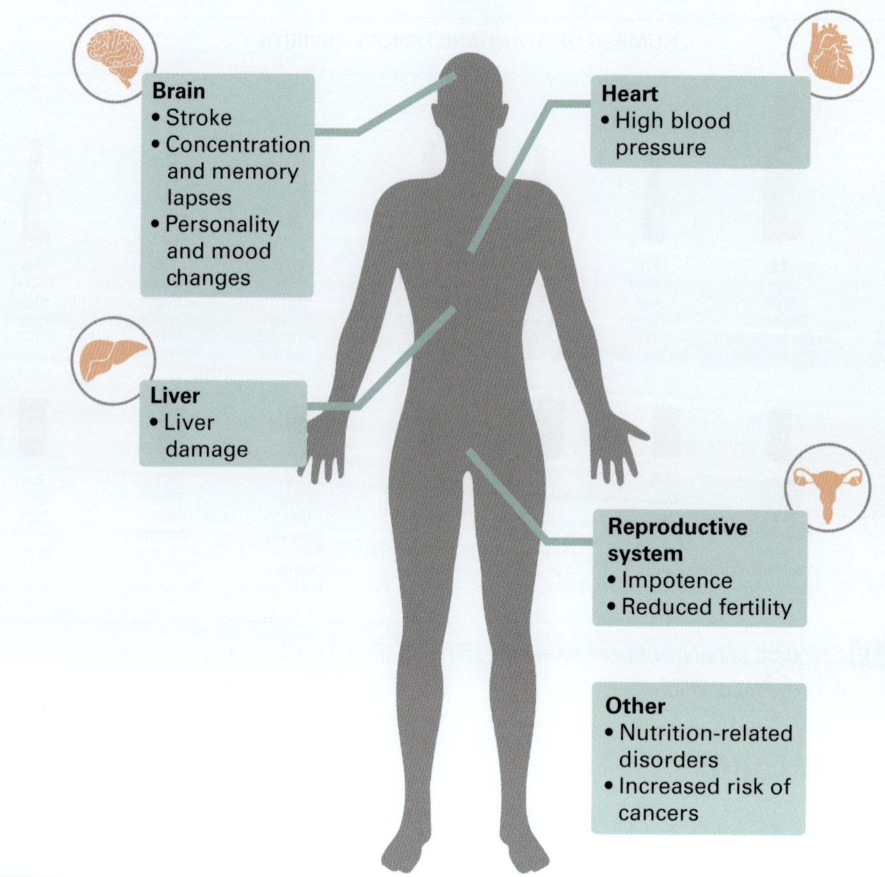

WB Worksheet 1.7

Figure 1.17 The long-term effects of alcohol

Alcohol and the law

Although alcohol is usually considered a socially acceptable legal drug, it is illegal for people to sell alcohol to those aged under 18, and for those under 18 to purchase it. In Australia, a zero blood alcohol concentration limit applies to all learner and probationary/provisional driver's licence holders, regardless of age.

INVESTIGATION

ALCOHOL AND THE BRAIN

Purpose

The human brain is the focal point of the human body, responsible for all of our body's functions. The brain operates like a control centre, sending and receiving messages via the central nervous system. The brain controls everything from our thoughts, emotions, speech, movement and memory to our vital organs, such as our kidneys and heart. In this investigation, you will explore the effects of alcohol on the brain.

Method

Frontal lobe
Responsible for your personality. Around the age of 25, the frontal lobes have become the centre for decision-making, emotional expression, memory, problem-solving and judgement.

Temporal lobe
A centre for information processing located behind the ear. The temporal lobes have several functions in the human body, including the processing of auditory information, memories, speech, language, emotions and visual perception.

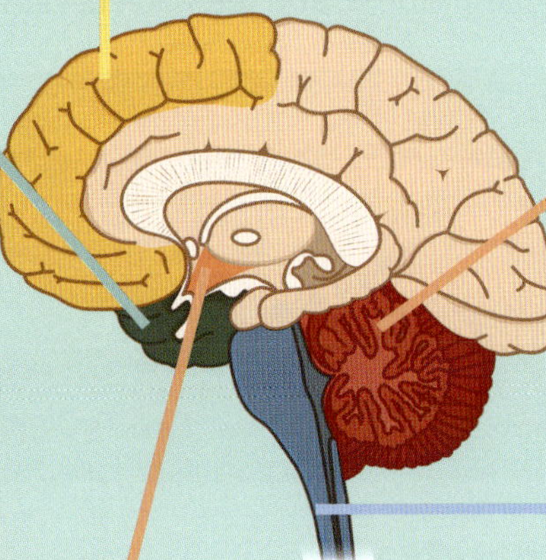

Cerebellum
Responsible for all physical movement, including balance, coordination, eye movement and motor learning (e.g. throwing a baseball or playing the drums).

Brain stem
Controls the flow of information between the brain and the rest of the body, and regulates important functions such as breathing and swallowing.

Hypothalamus
Responsible for keeping the body in a healthy and balanced state otherwise known as homeostasis. Essential functions include the release of hormones to help regulate body temperature, blood pressure and heart rate, thirst, emotions, hunger and sexual development.

Figure 1.18 Functions of areas of the brain

1 Using the internet, research the impact of alcohol on each of the following areas:
   - Frontal lobe
   - Cerebellum
   - Brain stem
   - Temporal lobe
   - Hypothalamus
2 Present your results in a report format.
3 Conduct further research to answer the following discussion questions.

Discussion

1 Compare the impact of alcohol on the developing brain with non-alcohol using teenagers.
2 Identify the part of the brain responsible for effective decision making. Infer how alcohol consumption would affect your ability to make decisions.
3 Recall the part of the brain responsible for motor learning. Describe how alcohol affects your ability to walk in a straight line.
4 Review the DrinkWise website to investigate how Australian drinking habits are changing. Do you think Australians have a positive or negative relationship with alcohol? Give reasons for your answer.
5 Predict what you believe the Australian drinking landscape will be in the future. Provide reasons.

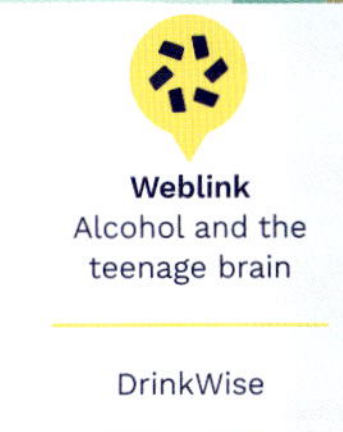

Weblink
Alcohol and the teenage brain

DrinkWise

FAST FACT

1 The human brain weighs around 1.5 kilograms.

2 The brain accounts for around 2 per cent of our total weight.

3 Male brains are marginally larger than female brains.

TOBACCO

Drug classification: stimulant

In Australia, tobacco smoking has been practised for more than 300 years. It was first introduced to First Nations communities in northern Australia by Indonesian fisherfolk in the early 1700s. After British colonisation in 1788, tobacco smoking was a popular pastime, and the habit became an accepted part of Australian society.

FAST FACT

Tobacco is known by a variety of names, including smokes, cigs and ciggies.

Video
Smoking: Why do people smoke? How can we further reduce smoking rates? Watch the video and start the discussion!

FAST FACT

1 Smoking has been linked to at least 19 forms of cancer.

2 Daily smoking rates for Australians aged 18 and over dropped from 22.4 per cent in 2001 to 13.8 per cent in 2017–18.

3 In 2016, 61 per cent of daily smokers had tried to quit or cut back their smoking over the past year.

What's in a cigarette?

The main ingredient in a cigarette is tobacco. Tobacco is made from the leaves of the tobacco plant, which is grown in warm climates around the world. The leaves are picked and dried, then processed by machines. Chemicals and artificial flavours are added to the dried tobacco by cigarette manufacturers.

Cigarettes contain approximately 600 ingredients. When smoked, these ingredients will produce more than 7000 chemicals. Around 70 of these chemicals are **carcinogens**, known to cause cancer, and can have deadly effects. Some of the main chemicals and substances found in cigarettes are shown in Table 1.4.

carcinogens substances that cause cancer

Table 1.4 Chemicals and substances in cigarettes

Nicotine	An addictive drug that makes people want to smoke more. Also used as an insecticide.
Ammonia	Commonly found in household cleaning products. Used to boost the effect of nicotine.
Carbon monoxide	Toxic, tasteless, odourless gas found in motor vehicle exhausts.
Methanol	A key component of rocket fuel.
Acetone	A solvent commonly used to remove nail polish.
Pesticides	Toxic chemicals used to kill insects.
Formaldehyde	A chemical found in a variety of products, from disinfectants to cosmetics. Used also to delay the decomposition of dead bodies!
Hydrogen cyanide	A toxic gas used in gas chambers.
Arsenic	An ingredient in rat poison.
Tar	Used to surface roads.
Butane	A highly flammable substance found in lighter fuel.
Radon	A radioactive gas.
Cadmium	An active component of battery acid.

Tobacco packaging in Australia

In order to reduce tobacco consumption, all tobacco products in Australia must be sold in standardised plain, logo-free, drab, dark brown packaging. The company brand name must be a certain size, in a certain font and in a certain place on the pack. No colours, logos or promotional text can be featured. Additionally, health warnings and other legally required information, such as toxic ingredients, must be identified on the packaging. The health warning must cover 75 per cent of the front of the pack and 90 per cent of the back.

Figure 1.19 Cigarette packs are required by law to display graphic anti-smoking messages, designed to 'scare' consumers into thinking about the implications of smoking.

Effects of smoking

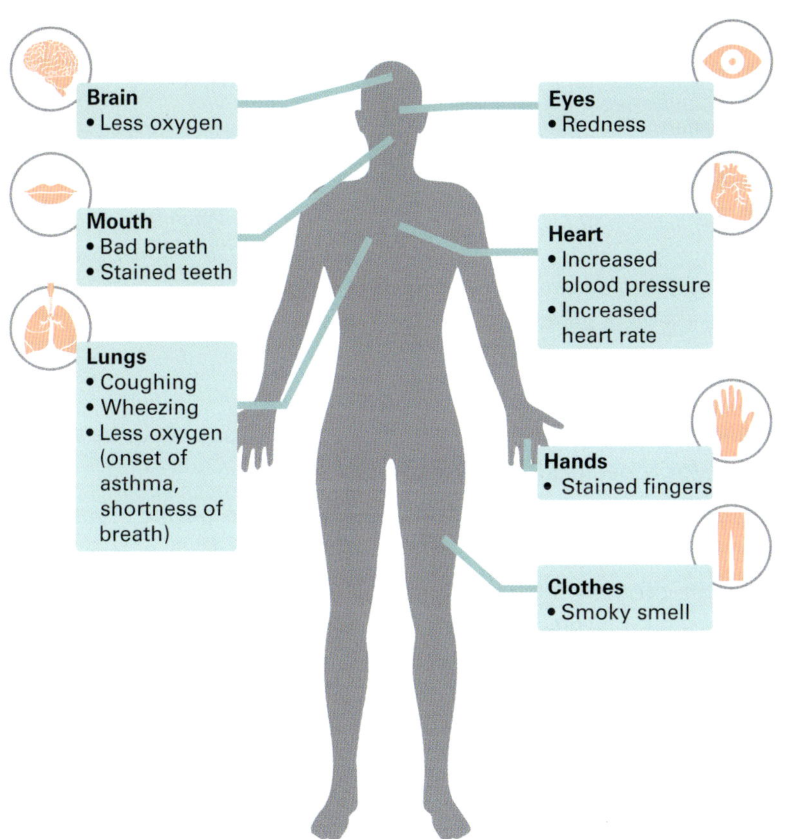

Figure 1.20 Short-term effects of smoking

Worksheet 1.8

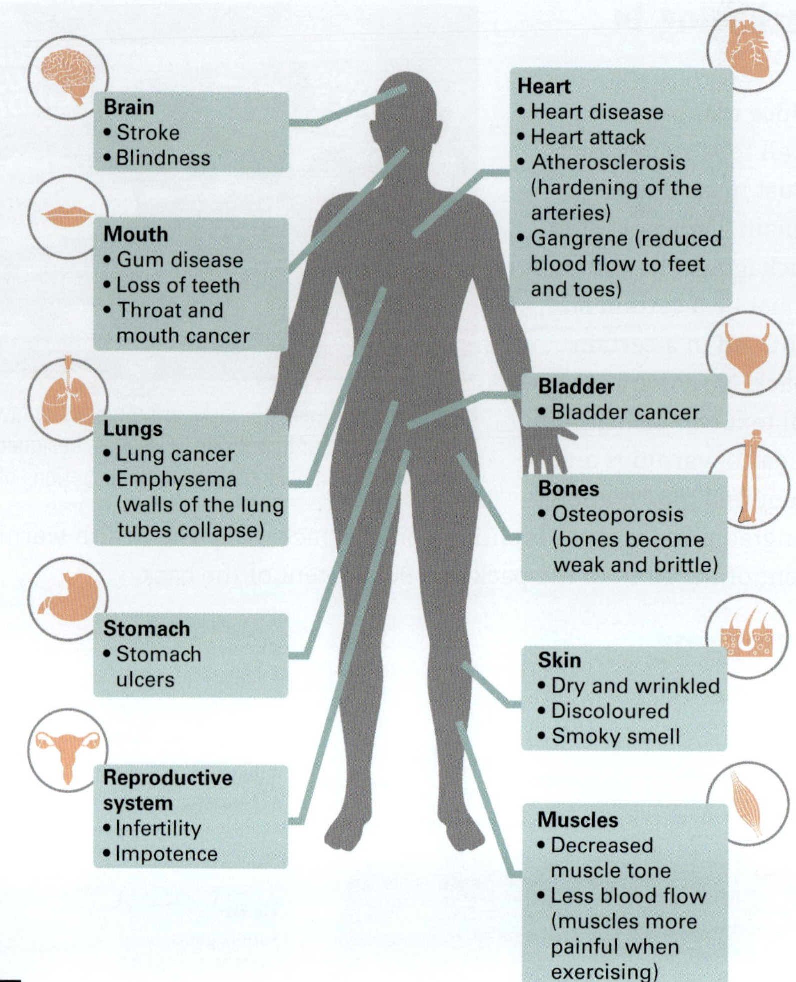

Weblink
Explore the Smokefree website to see the damaging effects of smoking on your body.

Figure 1.21 Long-term effects of smoking

Tobacco and the law

Worksheet 1.9
Worksheet 1.10

Although tobacco is a legal drug, it is illegal for tobacco products to be sold to those aged under 18, and for those under 18 to purchase them.

In Australia, there are severe restrictions on smoking in public areas, such as shopping centres, bars and restaurants, and within the workplace. It is also against the law to smoke in a car carrying a person under 18 years of age.

Smoking is now prohibited by almost all airlines around the world. People caught smoking on an aeroplane may face a hefty fine or jail term.

Vaping and e-cigarettes

Vaping refers to the inhaling of a vapour produced by an electronic device or e-cigarette. Unlike an actual cigarette, an e-cigarette does not produce tobacco smoke, but rather a vapour, which is often mistaken for water vapour. These battery operated devices heat the fluid-filled cartridges, and vaporised doses are then inhaled by the user in an act known as vaping. The vapour released by these electronic devices typically contains nicotine, flavourings and other toxic chemicals.

> **FAST FACT**
> E-cigarettes are known by a variety of names, including e-cigs, vape pens, hookah pens, e-hookahs, vapes and mods.

While vaping is a relatively new trend, the concept of vaping has been around for a very long time. Shisha, a traditional smoking device, was introduced to India

9780170463096

thousands of years ago, and even ancient Egyptians used hot stones to vape herbs. These ancient methods have led to the vaping methods we have today. In 1927, Joseph Robinson was the first person to initiate the idea of an e-cigarette to help inhale vapours, in what was considered a safer alternative to cigarette smoking. In 1963, Herbert Gilbert invented the first 'smokeless' e-cigarette, but no businesses were interested in manufacturing the product. In 2004, electronic cigarettes or e-cigarettes were released into the Chinese market. A Chinese pharmacist, Hon Lik, was credited with inventing a safer and more environmentally friendly method of inhaling nicotine in an attempt to reduce the harmful effects associated with tobacco smoking. Global internet sales saw the e-cigarette grow in popularity.

Figure 1.22 Vaping products

Warning: The lithium-ion batteries within e-cigarettes or 'vapes' have been known to explode resulting in serious injury and even death. While the explosions are rare, they are very dangerous.

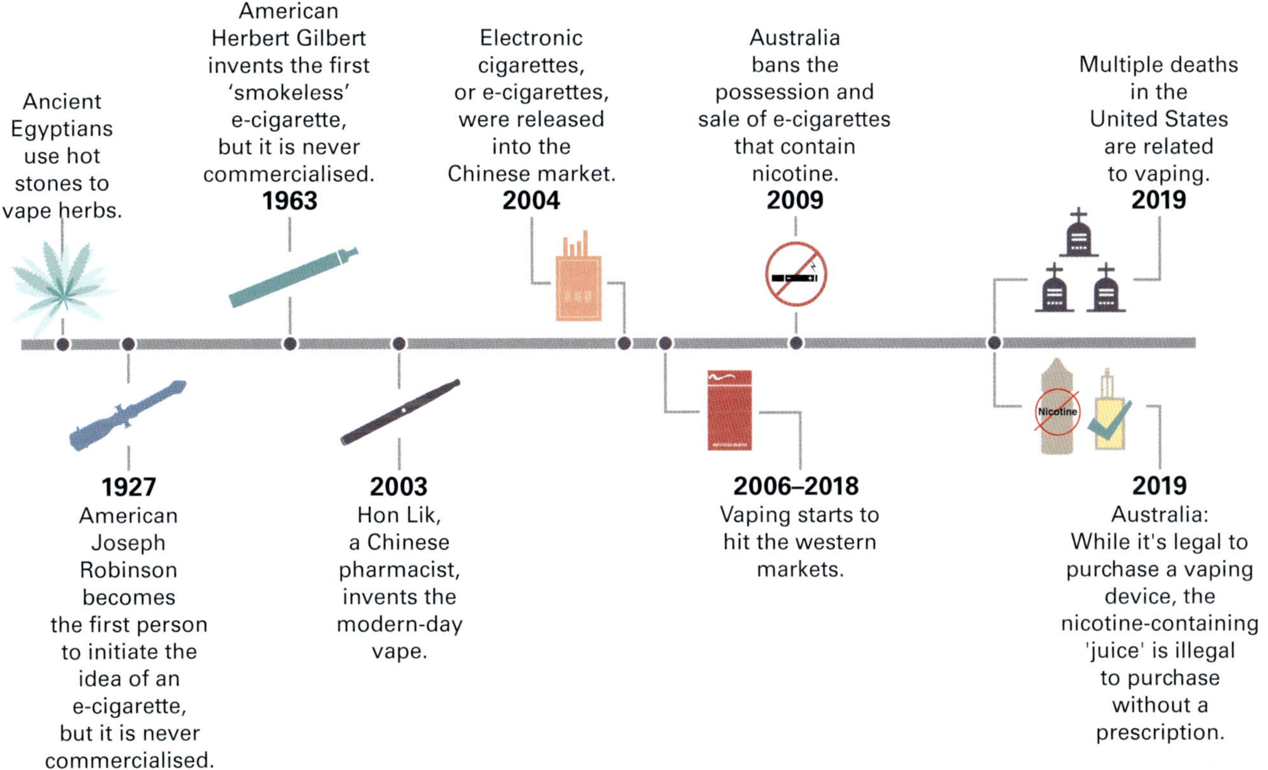

Figure 1.23 Timeline showing the development of e-cigarettes

Effects of vaping

Vaping has not been around long enough for researchers to know all the harmful effects on the body. However, there is mounting evidence to suggest that the toxic chemicals in the vapour have been linked to cancer, heart disease and serious lung damage, which has resulted in death.

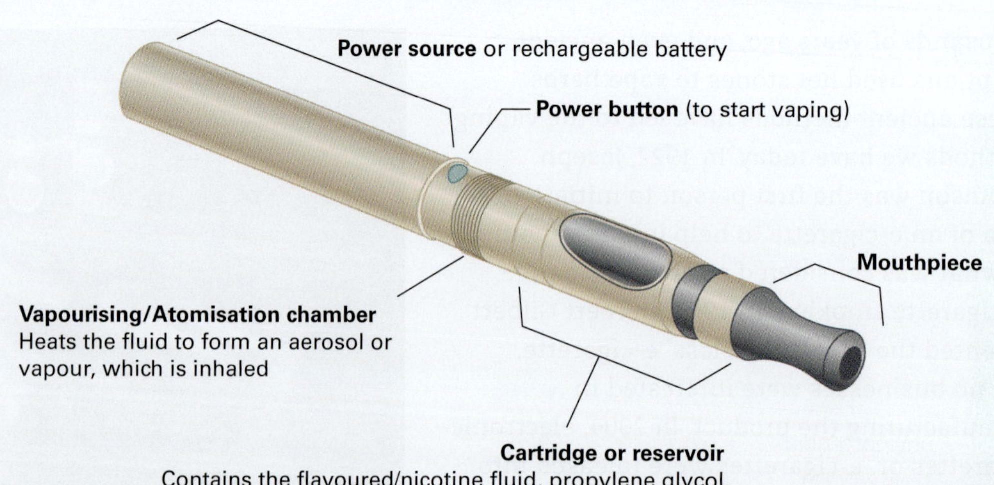

Figure 1.24 Parts of an e-cigarette

CASE STUDY ➡ UP IN SMOKE

Identify

It's the $20-billion industry trying to take out the $900-billion tobacco heavyweight but is vaping really a healthier alternative to smoking?

Understand

The number of people who smoke cigarettes in NSW has dropped to record levels, although experts have warned rising rates of vaping among under-25s could undo decades of work to reduce smoking rates.

Eleven per cent of people aged 16 to 24 reported being a current user of e-cigarettes, or "vapes", more than double the number in 2020, data from the NSW Population Health Survey, published today, showed.

In the same period, daily rates of smoking cigarettes among all people aged 16 and over decreased from 9.2 per cent to 8.2 per cent.

Anita Dessaix, chair of the Cancer Council's public health committee, said she was concerned e-cigarettes were making smoking a habit for the next generation.

"The biggest increase in vaping we are seeing is in that 16- to 24-year-old age group. E-cigarettes were being positioned as a way for people who were already smoking to quit, but there hasn't been a great increase in uptake among the middle-aged," she said.

"It does feel like the genie is starting to get out of the bottle with vaping, and we are not going to be able to get on top of the trend with the level of addiction and the sheer quantum of product available."

A recent ANU review of international vaping research found young non-smokers who vape were around three times more likely to take up smoking than those who did not vape.

NSW Chief Health Officer Kerry Chant agreed the uptick in vaping could undermine decades of tobacco control in Australia.

"There's strong evidence of a smoking 'gateway effect'," she said. The Cancer Council's NSW Smoking & Health Survey 2021, also released today, found 23 per cent of people who had vaped did so to help them quit or cut down on cigarettes. More common reasons given for vaping were curiosity (36 per cent) and to socialise or because someone offered it (28 per cent).

The survey only interviewed people aged 18 and over, meaning the results did not include contributions from younger vape users. Health authorities are concerned by rising rates of vaping among high school students.

Under 40s were significantly more likely to report using disposable vapes compared to people in their 50s and 60s (50 per cent compared to 32 per cent). Dessaix said this was likely due to the types of vapes sold at convenience stores, where younger people are more likely to purchase.

"What is available and what is easily accessible are those bright, colourful, fruit or candy flavoured vapes. If you're at school, they can look like a highlighter and be hidden in pencil cases," she said.

Across all adult vape users, the most common place to buy e-cigarettes was the internet (17 per cent of users), with more than half of those users buying from overseas websites. The number of people buying vaping products at convenience stores has increased five-fold since 2019 to 5 per cent of users.

A 2020 study of 52 e-liquids by Curtin University and the Telethon Kids Institute found all contained at least one chemical that had unknown effects on respiratory health, with some using up to 18 chemicals in this category.

"Vapes can contain many harmful chemicals and toxins, even if they are nicotine free. We know vapes can harm your health in the short term, but the long-term effects are largely unknown," Chant said, adding her message to young people who vape or smoke cigarettes was to quit today.

About 23 per cent of NSW residents now consider themselves a "former smoker" of cigarettes, a figure which can be extrapolated to about 1.5 million people.

The Cancer Council's survey showed 41 per cent of people in NSW who smoke were thinking about quitting in the next six months.

"Quitting smoking is one of the most important things you can do for your health. It will reduce your risk of sixteen different types of cancer, coronary heart disease, stroke and other debilitating conditions," Chant said.

Source: 'Number of young people vaping doubles in a year as smoking rates' drop by Mary Ward, 31 May 2022, https://www.smh.com.au/national/nsw/number-of-young-people-vaping-doubles-in-a-year-as-smoking-rates-drop-20220531-p5apur.html.

Discuss

1 In pairs, complete the PMI chart below relating to the article above. A PMI chart will help you examine the pluses, minuses and interesting facts associated with the article.

Pluses	Minuses	Interesting facts

2 Present your findings to the rest of your class.

CAFFEINE

Drug classification: stimulant

Caffeine is a naturally occurring compound found in the leaves, seeds and fruits of a variety of plants, including cocoa, coffee beans and tea leaves. Drinking caffeine triggers the release of adrenaline, a hormone that acts on the central nervous system by speeding up messages that are sent to and from the brain. In small amounts, caffeine can make you feel alert, more focused and able to think and react more quickly. However, larger amounts of caffeine may lead to heightened irritability, anxiety and difficulty sleeping.

Shutterstock.com/philippou

Figure 1.25 Coffee culture is booming in Australia.

Video
Caffeine: Why is coffee so popular? Do you think there should be restrictions on caffeine for teens and children? Watch the video and join the discussion!

FAST FACT

- Australia has a booming coffee culture. Seventy-five per cent of all Australians enjoy at least one coffee a day, and 28 per cent have three or more cups a day! Apparently one in four Australians (27 per cent) say they cannot survive the day without coffee.
- Australians spend more than $500 million a year on energy drinks. Energy drinks can contain as much as 160 milligrams of caffeine per can.
- Children and teenagers should not consume more than 2.5 milligrams of caffeine per kilogram of body weight per day. For a teenager weighing 50 kilograms, that means no more than 125 milligrams of caffeine per day. Healthy adults should consume no more than 400 milligrams of caffeine a day.

Worksheet 1.11

FAST FACT

The chemical name for caffeine is 1,3,7-trimethylxanthine. This chemical is also used as a pesticide to kill frogs, but is best known as the world's most popular drug.

Caffeine has been around for thousands of years. Even though the coffee 'tree' originated from Ethiopia, it was the Arabs who first cultivated the plant and used it as a drink. Although coffee was well established in the Islamic world in the 16th century, it was not until the 17th and 18th centuries that it became popular in European coffee shops. Today, 120 000 tonnes of caffeine are produced globally each year. Caffeine is found in a variety of food and beverage products, including:

- coffee
- tea
- chocolate
- energy drinks
- energy bars
- over-the-counter medications (such as cough medicines)
- cola drinks
- chocolate milk.

Worksheet 1.12

FAST FACT

A traditional Ethiopian story: One day, a farmer moved his herd of goats to a new pasture and noticed they were becoming restless and irritable. After watching the goats closely, the farmer noticed they were grazing on small seeds. These seeds were later dried and called 'coffee beans'.

Energy drinks

electrolytes
inorganic compounds used to create electrical energy for a variety of bodily functions; an example is salt

Energy drinks are non-alcoholic carbonated drinks that contain substances known to boost energy levels, such as caffeine or guarana, a herbal source of caffeine. The energy drink industry in Australia is booming. There are hundreds of different energy drinks available, ranging from energy shots to massive 500 millilitre cans! Australia has even produced an energy drink in a powdered form.

Energy drinks should not be confused with sports drinks, which are designed to rehydrate and replace **electrolytes** lost during physical activity. Sports drinks provide carbohydrates that the body requires to create energy for muscular contraction. Energy drinks contain higher concentrations of caffeine, producing a sense of alertness and focus. Some energy drinks contain twice as much caffeine as many soft drinks.

FAST FACT

Caffeine: The legal limit of caffeine in a 250 millilitre energy drink in Australia is equivalent to 1 cup of coffee (80mg).

9780170463096

What's in energy drinks?

- Caffeine: a 250 millilitre can of energy drink contains 80 milligrams of caffeine. A 500 millilitre can contains around 160 milligrams of caffeine. These levels are well over the recommended limits for a child.
- Herbal extracts: guarana and ginseng
- Protein: taurine, an amino acid, is added to energy drinks
- Sugar: typically around 13 teaspoons per can
- Vitamin B

FAST FACT

Thirty-one per cent of 12–19 year olds regularly consume energy drinks.

The harmful effects of energy drinks

The risks associated with the long-term use of energy drinks are not yet known. However, recent data collected by the Australian Poisons Centre highlights the most common side effects associated with over-consumption of energy drinks. These include:

- palpitations/faster than normal heartbeat
- tremors
- agitation
- upset stomach
- chest pain
- dizziness
- tingling/numbing skin
- difficulty sleeping
- breathing problems
- headaches.

iStock.com/skodonnell

Figure 1.26 Energy drinks often contain more caffeine than the daily recommendation for a child.

CASE STUDY ENERGY DRINK DEATH

Identify

The death of a 35-year-old man after consuming energy drinks has prompted a Perth mother to educate others about the dangers.

Understand

Mick Clarke died suddenly earlier this year. The truck driver was a regular runner and never smoked. However, he drank about four caffeinated energy drinks a day as well as coffee.

The coroner said the cause of death was caffeine toxicity in a man with myocardial scarring – damage to his heart from a heart attack he did not know he had suffered.

His mother, Shani Clarke, said he had complained of indigestion in the weeks before his death. 'The coroner said the indigestion could have been mistaken for a mild heart attack,' she said …

Insomnia is one of the better known side effects of excess caffeine intake.

Others include increased anxiety, panic attacks, high blood pressure, bowel irritability and cardiac arrhythmia, according to WA Health's deputy chief health officer Andy Robertson …

The drinks' marketing suggests they improve performance, endurance and mental concentration, although none of these have been substantiated clinically. However, they can cause serious health effects.

Dr Robertson said it was recommended not to consume more than 400 milligrams of caffeine a day. A 500mL can of a typical energy drink carries about 160mg …

Ms Clarke has started a Facebook page called Caffeine Toxicity Death Awareness to highlight the dangers associated with the drinks … Ms Clarke said her aim now was to have the drinks banned for people under the age of 18 … Dr Robertson agrees the drinks have a far greater effect on younger people.

'There's certainly a number of countries [that] have already banned them and certainly … adolescents and the children are far more susceptible to their effects, they haven't developed any tolerance to caffeine,' he said.

Source: 'Man's death after too many energy drinks prompts mother to highlight dangers', Natasha Harradine, 2 September 2014, *ABC News*. Reproduced by permission of the Australian Broadcasting Corporation – Library Sales. Natasha Harradine © 2014 ABC

Discuss

1. How many cans of energy drink did Mick Clarke consume on a daily basis?
2. In addition to the energy drinks, what else was Mick consuming?
3. Identify the symptoms Mick suffered as a result of the drinks he was consuming.
4. Describe the side effects of excessive caffeine intake.
5. Analyse the following scenario and propose a response. Use facts to support your answer.
   Your friend plays netball and her team is playing in the end of season Grand Final. She is excited and has asked you to come along for support. You arrive courtside and see her warming-up prior to her game. You notice she is running back to her bag to consume an energy drink. You watch her finish the drink, reach into her bag and grab another one. She sees you, waves and shouts "Hey … need all the energy I can", then runs back to continue her warm-up.' You are very surprised and worried. What will you say?

Energy drinks and alcohol consumption

Recent research suggests the combination of alcohol and energy drinks could be more harmful than drinking alcohol alone. There has been an alarming increase in the number of people choosing to drink energy drinks with 'shots' of alcohol. This polydrug use combines a depressant (alcohol) with a stimulant (caffeine-laden energy drink). This blend will allow the person to feel the effects of alcohol while remaining more alert and awake, with the energy drink masking the tiredness and relaxed state associated with alcohol consumption.

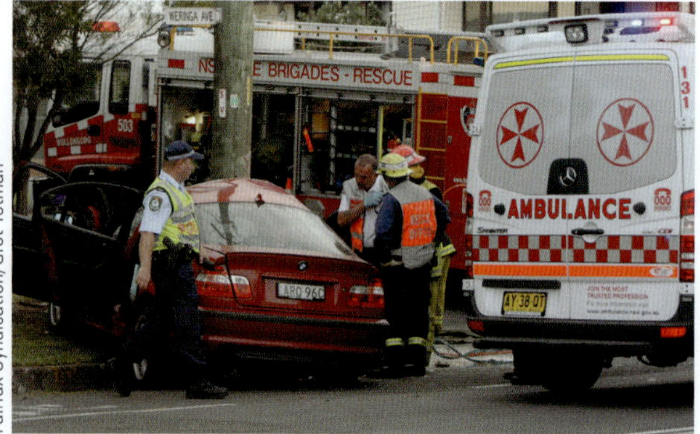

Fairfax Syndication/Gret Totman

Figure 1.27 The combination of alcohol and energy drinks can have fatal consequences.

People who combine alcohol with energy drinks often falsely perceive themselves to be more confident and alert, and subsequently take more risks, including driving while under the influence of alcohol. They may increase their chances of experiencing alcohol-related accidents.

Table 1.5 Caffeine content in food and drink

Drink/product	Amount	Caffeine content
Brewed black tea (medium strength)	mg/100ml	22.5
Coffee	mg/100ml	
⇨ Cappuccino		101.9
⇨ Flat white		86.9
⇨ Long black		74.7
⇨ Espresso style		194.0
Cola	mg/100ml	
Coca Cola		9.7
Diet Coke		9.7
Coke Zero		9.6
Energy drink (Red Bull)	mg/100ml	32.0
Milk chocolate bar	mg/100ml	20.0
Dark chocolate bar	mg/100ml	59.0

Source: Australian Drug Foundation. Adapted from Food Regulation Standing Committee, Caffeine Working Group. (2013). The regulation of caffeine in foods.

From top to bottom: Shutterstock.com/Nitr; Shutterstock.com/ikontee; iStock.com/gvictoria; Shutterstock.com/Baranov E; Shutterstock.com/Andrjuss; Shutterstock.com/Andrea1971

FACE TO FACE Calculating your caffeine intake

It is recommended that children and teenagers consume no more than 2.5 milligrams of caffeine per kilogram of body weight per day. Based on this recommendation, calculate how much caffeine you are allowed to consume on a daily basis. Now, using the information in Table 1.5, work out which products you can consume that will keep you within your recommended daily allowance. Discuss your findings with the rest of the class.

Effects of caffeine

The effects of caffeine can be felt in as little as five minutes after consumption, and can last for up to 12 hours. As with all drugs, the side effects can differ among individuals. Symptoms are more likely to increase if consumption increases.

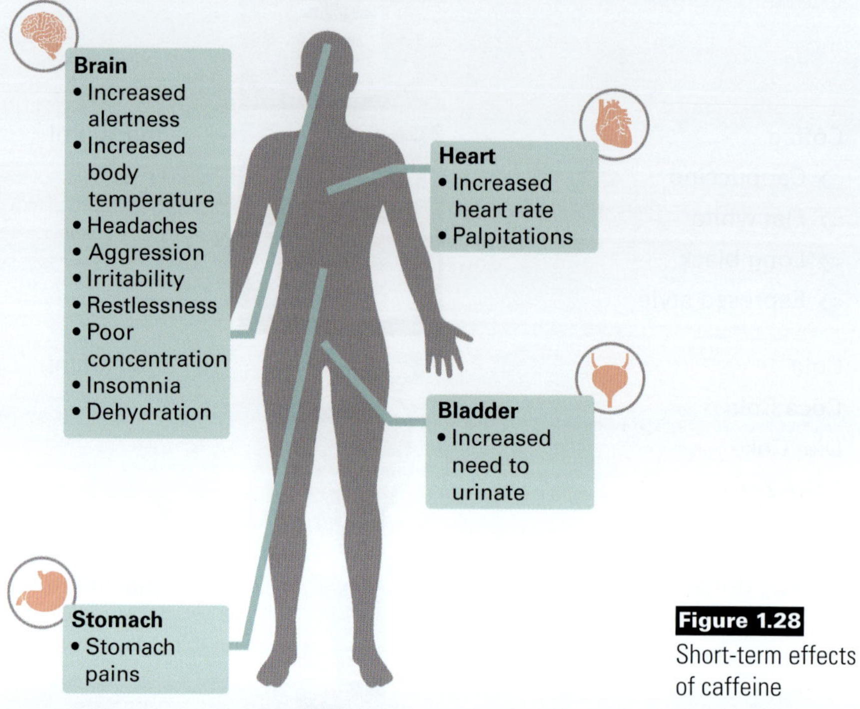

Figure 1.28 Short-term effects of caffeine

Caffeine and the law

In Australia, it is legal to purchase and sell caffeinated products. There is no legal limit on the consumption of caffeine, and it is considered socially acceptable.

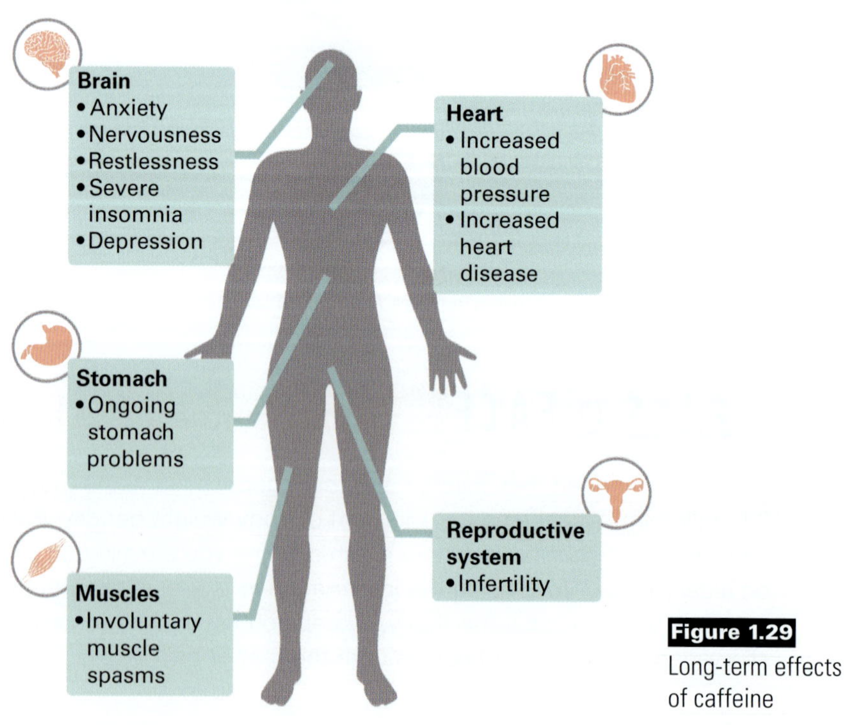

Figure 1.29 Long-term effects of caffeine

CANNABIS

Drug classification: depressant (small doses); hallucinogen (larger doses)

Cannabis is an illegal drug made from the leaves and dried flowers of the cannabis plant. There are two species of cannabis plant, *cannabis sativa and cannabis indica*. Each species produces different effects on the body. The leaves of the cannabis plant have very distinct characteristics, with between five and seven arrow-shaped leaflets attached to a centre point. The greyish-green dried plant matter can either be smoked or prepared as an edible ingredient.

The main active ingredient in cannabis is a chemical called delta-9-tetrahydrocannabinol, otherwise known as THC. The concentration of THC varies depending on the part of the plant being used and its growing conditions. This ingredient causes the 'high' associated with cannabis use, leaving the user with feelings of happiness and euphoria.

Cannabis has been around for thousands of years. It is the only plant in the world cultivated for its fibre, used in the manufacture of cloth and also used for its psychoactive properties. Early Chinese records dating back to 2737 BCE refer to cannabis being used to treat conditions such as rheumatism and malaria. Originally, there was mention of the intoxicating properties of cannabis, but the medicinal properties were considered more important. Recreationally, the drug was used in India and by Muslims in place of alcohol, which is forbidden by the Qur'an.

Worksheet 1.13

FAST FACT

Cannabis is known by a variety of names, including marijuana, grass, pot, dope, hash, weed, ganja, head, bud, doobie, mary jane and bhang.

Worksheet 1.14

FAST FACT

Cannabis is the most common illegal drug in Australia. One in 12 students aged 12–17 have used cannabis in the past four weeks.

Source: 'Cannabis: Factsheet', Positive Choices, https://positivechoices.org.au/teachers/cannabis-factsheet

Effects of cannabis

The effects of cannabis usually occur within the first few minutes of the drug being smoked, and may last for up to two or three hours, depending on the concentration of THC. If cannabis is eaten, the effects are slower to occur and may last longer than when smoked.

Over time, the regular use of cannabis may result in a number of health-related problems.

Worksheet 1.15

Worksheet 1.16

Medicinal cannabis

Medicinal cannabis refers to the legal use of high quality and regulated cannabis products prescribed by doctors to ease the symptoms associated with a medical condition. Unlike recreational cannabis, medicinal cannabis is taken by those suffering either chronic or terminal illnesses to alleviate debilitating side effects. It is prescribed where conventional medicine has failed to be effective, and is not a cure. Cannabis has been used for medicinal purposes for thousands of years to treat various conditions. The cannabis plant contains a vast number of unique compounds, known as cannabinoids. Over 100 different cannabinoids have been identified. The main two active ingredients are Delta-9-tetrahydrocannabinol (THC) and Cannabidiol (CBD).

Research suggests these cannabinoids may help to:

- relieve anxiety
- reduce inflammation
- alleviate pain
- control vomiting and nausea
- relax muscle groups
- increase appetite

iStock.com/sumografika

Figure 1.30 The dried leaves and flowers of the cannabis plant are known as marijuana.

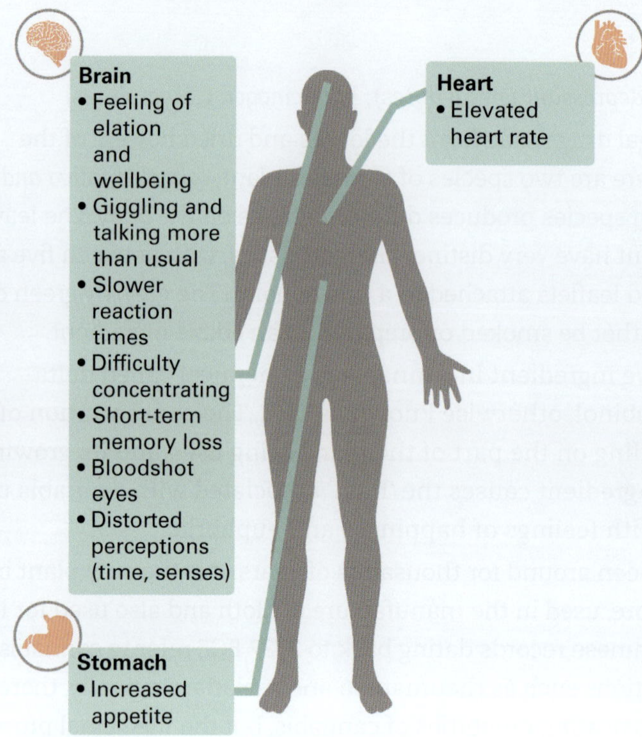

Figure 1.31 Short-term effects of cannabis use

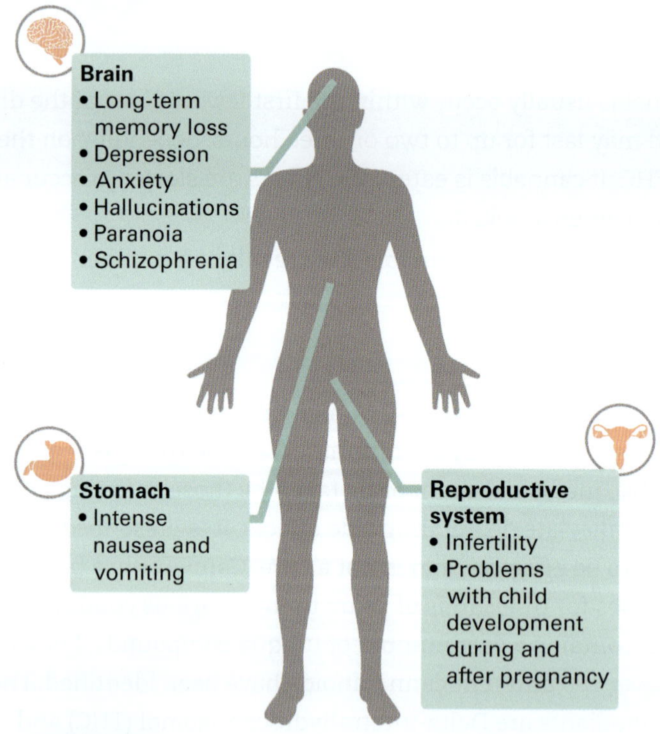

Figure 1.32 Long-term effects of cannabis use

- improve weight gain
- treat cancer.

In 2016, the use of medicinal cannabis was legalised in Australia. While there has been much debate over the legalisation of medicinal cannabis, the Australian government openly supports further research into the medicinal properties of cannabis and its regulation in order to safeguard patient access.

INVESTIGATION

MEDICINAL CANNABIS

Purpose

To investigate the effectiveness of medicinal cannabis to treat endometriosis and other conditions.

Method

1 Read the following extract from an article discussing treatment of endometriosis using medicinal cannabis, then conduct further research to investigate the discussion questions.

Endometriosis is a chronic, inflammatory condition where tissue similar to the lining of the uterus is found outside of the womb. It affects around one in ten women of reproductive age, causing pain, infertility and gastrointestinal symptoms. Women often report difficulty getting their pain and other symptoms under control, despite medication or even surgery.

Our research, published today, found one in ten Australian women with endometriosis reported using cannabis to manage their pain and other symptoms. We surveyed 484 women with surgically diagnosed endometriosis about the self-management strategies they used.

Of the respondents, who were aged 18 to 45, 76 per cent reported using self-management techniques in the past six months. This included the use of heat packs (70%), dietary changes (44%), exercise (42%), yoga or Pilates (35%) and cannabis (13%).

Out of all of the self-management techniques, cannabis was rated as the most effective for managing pain …

Emerging research shows medicinal cannabis can help manage a number of conditions, including chronic pain in adults, the spasticity of multiple sclerosis, intractable epilepsy (where seizures can't be controlled with medication) and chemotherapy-induced nausea and vomiting.

Source: '1 in 10 women with endometriosis report using cannabis to ease their pain', by Justin Sinclair (Research Fellow, NICM Health Research Institute, Western Sydney University) and Mike Armour (Post-doctoral research fellow, Western Sydney University), *The Conversation*, 12 November 2019. Read the full article on *The Conversation*: https://theconversation.com

Weblink
Read the rest of the survey research findings

Discussion

1 What is endometriosis?
2 Describe the symptoms of endometriosis.
3 How many Australians are affected by this condition?
4 Research suggests that medicinal cannabis can help manage a number of conditions. Identify three other conditions mentioned.
5 Conduct online research to find five interesting facts about one of the aforementioned conditions.
6 Find one article discussing the effectiveness of using medicinal cannabis for the condition and summarise the findings.
7 Considering the extract above and your own research, discuss the following statement: 'Medicinal cannabis is an effective and safe treatment option.'

Cannabis and the law

Although medicinal cannabis has been legalised, it is illegal to use, sell or give cannabis to someone else in Australia. Serious penalties such as substantial fines or prison terms apply to those convicted of supplying cannabis or being in possession of items used to smoke cannabis, such as pipes and bongs.

Worksheet 1.17

ECSTASY

Drug classification: stimulant, hallucinogen (rare)

Ecstasy, or molly, is the common name for the synthetic drug MDMA (methylenedioxymethamphetamine). Primarily a stimulant, ecstasy is an illegal psychoactive drug with hallucinogenic properties. People taking ecstasy may feel physically energised and emotionally relaxed. Depending on the contents of the drug, there may also be some hallucinogenic effects, including distortions in reality.

MDMA was originally manufactured in Germany in the early 1900s. In the 1970s and early 1980s, MDMA was used experimentally by psychotherapists to help people understand their feelings by promoting deep inner thinking and enhancing communication. MDMA has been used in nightclubs and on the 'party' scene since the 1980s.

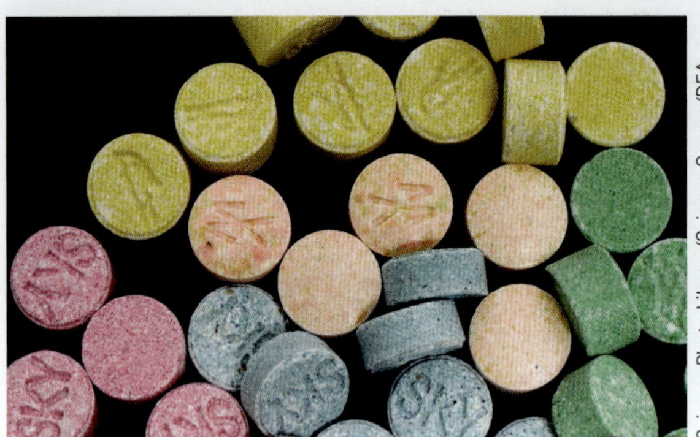

Science Photo Library/Science Source/DEA

Figure 1.33 Ecstasy tablets are sold in a variety of forms, usually with a distinctive design.

> **FAST FACT**
>
> Four per cent of young people aged 12–17 years have used ecstasy (MDMA) in the past 12 months.
>
> Source: 'Young people and drug use', Alcohol and Drug Foundation, https://adf.org.au/talking-about-drugs/parenting/talking-young-people/the-other-talk/drugs-young-people/

> **FAST FACT**
>
> Other names for ecstasy include Molly, E, pingers, the love drug, pills, eccy, candy and e-bomb.

Forms of ecstasy

Ecstasy can be supplied as a tablet, capsule or a powder. Tablets (or pills) are the most common forms of MDMA. They come in a variety of colours, shapes and sizes, and usually display popular imprinted commercial logos or graphic designs. Ecstasy is primarily swallowed, but can also be injected or snorted.

It is important to note that not all ecstasy tablets contain MDMA. Although MDMA is usually the main ingredient in ecstasy, it is becoming difficult to source, so alternative ingredients are being used to mimic its effects. These alternatives include caffeine, amphetamine, ephedrine, methamphetamine and talcum powder. Some ecstasy tablets may contain no MDMA at all. The majority of the ecstasy sold in Australia is amphetamine-based.

Effects of ecstasy

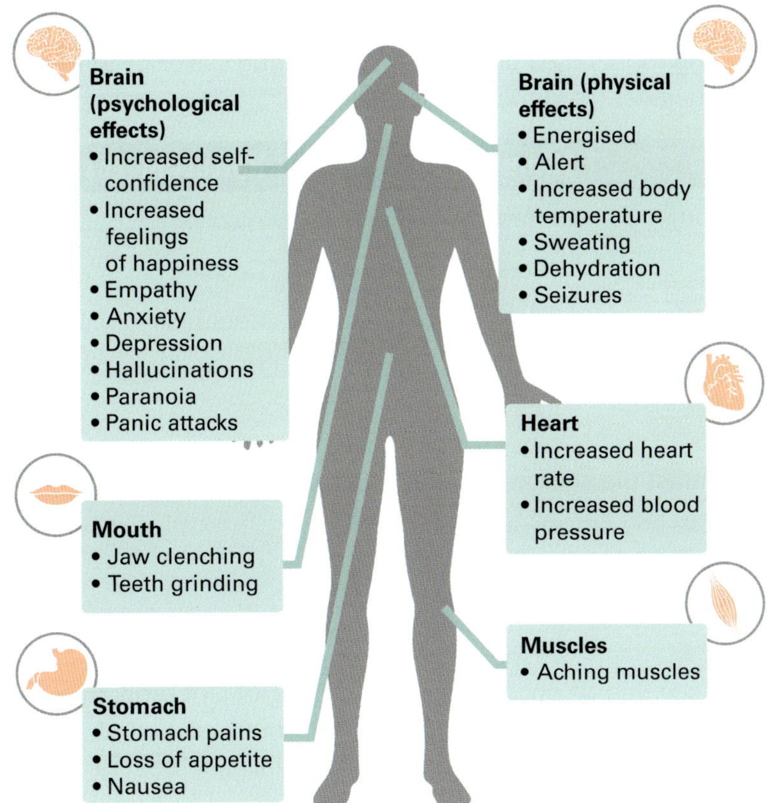

Figure 1.34 Short-term effects of ecstasy

Long-term effects of ecstasy

Unfortunately, little is known about the long-term physiological effects of ecstasy use. Some evidence suggests that long-term use can cause damage to organs such as the liver, heart and brain. In addition, research suggests that those who regularly use ecstasy are exposed to an array of mental health issues and may experience irrational emotional behaviour, poor memory and concentration, anxiety, paranoia, irritability, depression and personality changes. Long-term use can lead to dependency.

Ecstasy and the law

It is illegal to use, sell or give ecstasy to someone else in Australia. Serious penalties such as substantial fines or prison terms apply to those convicted of supplying ecstasy.

Worksheet 1.18

FACE TO FACE Debate on drugs

In groups of four, formulate an argument on one of the following drug-related topics. You will be asked to debate your argument either for or against the motion selected, against another team of four.

- Cannabis is an evil weed leading to harder drugs.
- Alcohol is the most damaging drug in society.
- Is vaping more dangerous than smoking?
- Should energy drinks be banned?

Weblink
Test your knowledge with the Mind Matters quizzes from the National Institute on Drug Abuse for Teens.

Quiz
How will drug use affect my body?

1 Explain the difference between a depressant and a stimulant. Give two examples of each.

2 Group the following list of drugs under their correct drug classification. Remember, a drug may be listed under more than one heading:

Depressants	Stimulants	Hallucinogens
Magic Mushrooms	Coffee	Alcohol
Cannabis	Ecstasy	Cocaine
Khat	LSD	Ketamine
Heroine	Inhalants	Nicotine
Codeine	Ephedrine	Ice

3 Compare and contrast the short-term effects of alcohol and caffeine.

4 Define polydrug use and provide an example.

5 Outline the four main types of alcohol.

1 The National Tobacco Campaign is an Australian Government program aimed at reducing smoking rates in Australia. Launched in 1997, this is one of Australia's longest-running public health campaigns. Investigate this campaign and then answer the following questions:

Weblink
The National Tobacco Campaign

a Explain why the National Tobacco Campaign important.

b Identify the percentage of all cancers in Australia caused by smoking tobacco.

c Explain, using the 'key findings' on tobacco use, whether there has been an increase or decrease in smoking rates in Australia.

d Take time to explore the National Tobacco Campaign website and watch the 'quit stories' on smoking. Using a transcript format, write a 'quit story' that presents the anti-smoking message to teenagers.

2 Watch the video about vaping, then answer the questions below:

a Describe the differences in vaping laws between Australia and the US.

b Why do you think there are different laws for different countries?

c Do you think there is a link between increased vaping among young people and vaping marketing?

1 Many young people believe cannabis is no big deal, and that cannabis should be legalised in Australia for both recreational and medicinal use. Although the media has been sending mixed messages about the use of cannabis, the risks associated with this drug are very real and should not be ignored.

Use the internet to research information on the effects of cannabis. Create a 20-second radio commercial giving listeners important health information relating to why cannabis could lead to lower academic grades at school.

2 In groups, debate whether cannabis should be legalised. Research information to prepare your argument for or against the legalisation of cannabis and draw a conclusion. Justify your group decision using facts from your research.

3 In Australia, cigarette advertising is banned and cigarette packaging must display anti-smoking messages. Should these rules also apply to e-cigarette use?

WHAT FACTORS INFLUENCE THE USE OF DRUGS AND ALCOHOL?

Worksheet 1.19
Worksheet 1.20
Worksheet 1.21

Young people choose to take drugs for a variety of reasons. They are influenced by three main factors: personal, environmental and social. Understanding these factors and the risks associated with drug use will help you to make responsible, safe and informed decisions.

PERSONAL FACTORS: DO YOU HAVE A FRIEND LIKE BEN?

Ben had been going out with Charlotte for nearly three years. Charlotte ended the relationship by text message. She gave no reason, only saying she didn't want to go out with Ben anymore. Ben pretended the break-up didn't bother him, but deep inside he was struggling to come to terms with it. After a while, Ben began to feel depressed and wanted to find a way to deal with his emotions. He started drinking energy drinks to pick him up and was soon drinking up to four energy drinks a day.

Stress and self-esteem are two personal factors that could lead to drug use.

Shutterstock.com/Oleg Golovnev

Figure 1.35 Stress and low self-esteem can influence the use of drugs and alcohol.

Shutterstock.com/LightField Studios

Figure 1.36 Peer pressure, role models and socio-economic background all influence drug use.

Alamy Stock Photo/Wavebreakmedia Ltd UC99

Figure 1.37 Family members, conflict and education can all influence drug use.

Stress

From time to time, everyone will experience stress. People handle stress in many different ways. In Ben's case, he was unable to deal with the break-up and needed to find a way to ease his emotions. Unfortunately, substance use does not address the underlying factors causing the stress. Ben's need for a caffeine rush may make the situation worse.

Self-esteem

Low self-esteem can significantly increase the likelihood of drug use in young people, as they are less likely to consider the consequences associated with their actions.

SOCIAL FACTORS: DO YOU HAVE A FRIEND LIKE KAITLIN?

Kaitlin is 16 years old and has smoked four cigarettes a day since she was 12. She first started smoking when she was offered a cigarette by one of her older brother's friends. Now, whenever Kaitlin hangs out with her friends, they smoke. She has also tried marijuana, for a laugh and because it made her feel part of the in-crowd. Kaitlin says she can stop smoking at any time, but she is yet to do so.

Peer pressure, role models and socio-economic background are all social factors that could influence young people to use drugs.

Peer pressure

In Kaitlin's case, she first smoked a cigarette because she wanted to be accepted by her brother's friend. She may also have felt pressure from her peers.

Role models

Some teenagers may have seen their role models on television or in movies smoking, looking cool and unconcerned. This 'glamorisation' of drug use could influence teenagers' decisions to use drugs.

Socio-economic background

Poverty, financial stress at home and mental and physical abuse may prompt teenagers to use drugs as a way of coping with certain situations. Often, drug accessibility is more prevalent in low socio-economic areas, as some people may sell drugs as a means of overcoming poverty.

ENVIRONMENTAL FACTORS: DO YOU HAVE A FRIEND LIKE PATRICK?

Every night, Patrick's father would come home from work and head straight for the fridge. He would drink an average of six bottles of beer a night, often asking Patrick to get another one for the 'old man'. After a couple of beers, Patrick found that his father grew funnier, and he loved listening to the stories he told. One day, when Patrick was 16, he asked his father if he could have a beer too. His father gave him a pat on his back and said, 'Go for it, son!'

There are a number of environmental factors that may influence teenagers to use drugs, including family members, family conflict and education.

Influence of family members

Family members who smoke or drink alcohol are more likely to influence young people living in the same household to do the same, as they start to see the drug use as a normal part of everyday life.

Family conflict and home-management issues

Family conflict, lack of support and direction, divorce or separation are all factors that may contribute to the increased risk of drug taking. Family transition and moving house may also add stress to young people.

Education

A student's performance and participation at school can be a major risk factor associated with drug use. Expulsion, truancy, boredom and poor academic achievement, coupled with a lack of motivation and commitment, can also increase the risk of drug use.

RISK FACTORS ASSOCIATED WITH DRUG USE

Anyone can become dependent on drugs. It can happen to people of any age, economic status or gender. There are, however, certain risk factors that can increase the likelihood of becoming dependent on drugs.

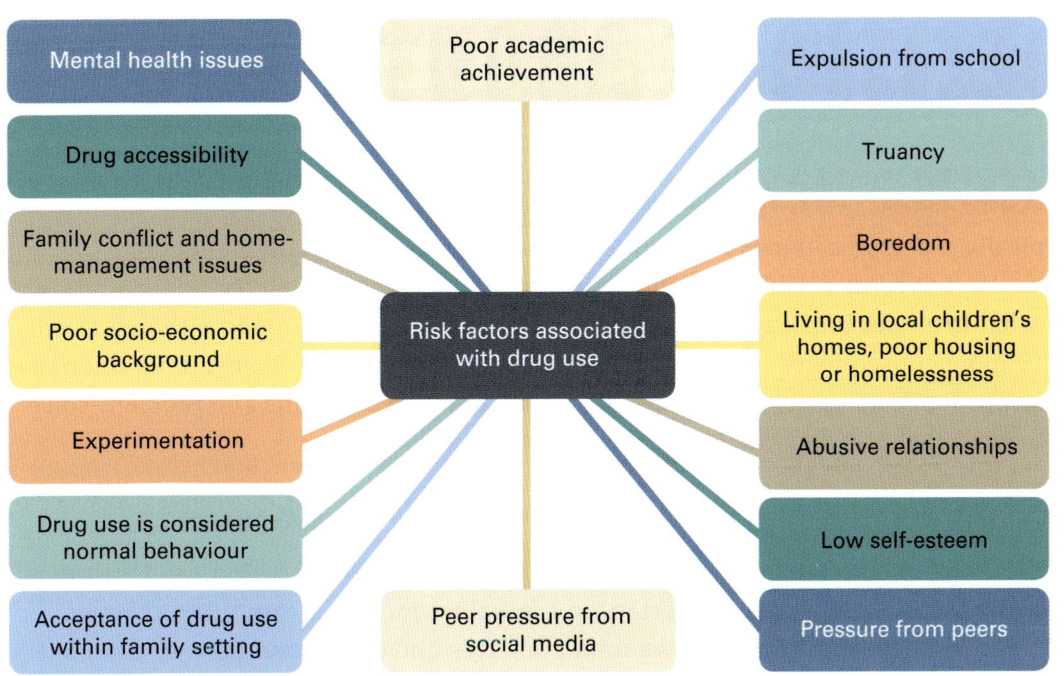

Figure 1.38 Risk factors that can increase likelihood of becoming drug dependent

SOCIAL MEDIA INFLUENCES ON DRUG USE

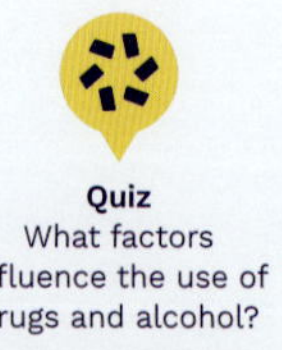

Quiz
What factors influence the use of drugs and alcohol?

Do you like to use social media and chat to friends online? SMS has fallen out of fashion in recent years, in favour of apps that allow friends to connect, such as Snapchat, WhatsApp, Tumblr, Kik, WeChat, Viber, GroupMe, Jott, Tango, Instagram and Facebook.

When friends post photos of themselves drinking and partying on social media, this can lead others to feel like this behaviour is socially acceptable and even desirable. The desire to fit in can lead to experimentation, which ultimately can lead to dependence.

1 There are many factors that can influence the use of drugs and alcohol. Under each of the following headings, identify the factors that may contribute to the use of drugs and alcohol:
- Personal
- Social
- Environmental

1 Ninety per cent of all addictions start when people are in their teenage years. Certain risk factors can increase a person's chances of taking drugs. A risk factor for one person may not be the same for another. Review the list of risk factors associated with drug use and identify the five risk factors that you think are the most likely causes of drug use. In pairs, discuss your lists and identify the similarities and differences.

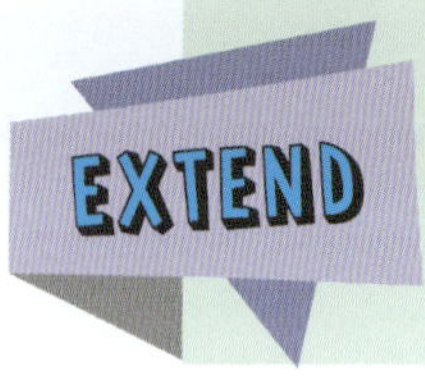

1 Review the three scenarios detailed in this section (Ben, Kaitlin and Patrick). What advice or help would you give each of these teens if you were friends with them?

HOW CAN I MAKE INFORMED DECISIONS ABOUT DRUGS?

DECISION-MAKING AND PROBLEM-SOLVING

Typically, a person can make around 2000 decisions a day. This might sound like a lot, but most of these decisions are very minor, such as which clothes to wear or what to eat for breakfast. More complex decisions may require more thought and have more consequences linked to them. If these decisions are rushed or emotionally driven, poor decisions can be made. The ability to make decisions and problem-solve are useful tools, as they can help us to make smart choices.

USING VALUES AND EMOTIONS TO MAKE DECISIONS

Values are fundamental behaviours and beliefs that inform our actions and attitudes. They help us decide what is important in life. Values mould us into the person we want to be; they determine how we treat ourselves and others and how we interact in our communities. Values are principles that help us determine what is right and what is wrong. They describe our personal qualities that enable us to guide our behaviour. When we make decisions, we consider our values first.

Emotions are strong **feelings** usually accompanied by specific physical changes including an increased heart or respiratory rate, shaking or crying. Emotions can be triggered by the people you are with or a situation you may find yourself in.

emotions strong feelings usually accompanied by specific physical changes

feelings a reaction or emotion such as anger or sadness

FACE TO FACE Values and decision-making

In pairs discuss the following questions:

1 Create a list of values that are important to you.
2 Why it is important to consider your values when making decisions?
3 What factors influence your emotions?
4 Read the following two scenarios and make a decision about what actions you would take if you were involved in the scenario. Consider your values and emotions:

Scenario	Identify the values that would be important in making a decision	Identify the factors that would influence your emotions
You are enjoying the day at the beach with friends. Your friend's brother asks if you want a beer.		
You are riding your dirt bike with your best friend. You stop for a break and she offers you a cigarette.		

UP AND MOVING Values influencing decision-making

As a class, consider the following statements and decide whether you agree or disagree with the following statements:

- Adolescents should encourage parents to quit smoking.
- Cannabis should be made legal.
- The legal drinking age should be increased from 18 years to 21 years.
- Schools should provide pain relief medication for those suffering from headaches at school.
- The legal age to drive should be 18 years.
- Passengers should be allowed to smoke on planes.
- Parents are to blame when adolescents abuse alcohol and other drugs.
- Prescription and over-the-counter medication is not harmful.

1 Discuss the importance of personal **values** when making decisions about alcohol- and drug-related situations.
2 Your personal values and beliefs may change as you get older. Identify factors that may contribute to this change.

values fundamental behaviours and beliefs that inform our actions and attitudes

SEVEN STEPS TO EFFECTIVE DECISION-MAKING

A step-by-step decision-making process can help you solve problems by considering all the relevant information and related consequences. This decision-making tool can help you choose the most appropriate path to take.

1

Identify the decision or define the problem

Clearly define the nature of the decision that must be made.

For example: You have just been watching the AFL Grand Final with friends. Everyone was ready for a great game. Unfortunately, the game didn't go well, and your team was losing considerably by half time. You are all disappointed and restless. Your friend's older brother has been sitting with two cartons of beer and a pile of snacks. Your friend suggests, 'Let's have a beer to commiserate.'

2

Identify any alternatives or options that could help you with the decision

Consider some alternative options and jot them all down.

For example: Suggest watching a movie or decide to head home; Suggest having one beer; You're sure one beer won't hurt.

3

Consider the pros and cons for each option listed in Step 2

Weigh up the evidence and consider the advantages and disadvantages for each option.

For example: Suggest having one beer. You're sure one beer won't hurt.

Advantage – It will be fun

Disadvantage – What if my parents find out?

4

Rank the advantages and disadvantages

Rank the advantages and disadvantages out of 10 in order of importance, with 10/10 being the most important and 1/10 being the least important.

For example: *It will be fun* 5/10

What if my parents find out? 9/10

5

After considering the best alternative, make a decision

Once all the alternatives as well as the advantages and disadvantages have been weighed up, select the best alternative for you.

For example: You decide that the game is not worth watching and you don't want to end up in a situation that you may later regret. You head home instead.

6

Take action

This is when you will implement the solution to your problem.

For example: You politely decline the beer and head home.

7

Review and evaluate your decision

Review and evaluate the decision made and its consequences and learn from your experience.

For example: Reflect on the decision you made and consider how else you could respond if you find yourself in a similar situation in the future.

Figure 1.39 The seven steps to effective decision-making

9780170463096

DEVELOPING SENSITIVITY AND EMPATHY

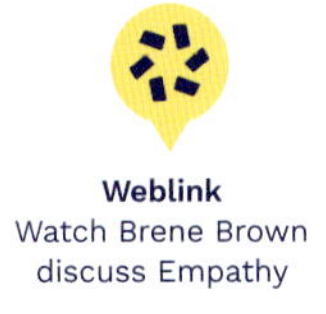

Weblink
Watch Brene Brown discuss Empathy

The term 'empathy' refers to being able to sense another person's emotions and feelings coupled with the ability to visualise what someone else may be thinking or feeling. Empathy and sensitivity are important skills to add to our toolkits. They help us to solve problems, consider things from another person's point of view (or perspective) as well as cope with our own and others' emotions while avoiding disagreements. These skills will help you.

FACE TO FACE

1 To help you develop sensitivity and empathy, in pairs, read the following scenario and choose the most appropriate option to respond with.

Scenario – Your brother comes home from school and starts telling you about his mate, Faf, who seems to be really cranky recently.

Which response would you choose?

Option A: 'Well, you're always cranky… so what?'

Option B: 'I'd say just leave her alone for a while, he'll get over it.'

Option C: ASK QUESTIONS.

'Right, well tell me how Faf has been acting'.

'What do you think has been going on with him lately.'

'What do you think could make him act this way?'

'Is there anything you can do to help him?'

'If you were Faf, what would you like others to do?

2 In pairs, create a drug-related scenario and provide three options. Develop a 'most appropriate option' that demonstrates empathy and sensitivity. Once complete, hand your scenario to the next pair to solve.

ASSERTIVE BEHAVIOURS

Being assertive means being able to communicate thoughts and feelings freely and confidently while also considering the thoughts and feelings of others. When communicating with others, you will have a choice of three types of communication styles.

Passive communication

Passive communicators are often hesitant and nervous. They tend to place others' needs before their own. They tend to allow others to decide the outcome of a situation, are regularly indecisive and lack self-respect. Assumption and silence are key indicators of a passive communicator.

Assertive communication

Assertive communicators are able to freely express opinions, feelings and thoughts in a positive, open, honest and respectful manner, without hurting others. Assertive communicators are confident in the way they speak and are receptive to the needs of others.

Aggressive communication

Aggressive communicators tend to express their thoughts and feelings in a defensive, demanding and hostile manner, often at the expense of others. While this dominant behaviour may result in short-term gains, relationships with others will suffer in the long term.

These styles are all situation specific. While assertive communication seems like the most appropriate and healthiest method, it may not always be the best choice. Some situations will require a more passive or aggressive approach. Being assertive is often seen as the balance point between passive and aggressive behaviour. It's important to become familiar with all three methods of communication so you can choose the most effective option for each situation and build respectful, lasting relationships.

FACE TO FACE Assertiveness

Complete the following think-pair-share activity. In pairs, discuss the following questions:

1 Why don't people always communicate in an appropriate manner?

2 What is your own personal 'pet peeve' when communicating with others? What really irritates you?

3 Do you have any 'bad' communication habits you would like to break?

4 How can you tell when you are communicating well with others?

5 Make a list of situations in which you would like to be more assertive.

6 Think of a situation where you wish you had been more assertive. Write a statement you could use in this situation.

Shutterstock.com/Justek16

Figure 1.40 You have the right to say no to anything that makes you uncomfortable.

Your rights

As an individual you have many rights. You have the right to:

- make your own decisions
- express your own thoughts and feelings
- ask for what you want
- say 'no' without feeling guilty
- be treated with respect
- maintain self-control
- listen to the views of others (whether you agree or disagree)
- change your mind
- take reasonable risks
- make mistakes as well as apologise
- choose not to be assertive
- identify your needs
- take time to stop and think.

Being assertive does not necessarily mean winning an argument. It does, however, mean you have the responsibility to express your thoughts and feelings in an appropriate manner.

FACE TO FACE Your rights

In pairs, discuss the following questions:

1 How many of the above rights do you feel you currently have?
2 Identify which of these rights are the most difficult to carry out.

Steps to being assertive

1 Keep to the point; avoid lengthy explanations.
2 Maintain eye contact with the person you are talking to.
3 Remain calm and avoid anger.
4 Be polite, yet firm.

HARM MINIMISATION

Harm minimisation focuses on ways of reducing the harmful effects of alcohol and drugs on individuals, families and communities.

In order to minimise the harmful effects of alcohol and other drugs, it is important to understand these facts:

- Alcohol and drugs are prevalent in society.
- It is impossible to permanently remove drugs from society.
- The removal of drugs may increase the risk of harm to society.

Harm minimisation aims to improve the health, social and economic situations for individuals and the communities in which they live. There are many strategies used to inform people about the risks associated with alcohol and drug use. In Australia, the federal and state governments have adopted a number of strategies aimed at addressing alcohol and drug-related issues.

Shock tactics

Over the years, the Australian Government has run many confronting campaigns designed to shock people into breaking habits that may have a deadly effect on their health. These campaigns have highlighted the effects of smoking, drink- or drug-driving and even binge drinking on individuals, families and their communities.

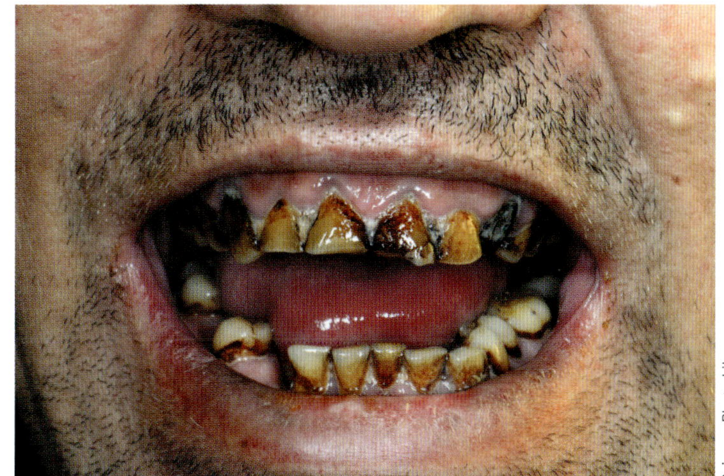

Science Photo Library

Figure 1.41 Graphic photos are used in government campaigns to stop people smoking.

PEER INFLUENCE

Deciding whether to take drugs is an individual choice. Dealing with life's pressures can be difficult at times, but seeking help and advice from friends and family can help you make the right decision and feel in control of your choices. While friends can have a positive influence, some may encourage you to do something you wouldn't normally do. This is known as 'peer influence' or 'peer pressure'. Don't be afraid to say no. A true friend will respect your

Shutterstock.com/Monkey Business Images

Figure 1.42 Peer groups can put pressure on individuals.

decisions, even if they choose to act differently.

Making informed decisions is a key factor in minimising harm and reducing associated risks. Before you make a decision, consider the consequences. It's important to remember that it's not all about you. Consider the impact of your decision on your friends and family, too. Anyone can reduce risky behaviour by adopting a safe attitude. A safe attitude can help you focus on your own safety and the health and wellbeing of those around you. Don't tell yourself 'it won't happen to me'; it's important to accept responsibility for your own health and safety.

Awareness of blood borne viruses (BBVs)

A blood borne virus (BBV) is a virus that is carried in the blood or in other bodily fluids including vaginal fluid, semen and breast milk. A BBV is passed on from one person to another via blood-to-blood contact. Risky behaviours such as unprotected sex or sharing injection needles associated with alcohol and other drug use are frequent causes of BBV transmission. The three most common BBVs include:

- HIV (human immunodeficiency virus)
- hepatitis B
- hepatitis C.

All too often people with a BBV may not even be aware that they are carrying a virus so it is important to consider ways to reduce your chances of contracting a BBV. COVID-19, a recently discovered coronavirus causing serious illness and death globally, is currently not considered a BBV. However, in order to reduce the risk of both COVID-19 and BBV transmission, it is recommended that you practise good personal hygiene and wash hands thoroughly.

Where can you go to seek help?

When you're a teenager, it may seem as though other teenagers are the only people who can really understand you. Talking to a friend may be easier than talking to an adult. However, it is important to speak to someone you can trust. It might be your favourite teacher, a close friend or a member of your family.

There are also many places where you can go to get help. Medical professionals and school counsellors can offer advice and can even refer you to a community drug program appropriate for your needs. If you do not want to talk to anyone at school, don't be afraid to pick up the phone and call a helpline such as Lifeline (13 11 14 or text: 0477 13 11 14), Kids Helpline (1800 55 1800) or Beyond Blue (1300 22 4636). You can also check them out online.

There are a number of alcohol and drug information services across Australia. The Australian Alcohol and Drug Foundation is committed to minimising drug- and alcohol-related problems in Australian communities. Its website also provides an in-depth list of support services for those with drug-related problems.

Figure 1.43a Lifeline can provide support if you or a friend need help.

Weblink
The Alcohol and Drug Foundation

Figure 1.43b Kids Helpline is Australia's only free (even from a mobile), confidential 24/7 online and phone counselling service for young people aged 5 to 25. You can reach the Kids Helpline phone counselling service on 1800 55 1800.

Weblink
Kids Helpline

ASKING FOR HELP

This activity has been developed in collaboration with Dr David Bakker of MoodMission

Identify

Sometimes it's hard to ask for help, but it can actually be the most helpful thing you can do.

Understand

Humans are social animals. We need connections to other people to feel okay. However, lots of things can get in the way of this, such as not trusting others, fear of burdening people with our problems, or feeling like there's no actual way others can help. When these barriers crop up, it's important to remind ourselves of the power of help – both giving and receiving.

Practise

1 Think of a problem you're currently facing. Maybe you're feeling stressed about schoolwork, or are having trouble with friends.
2 Now think of one thing you could ask for that would help with the problem. It might not be something specific – it could just be the opportunity to talk to someone, and get your worries off your chest.
3 Who could help you? It might be a friend or family member, for example.
4 Now that you have identified a problem, something you could ask for, and the person you could ask to help, make a plan to do it! Will you send them a message, or talk to them face to face?

Reflect

There are often lots of barriers that prevent us from asking for help. Sometimes we're afraid we'll be a 'burden', or that we'll be rejected, or that no one else can help. But it's always worth trying. You might be surprised at how much people want to help.

BASIC FIRST AID

Would you be able to assist a friend if they needed first aid? If you think your friend has taken something they shouldn't have and is having a bad reaction, the following tips may help:

- If you're not sure how serious the problem is, **always call 000**. It is better to get immediate medical assistance than to delay. Never avoid calling an ambulance because you don't want to get the police involved or you don't want to get into trouble. It's not about you, it's about your friend. Any delay could have severe consequences. Monitor your friend carefully while you wait for help to arrive.
- If your friend is unconscious but breathing, place them on their side, in the recovery position, making sure their head is tilted back so they can breathe and avoid choking on any vomit.
- If your friend has stopped breathing, start CPR (cardiopulmonary resuscitation). Don't panic! If you call 000, the operator will talk you through this process while the ambulance is on its way. You could save your friend's life.

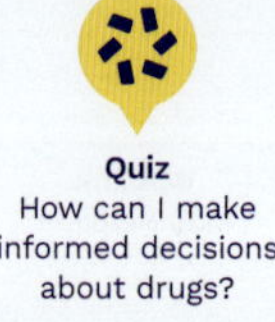

Quiz
How can I make informed decisions about drugs?

1 Identify the seven steps for effective decision-making.
2 Describe three forms of communication style.
3 Differentiate between a 'value' and an 'emotion'.
4 Recall three support services you could contact if you need help.

1 Read the following four role-play situations with a partner. Provide the 'opening line' for your partner to respond to. The opening lines are written in brackets after each situation.
 a Responding to someone who has asked if you want to buy some cannabis.
 ('Hey, it's super high quality, and cheap! You know you want it!')
 b Responding to a friend who wants to borrow money to buy some alcohol for a party that night.
 ('Come on, loan me $50. You owe me a favour anyway!')
 c Confronting a friend who has been self-administering prescription pain killers for a painful knee.
 ('Give me a break, I'm playing in the basketball semi-finals tomorrow night and I need to get this knee in shape!')
 d Deciding whether to call an ambulance for a friend who has collapsed after binge drinking.
 ('No! We can't call an ambulance. My dad is going to be so mad at me if he finds out I've been drinking!')
2 A powerful, yet subtle, anti-smoking campaign highlights the link between smoking and the risk of stroke. Visit the Quit website and watch the clips in the Smokes Lead to Strokes section. This particular campaign uses interviews with survivors of strokes. Other campaigns may use shock tactics instead, although many people say shock tactics don't work.
 In pairs, discuss the following questions:
 a What is a stroke and how can smoking increase your risk of suffering a stroke? How did you feel after watching the interviews?

Weblink
Quit: Smokes lead to strokes

b Do you think this is an effective campaign or do you think shock tactics would work better?

c Are shock tactics a good way to inform young people about the dangers of drug use?

1 Research two online drug and alcohol support services available in your local community. Use the information you find to justify which one is your preferred service. Consider criteria such as accessibility, needs and cost.

HOW WILL DRUG USE AFFECT MY PERFORMANCE IN SPORT?

Worksheet 1.22

Worksheet 1.23

ergogenic aids any substance or factor that may improve sporting performance

performance-enhancing substances substances taken by athletes to improve sporting performance

More than 6 billion people around the world play sport regularly. Many people play sport to stay healthy, socialise and relax, as well as to have fun. For others, sport is a business, and some athletes will risk anything in order to win. **Ergogenic aids** are substances used to improve performance and recovery times. These can include both legal and illegal **performance-enhancing substances**. Testing for illegal performance-enhancing drugs has improved over the years, so to avoid the risk of a lifetime ban from the sport, coaches are turning to legal alternatives.

Athletes put themselves under enormous pressure to be the best, with most expecting to make a lot of money if they are successful. The overwhelming desire for fame and fortune can cause some athletes to make the wrong decision. Many athletes' careers are relatively short, so they need to reach peak performance quickly. This can motivate some to seek alternative aids to help them to succeed.

There are a number of reasons why athletes choose to use both legal and illegal ergogenic aids. These include peer pressure and the pressure from other athletes to excel. Athletes use performance-enhancing substances for many reasons, including to:

- improve performance and 'win'
- control appetite and lose weight
- manage an injury
- improve physical appearance
- improve recovery rate
- build confidence
- enhance self-esteem.

FAST FACT

If one athlete on an Olympic team (e.g. 4 × 100 metre relay) is found guilty of using performance-enhancing drugs, the entire team may be disqualified and forced to return any medals they may have won.

Source: Did You Know About Drug Use in Sports? © 2020 ProCon.org

CASE STUDY

SUN YANG BANNED FOR EIGHT YEARS FOR BREAKING ANTI-DOPING RULES

Identify

Chinese swimming star Sun Yang has been banned from swimming until February 2028 for refusing to cooperate with sample collectors. The ban means Sun will be unable to defend his 200-metre freestyle title at the Tokyo 2020 Olympic Games.

Understand

Chinese Olympic gold medallist Sun Yang has been banned from swimming for eight years for breaking anti-doping rules, but the swimmer has indicated he will appeal the Court of Arbitration for Sport's (CAS) ruling.

The CAS found the three-time Olympic champion guilty of refusing to cooperate with sample collectors during a visit to his home in September 2018 that turned confrontational.

In a rare hearing in open court in November, evidence was presented of how a security guard instructed by Sun's mother used a hammer to smash the casing around a vial of Sun's blood.

Figure 1.44 Mack Horton refused to stand on the podium with Chinese Swimmer Sun Yang.

FINA's doping panel had cleared Sun of any wrongdoing in the incident but the World Anti-Doping Agency (WADA) appealed that decision.

Discuss

1 What do you think is the most appropriate punishment for someone who is caught refusing to provide blood and urine samples for drug testing?

2 In 2019, prior to Sun's ban, Australian swimmer Mack Horton refused to stand on the podium with Sun Yang, after he finished in second place in the 400-m final at the world championships. While Sun Yang had not officially been caught using performance enhancing drugs, there were rumours that he was cheating. Mack Horton staged a silent protest after the presentation ceremony and called Sun Yang a drug cheat.

Do you agree with Mack Horton's decision to not take to the podium? Provide reasons for your decision.

Source: Sun Yang banned for eight years for breaking anti-doping rules', AP/ABC, https://www.abc.net.au/news/2020-02-28/sun-yang-banned-from-swimming-for-eight-years/12012900'

LEGAL PERFORMANCE-ENHANCING DRUGS IN SPORT

There are many different ways athletes can legally enhance their performance. Some legal substances and practices include:

- bicarbonate of soda
- sports drinks

- caffeine
- creatine supplements
- sports gels
- carbohydrate loading.

It is not necessary to 'cheat' to enhance sporting performance. Many of these legal performance-enhancing products can be found on supermarket shelves. However, each legal ergogenic aid has its own advantages and disadvantages, as seen in Table 1.6.

Ergogenic aids used to enhance performance can be mechanical (practical), pharmacological (drugs), nutritional (diet), physiological and psychological (mind).

dLibrary

Figure 1.45 Some of the supplements readily available to athletes to enhance sporting performance

Table 1.6 Advantages and disadvantages of legal ergogenic aids

Drug	Advantages	Disadvantages	Sports/athletes most associated with its use
Sodium bicarbonate	Creates a buffer against the build-up of lactate in the muscle, delaying muscle fatigue	Vomiting Stomach problems Diarrhoea	Sprint cycling Rowing
Sports drinks (e.g. Powerade/ Gatorade)	Rehydration Immediate source of energy	No known problems associated with the consumption of sports drinks	All sporting activities
Caffeine (coffee/energy drinks)	Increases alertness, reaction times and arousal levels	Increased urine production Irritability Lack of sleep	Tennis Volleyball
Creatine supplements	Increased training volume and decreased recovery time	Stomach problems Muscle cramping Increased water retention	Sprinters Javelin/shot-put
Sports gels	Concentrated form of carbohydrates for energy production Easy to carry	Stomach problems Psychological dependency	Endurance athletes
Carbohydrate loading (e.g. pasta/rice)	Diet of starchy foods designed to increase carbohydrate reserves in muscles	Increases water absorption, leading to weight gain	Endurance athletes

INVESTIGATION  ERGOGENIC AIDS

Purpose

To investigate the advantages and disadvantages of five categories of legal ergogenic aids (mechanical, pharmacological, nutritional, physiological and psychological) and their effect on an athlete's health and performance in sport.

Method

1 Using the internet, research the five categories of ergogenic aids:
- Mechanical
- Pharmacological
- Nutritional
- Physiological
- Psychological

2 Identify one ergogenic aid of interest from each category and explain one advantage and one disadvantage of each aid.

3 Create a diagram of each aid.

Discussion

1 How can the use of ergogenic aids affect sporting performance?

2 Why might an athlete choose to use ergogenic aids to improve performance? Discuss possible motivating factors.

3 Would these motivating factors vary from sport to sport?

4 Consider your own sporting performance. Do you currently use any products to enhance your own sporting performance?

HOW DOES THE MEDIA INFLUENCE DRUG USE?

body image the way a person feels about their own body

Have you ever tried to model yourself on your favourite sportsperson? The media plays an important role in creating what is considered to be the 'desirable' **body image** for men and women. Television, the internet, magazines, movies and newspapers are all influential in creating perceptions of the 'ideal' man or woman. Many famous athletes have used their bodies to advertise well-known products. Can you think of any athletes and the products they advertise?

Unfortunately, many people resort to exercise and dieting in order to transform their body into what they perceive to be the ideal. Eating disorders are very common among female athletes, especially those in competitive sports. Some sports, such as gymnastics, trampoline, beach volleyball and diving, require perfect body presentation. Unfortunately, some athletes resort to taking appetite suppressants to keep their weight under control.

Appetite suppressants

Appetite suppressants (or diet pills) are readily available in Australia from a doctor or pharmacy. As their name suggests, these drugs are used to reduce feelings of hunger, but their dangers far outweigh the benefits of any potential weight loss.

1

FAST FACT

The dangers of appetite suppressants include:

- increased risk of heart attack or stroke. Diet pills are stimulants, known to increase the risk of heart-related problems or blood clotting
- dependency. Diet pills often contain a cocktail of highly addictive drugs, including amphetamines, anti-anxiety drugs and anti-depressants
- multiple side effects, including constipation, headaches, stomach upsets and mood swings
- misleading claims. All too often, claims that diet pills promote weight loss are misleading. Many diet pills contain a combination of caffeine, a stimulant, and a diuretic, which promotes fluid loss. You may seem slimmer on the scales, but this is the result of water loss, not fat loss.

HEALTHY PERFORMANCE-ENHANCING STRATEGIES

Here are a few tips to help you improve your performance and give you that 'winning edge' without using performance-enhancing supplements.

- Eat a nutritious, balanced diet, including carbohydrates, protein, fruit and vegetables.
- Include protein for recovery. Protein is needed for growth and muscle repair. Sources of protein include fish, red meat, chicken and beans.
- Include aerobic exercise, such as running, walking, swimming or any other active sport as part of your fitness routine.
- Avoid power lifting. Lifting weights at a young age can cause problems in your development. Wait until you are at least 16 years old before commencing a weight-training program.

Shutterstock.com/Doug James

Figure 1.46 Aerobic exercise should be a key part of your fitness routine.

An athlete training to compete in a marathon should consume a well-balanced diet with plenty of carbohydrates, fresh fruit and vegetables, lean protein and healthy fats.

Getty Images/Clive Brunskill

Figure 1.47 Sinead Diver finished 10th at the Tokyo Olympics in 2021, finishing 4 minutes behind the winner, Kenyan Peres Jepchirchir.

1. Define the term 'ergogenic aid'.
2. Identify three legal ergogenic aids and outline how they assist sporting performance.

1. Discuss how athletes justify using illegal performance enhancing drugs
2. Review Table 1.6 and recommend a legal ergogenic aid for each of the following scenarios:
   - **a** A tennis player looking to increase their energy levels during a game
   - **b** A marathon runner on the day of a race
   - **c** A rower looking to improve in the lead-up to a race

1. Use the internet to research an athlete who has taken illegal ergogenic aids and been exposed as a drug cheat. Answer the following questions:
   - **a** Who was the athlete and what was their chosen sport?
   - **b** What illegal substance did the athlete take and what category of ergogenic aid does this substance fall under?
   - **c** How was the substance detected?
   - **d** What was the outcome of the detection?

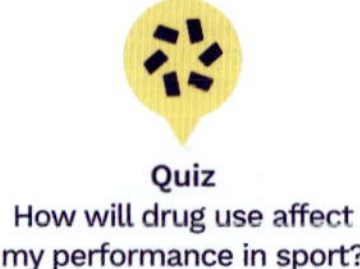

Quiz
How will drug use affect my performance in sport?

CHAPTER 1 REVIEW

1. What is the meaning of the term 'drug use'?
2. What is the difference between over-the-counter drugs and prescription medication?
3. Why is it dangerous to take a prescription drug for any reason other than its intended purpose?
4. What is meant by the term 'bush medicine'?
5. Explain the three main classifications of drugs.
6. Name three short-term and three long-term effects of alcohol on the body.
7. How many grams of alcohol are in a standard drink, and why is it important to be aware of the alcoholic content of a standard drink?
8. Name three short-term and three long-term effects of smoking on the body.
9. Why is vaping considered to be harmful to your health?
10. Alcohol is considered to be the most harmful legal drug available. Explain five reasons why this is the case.
11. Energy drinks can be potentially dangerous when consumed in large quantities. Identify the main drug found in these drinks and list five common side effects of the over-consumption of energy drinks.
12. Discuss the dangers of ecstasy use.
13. Explain the social factors that could influence a person's decision to use drugs.
14. In Australia, the harm caused by drug use in communities can be very costly. In what ways does drug use affect Australian communities?
15. Discuss the meaning of being assertive. Do you think someone who lacks confidence or is shy can still be assertive? Explain your answer.
16. The Australian Government has released many confronting campaigns designed to 'shock' people into breaking a habit that may have a deadly effect on their health. Identify one such campaign and explain how it aims to raise awareness among the Australian population.
17. You find your friend on the floor, unconscious but breathing. What would you do?
18. There are numerous ways of enhancing sporting performance without resorting to banned substances. Discuss the various legal aids that can be used to enhance performance.

EAT WELL, LIVE WELL

IN THIS CHAPTER

You will learn about the importance of good nutrition at all stages of your life.

By the end of this chapter, you should be able to:

- understand energy and nutrition requirements for healthy living and performance
- consider and manage factors that influence eating habits
- make informed and justifiable decisions about eating
- analyse eating habits and propose strategies for improvement
- demonstrate an understanding of the trends and consequences of eating habits in Australia and the cultural and contextual factors that shape these trends.

Shutterstock.com/Nicetoseeya

HOW CAN I EAT A HEALTHY, BALANCED DIET?

Quiz
Pre-chapter

Before you start, take the pre-chapter quiz to find out how much you already know.

diet the usual food and drink consumed by a person; not to be confused with 'dieting', which is the practice of eating food according to a regulated or restricted system to cause a change in body weight

When considering **diet**, nutrition, food choices and eating habits, it is important to find the most appropriate and relevant information, from the most trusted sources available. This isn't always easy, as information about diet can come from many different sources, and not all of these sources have your health interests at heart.

Shutterstock.com/Robyn Mackenzie

Worksheet 2.1

Figure 2.1 Examples of healthy foods from different food groups

HOW TO STAY INFORMED ABOUT NUTRITION

How do you find nutritional information you can trust? You can usually rely on well-researched, credible information from government or educational organisations such as the Department of Health or the National Health and Medical Research Council (NHMRC). Organisations such as these are the leading experts in health research, promotion and standards and, perhaps most importantly, they're not selling anything.

Fortunately, there is a lot of clear dietary and nutritional information, advice and support available to the Australian community. That information is the focus of this chapter.

FACE TO FACE Nutrition

1 What does 'nutrition' mean to you? Provide a definition based on your own understanding.
2 Compare and contrast your definition with your classmates. You may notice some similarities and differences between your definitions. Write down two differences that you noticed between your definition of nutrition and your classmates' definitions.
3 Explain how people develop different ideas about what nutrition means. What influences people to have these different ideas?
4 Compare your definition of nutrition to the dictionary definition (you may need your teacher to help with this, or perhaps search online). Evaluate how similar your definition was to the one in the dictionary.

Nutrition is the process of obtaining the food necessary for health and growth. There are two documents that all Australians should use to guide their daily food and drink choices: the Australian Guide to Healthy Eating and the Australian Dietary Guidelines. There are five guidelines in the Australian Dietary Guidelines, but this chapter will focus only on the first three. The fourth and fifth guidelines refer to breastfeeding infants and the preparation and storage of food.

Video
Australians' diets: What do we mean by 'healthy eating'? What changes could we make to our diets to be healthier? Watch the video and join the discussion!

THE AUSTRALIAN GUIDE TO HEALTHY EATING

The Australian Guide to Healthy Eating separates food into five main groups:

- grain foods
- vegetable and **legumes**/beans
- milk, yoghurt and cheese
- fruit
- meat, fish, eggs and nuts.

legume seed or pod that is eaten in either the green immature form (e.g. peas and beans) or the mature form as dried peas, beans, lentils and chickpeas

FACE TO FACE What do you already know?

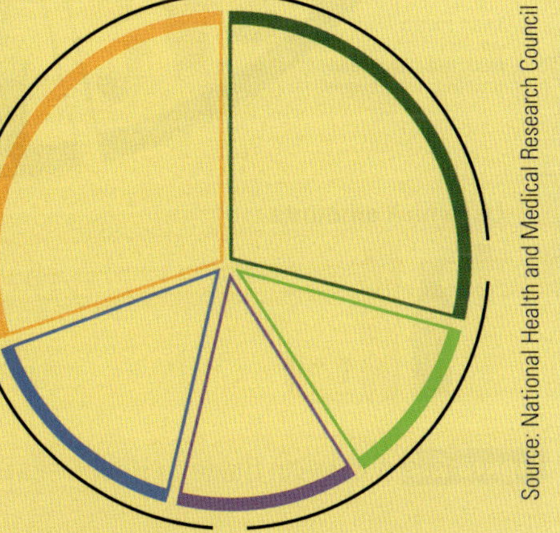

Source: National Health and Medical Research Council

Before looking at the Australian Guide to Healthy Eating, consider what you already know about food groups. Using the five groups listed, try to match them to the five segments in this diagram. You can create your own in your exercise book or access a blank template online in Nelson MindTap. Identify how the segments differ in size. This represents how much of each food group should be eaten – the larger segments represent the food groups you should eat more of. Determine which food groups you think are the most important part of a healthy diet, and assign them to the largest segments. Once you have matched all five groups, check with a classmate and compare and contrast for similarities and differences.

Worksheet 2.1

Scaffold
Australian Guide to Healthy Eating: What do you already know?

After completing the activity, examine the actual Australian Guide to Healthy Eating and judge how close you came to matching the food groups to the correct segments. Correct your answers if you need to.

Figure 2.2 Australian Guide to Healthy Eating

There are several other pieces of information included in the Australian Guide to Healthy Eating that will help guide you in your daily food choices. You already know that the food groups in the larger segments, such as the vegetable and legumes and the grains food groups, should make up more of your daily food intake than the smaller food groups, such as fruit, milk, yoghurt and cheese, and meat, fish, eggs and nuts. What else can you infer from the diagram? Consider these points:

- The examples of foods in each segment are healthy choices, e.g. lean meats, wholegrain breads, high-fibre cereals.
- There is a large variety of foods in each food group. Eating a wide range of vegetables provides a better balance of nutrients than eating the same few vegetables all the time. This also applies to other food groups, such as meats, grains, fruit and dairy products.
- Plenty of water should be consumed.
- Oils are included, but should only be consumed in small amounts.
- **Discretionary foods**, which are those high in sugar and fat, and highly processed foods and drinks, should be consumed in small amounts, and only occasionally.

discretionary foods foods and drinks that do not fit into the five food groups because they are not necessary for a healthy diet. These foods can be too high in saturated fat and/or added sugars, added salt or alcohol, low in fibre and contain too many kilojoules (energy). Many tend to have low levels of essential nutrients, so are often referred to as 'energy-dense' but 'nutrient-poor' foods.

FACE TO FACE Monitor your intake

How well do your daily eating choices match up with the recommendations made by the Australian Guide to Healthy Eating? Using the blank template, identify all the food and drinks you consumed yesterday and represent them with a drawing or image from the web. You can draw your own in your exercise book or access a blank template online in Nelson MindTap. Make sure you represent all ingredients in each meal, and categorise them into the correct location. For example, if you had a chicken and salad sandwich for lunch, depending on the exact ingredients, you might need to illustrate chicken, margarine, bread and the various salad ingredients. Draw in the approximate amount of glasses of water (or other drinks) consumed, as well as any discretionary foods.

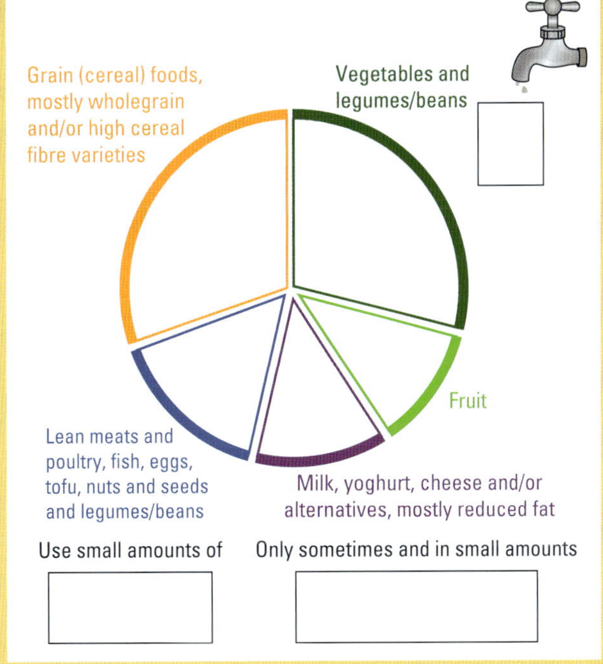

Source: Based on material provided by the National Health and Medical Research Council

Figure 2.3 Australian Guide to Healthy Eating template

1 Critique your finished diagram and determine if you have the right proportions of foods in each segment. You should have more drawings in the grains and vegetables segments than any of the others.

Identify any segments that you think had too many items in them, and also note any segments that didn't have enough.

Segments with too many food items	Segments with too few food items

Scaffold Australian Guide to Healthy Eating: Yesterday I ate ...

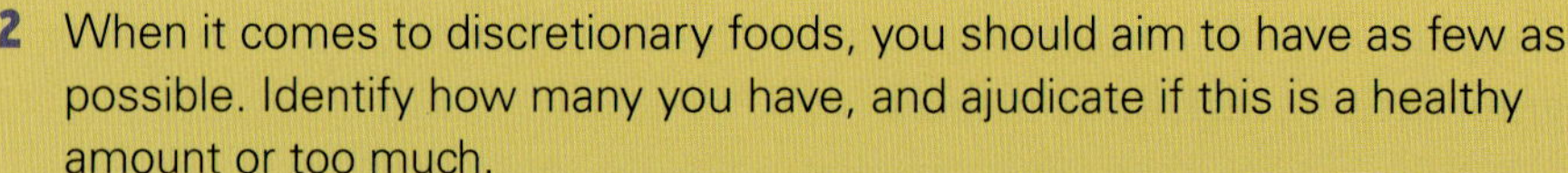

2 When it comes to discretionary foods, you should aim to have as few as possible. Identify how many you have, and ajudicate if this is a healthy amount or too much.

3 Analyse the variety of foods you included in your diagram. Appraise the range of different meats, vegetables, fruits and sources of dairy, grains and cereals you have. More is better! Speculate how you could try to include more variety in your daily diet.

Worksheet 2.3

EATING ISN'T JUST ABOUT ENERGY

Biologically, people really only need to eat and drink for two reasons:

1 To obtain the energy needed to survive, grow and function properly.

2 To obtain the nutrients needed to survive, grow and function properly.

satiety the feeling of being well fed, full and gratified, after a satisfying meal

There are other, more complex reasons why people eat, such as **satiety** or satisfaction, cultural or religious occasions, emotional fulfilment or as a part of socialising. We will explore some of these later in the chapter.

The foods you eat may contain large or small amounts of energy, as well as large or small amounts of various nutrients. Your body needs the following nutrients on a regular basis:

- a range of vitamins
- a range of minerals
- fibre
- water
- protein
- fats
- carbohydrates.

Your top priority should always be providing your body with what it needs to function at its best. Unfortunately, many people believe this means consuming large amounts of energy-rich food and drinks. This belief is partly due to the marketing of high-energy food and drink, which encourages people to include more of these products in their diet. This energy-driven focus usually comes at the expense of consuming a complete range of nutrients, such as vitamins, minerals, fibre and water.

VARIETY, BALANCE AND MODERATION

The Australian Guide to Healthy Eating shows the wide range of food that is needed in order to provide all the nutrients the body requires to function properly.

No single food item can provide the body with all of the required nutrients. Most food items provide a good amount of some nutrients, but very little or none of others. That is why it's important to consume as wide a variety of foods as possible to make sure you get all the nutrients you need. Some foods contain very few of the necessary nutrients, but high amounts of 'energy', often in the form of simple sugars. Eating too much or not enough of any food (whether you consider it a 'health food' or 'junk food') can be harmful to your health.

Some nutrients can actually be harmful if excess quantities are consumed, or if the body is not able to store them. This means that you can't just eat large amounts of food rich in a certain nutrient on one day, such as broccoli and oranges for vitamin C, and then not need any more vitamin C for the rest of the week. You need small amounts of each and every nutrient, every day. Perhaps the only exception to this rule is energy, which your body can store for long periods of time.

iStock.com/mediaphotos

Figure 2.4 Energy-dense, nutrient-poor foods can be very tempting.

In terms of variety, balance and **moderation**, the best advice is to eat small quantities of each of a wide range of nutrient-rich foods, every day. Try to avoid energy-dense, nutrient-poor foods. These are usually considered 'discretionary foods', which yield high amounts of energy and few nutrients. The surplus energy provided by these foods is often stored as dangerous body fat, which stops your body from functioning properly and causes a wide range of diseases and health conditions. The high levels of **kilojoules**, saturated fat, added sugars and salt contained in these foods can increase the risk of obesity and chronic disease such as heart disease, stroke, type 2 diabetes and some forms of cancer.

moderation within a limit, not extreme or excessive

kilojoules a measure of the energy in food, used by the scientific community. It is the accepted standard in Australia, and is found on the packaging and labels of food and drink products.

GOOD FOOD MEANS GOOD MOOD

Identify

You are what you eat, and you're not just your body – you're also your mind. What we eat can affect our mood.

This activity has been developed in collaboration with Dr David Bakker of MoodMission

Understand

Unhealthy or sugar-rich foods can mess with the body's energy supply, creating highs and lows in our blood glucose instead of a nice, consistent level. This is why you might get a burst of energy after having something sugary, but then crash soon after. Eating well can help make us feel good by providing our brains with a consistent supply of energy. Just like any machine, if our brain doesn't have enough fuel, it can start to malfunction. The prefrontal cortex is one of the most important parts of our brains for planning, concentrating and exercising self-control. It is also sensitive to blood glucose levels, so if we don't eat healthy foods that give us a steady energy supply, our ability to pay attention and control emotional impulses can quickly deteriorate.

Practise

1 When do you eat? It is important to eat food throughout the day so your body has a constant, reliable source of energy. Identify which of the following meals you usually eat:
   - Breakfast
   - Morning tea
   - Lunch
   - Afternoon tea/snack when you get home from school
   - Dinner
2 Some students do not have breakfast, lunch and dinner either by choice or their personal circumstances. Discuss the effect this might have on their development in terms of growth, immunity, concentration, etc.
3 Don't like breakfast? Maybe it's a case of finding the right sort of food – something that you will enjoy in the morning. Review the options laid out in this chapter and then list five breakfast ideas. Which choices do you think would be the best to give your body and brain the nutrition and energy they need?
4 Skip lunch? Maybe you need to take more food to school so you have a wider variety to choose from, or so you don't eat it all at recess. Identify some nutritional lunch options.
5 Skip dinner or have it alone? Having dinner together with family or friends can be a great way of maintaining regularity in meals and debriefing about your day.
6 One of the main symptoms of anxiety is having an upset stomach. If you find yourself unable to eat because of stomach pain, consider how anxious you are feeling. If you think anxiety is the culprit, try some of the strategies outlined in this book or talk to your parent or caregiver about seeing a doctor and getting help.

Reflect

Propose what you think are the main barriers to you eating healthily and regularly. Is it availability of good foods? The temptation of bad foods? Feeling too busy to take time to eat well? Or maybe anxiety is reducing your appetite? If any of these apply, what's something you could do to overcome this barrier?

HOW MUCH IS ENOUGH?

Every day, from the moment you wake up in the morning until the time you go to bed, you are faced with many decisions around what to eat and drink. Some people even interrupt their sleep to eat or drink! Eating and drinking are things we do several times a day, every single day, so it's no wonder that the decisions we make and, more importantly, the habits that are established can have a dramatic effect on our health (both short term and long term), and influence how well we cope with daily life.

Here are some of the daily decisions we face around eating and drinking:

- what to eat or drink
- how much to eat or drink
- when to eat or drink
- how often to eat or drink.

The Australian Guide to Healthy Eating is clear about the proportion of daily food and drink required from the five main food groups, as well as oils and discretionary foods. But exactly how much food is that?

Food intake is planned and measured using **serves** and **portions**. You will see the words 'serves' or 'serving size' in lots of nutritional advice, including on the labels of packaged food (see page 87). Portions refer to the amount of food you actually eat – for

serve the recommended set amount of a certain food. This should be used along with the 'serves per day' information to work out the total amount of food required each day from each of the five food groups.

portion the amount individuals actually eat, depending on energy requirements and level of hunger

instance, a tall male who is very active will probably eat a larger portion than a small female, even if she is also fairly active. If you choose to eat portions that are smaller than the recommended serving size, you won't be getting the recommended amount of that particular food group, so you may need to eat from that food group more often. If you eat portion sizes that are larger than the serving size, you will need to eat from that food group less often.

Weblink
Go to the Dietitians Association of Australia website and complete the Healthy Eating Quiz.

Standard serving sizes of foods from the five food groups

frozen vegetables
½ cup
½ medium
1 cup
tomatoes
beetroot
½ cup

Vegetables and legumes/beans

1 medium
peaches
1 cup
2 small

Fruit

1 slice
Noodles
brown rice
½ cup cooked
⅔ cup
½ cup cooked

Grain (cereal) foods, mostly wholegrain and/or high cereal fibre varieties

65 g
80 g
tuna
100 g
2 large
1 cup
baked beans

Lean meat and poultry, fish, eggs, tofu, nuts and seeds, and legumes/beans

low fat milk
1 cup
2 slice
¾ cup
soy drink
1 cup

Milk, yoghurt, cheese and/or alternatives, mostly reduced fat

Source: Based on material provided by the National Health and Medical Research Council

Figure 2.5 Standard serving sizes of foods from the five food groups

One of the hardest parts of maintaining a healthy diet is managing the size of the portions we eat. We know that the more food there is on a plate or in a package in front of us, the more we eat.

Worksheet 2.4

Table 2.1 Standard serves

Vegetables (75 g or 100–350 kJ)		Fruit (150 g or 350 kJ)		Grains or cereals (500 kJ)		Meat, eggs, nuts, seeds (500–600 kJ)		Dairy (500–600 kJ)	
½ cup	Cooked green or orange vegetables	1 medium	Apple, banana, orange or pear	1 slice	Bread	65 g	Cooked lean meats	1 cup	Fresh or long-life milk
½ cup	Cooked, dried or canned beans, peas or lentils	2 small	Apricots, kiwi fruits or plums	½ medium	Roll or flat bread	80 g	Cooked lean poultry	½ cup	Evaporated milk
1 cup	Green leafy or raw salad	1 cup	Diced or canned fruit	½ cup	Cooked rice, pasta, barley, noodles, polenta, bulgur or quinoa	100 g	Cooked fish fillets	2 slices	Hard cheese
½ cup	Sweet corn	**Or only occasionally:**		½ cup	Cooked porridge	2 large	Eggs	½ cup	Ricotta
½ medium	Potato	125 mL	Fruit juice	2/3 cup	Wheat cereal flakes	1 cup	Cooked or canned legumes	¾ cup	Yoghurt
½ medium	Tomato	30 g	Dried fruit	¼ cup	Muesli	170 g	Tofu	1 cup	Soy or rice milk
				3	Crispbreads	30 g	Nuts, seeds or peanut butter		
				1	Crumpet				
				1 small	English muffin or scone				

Here is a handy visual guide for different types of food:

'Handy' portion guide
Using visual cues is an easy way to get to know your portion/serve sizes.

Visual cue	Approximate portion size or serve size
Your fist 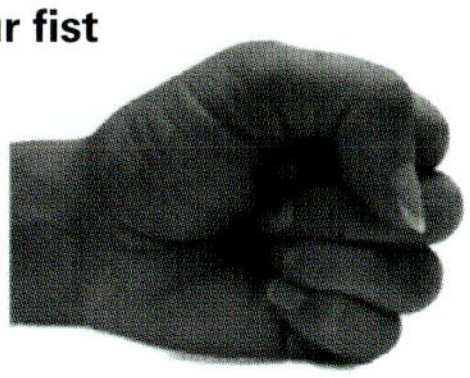	• 1 cup of raw salad vegetables • 1 piece of medium fresh fruit • 1 cup diced or canned fruit • 1 cup of cooked or canned legumes/beans
Cupped hand 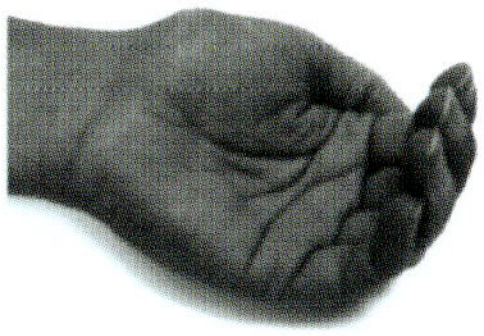	• 1 small piece of fruit • ½ cup cooked vegetables or legumes/beans • ½ cup cooked porridge • ½ cup of cooked rice, pasta, noodles, barley, buckwheat, semolina, polenta, bulgur or quinoa • ½ medium potato
Palm 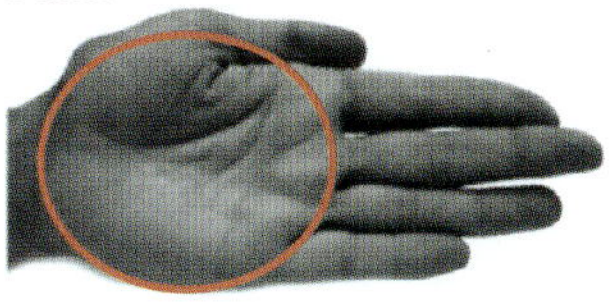	• 100g raw meat or poultry • 100g cooked fish • 1 slice of bread
Thumb 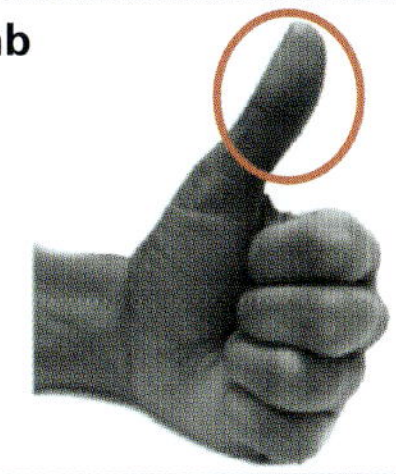	• 1 Tbsp. salad dressing • 1 Tbsp oil • 1 Tbsp peanut butter • 20g hard cheese (½ serve)
Thumb tip 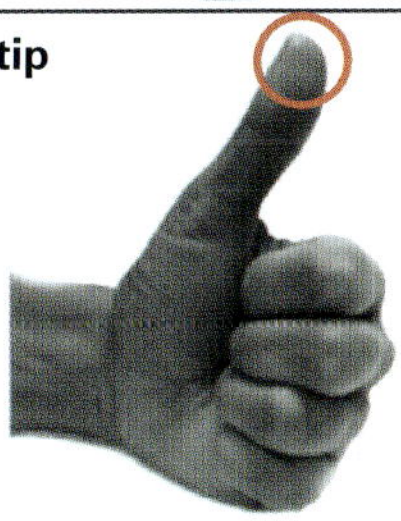	• 1 tsp sugar • 1 tsp oil • 1 tsp margarine or butter • 1 tsp mayonnaise

Figure 2.6 A 'handy' method to estimate portion sizes

Use Figure 2.6 to approximate how much of each type of food you eat in a sitting. Practise today with your lunch or dinner to see if you are getting your portion sizes correct.

There is a lot of official advice and many guidelines and recommendations about what makes a healthy diet and why you should be trying to eat healthily. Managing all this information can be difficult!

Worksheet 2.5

FOOD CHOICES

Here are some quick and easy tips and tricks to use whenever you are deciding which food and drinks to consume.

- Use low-fat versions of dairy products.
- Choose unsaturated fats over saturated fats. This information is included on all food packaging.
- Eat a nutritious breakfast, as this will make you less likely to snack on unhealthy foods later in the day.
- If having fast food, choose bread-based foods such as wraps, sandwiches or kebabs instead of pastry or deep-fried options. Limit your use of sauces, and only 'upsize' if it is a salad option.
- Choose lean meats, not over-processed or **cured** meats.
- Trim skin and fat from meat and poultry.
- Sip water throughout the day, and have water before and during your main meals.
- Choose a range of different coloured vegetables to ensure a variety of vitamins and minerals.
- Choose wholemeal bread and wholegrain cereals.
- Choose regulated sizes when snacking on discretionary foods. For example, have an ice-cream on a stick instead of dishing up a bowl of ice-cream out of a tub.
- Eat slowly, without distractions like TV, and give your body time to respond. Put your cutlery down between mouthfuls while chewing, and sip water in between bites of food to slow your pace. Concentrate on how your meal looks, smells, tastes and feels in your mouth before you swallow. You will enjoy food more and end up eating smaller portions.

cured food that has been preserved by salting, drying or smoking

Shutterstock.com/sarsmis

Figure 2.7 This meal just looks healthy!

Shutterstock.com/Gayvoronskaya_Yana

Figure 2.8 The more colours on the plate, the wider the range of vitamins and minerals.

Shutterstock.com/Robyn Mackenzie

Figure 2.9 Eating ice-cream on a stick instead of dishing up a bowl of ice-cream is a clever way to minimise your portion size.

Weblink
Visit Nutrition Australia or Live Lighter for some tasty and healthy food suggestions.

HEALTHY OPTIONS

In this section you'll find some good examples of meals and snacks. You will then have an opportunity to plan your own meals.

Sample meal plans and snack ideas

Table 2.2 contains samples of healthy meals and snacks for an adult. Remember, you should be eating less than these adult serves.

9780170463096

Table 2.2 Suggested healthy menu for an adult for one day

Breakfast	Wholegrain breakfast cereal with reduced-fat milk OR Wholemeal toast with baked beans and grilled tomato Glass of milk OR reduced-fat yoghurt		Shutterstock.com/MaraZe; Shutterstock.com/Joe Gough
Morning break	Apple Coffee with milk		Shutterstock.com/Ramon L. Farinos
Lunch	Sandwich with salad and chicken OR Roast beef, salad and cheese sandwich		Shutterstock.com/Yeko Photo Studio
Afternoon break	Coffee with milk Unsalted mixed nuts		iStock.com/kaanates
Evening meal	Pasta with lean beef mince and red kidney beans Green salad with olive oil and vinegar dressing OR Grilled fish on rice with lemon juice and vegetables		Shutterstock.com/jabiru; Shutterstock.com/farbled
Evening snack	Fruit salad and reduced-fat yoghurt		Shutterstock.com/Gaak
Drink plenty of water throughout the day			

Weblink
The Eat This Much website can automatically generate a range of meal options based on the amount of energy you think you need for the day. You will need to convert kilojoules to calories first.

Weblink
The Calorie King website is a very handy resource for determining the nutritional and energy information in common foods, and the breakdown of energy from fats, protein and carbohydrates.

Quiz
How can I eat a healthy, balanced diet?

1 Define nutrition.
2 List the five main food groups according to the Australian Guide to Healthy Eating. Identify three food items that fall into each group.
3 When should you eat discretionary foods?

1 a Brainstorm a week's worth of school lunches and snacks (five days). There needs to be a snack for the morning and one for the afternoon. Consider everything you know about what makes a healthy diet. Remember: variety, balance and moderation are the keys.

b Once you have created your meal and snack plan, make a shopping list for all the ingredients you will need.

c Now you can put your plan into action! Take your menu and shopping list home and discuss the task with your parents or caregivers. Join them on their next grocery-shopping trip, and make sure you don't forget anything. You might also need a parent or caregiver to help you prepare some of your lunches or snacks.

Scaffold Meal and snack plan

d One week later ...

After you completed your planning and shopping activities, reflect back on the week and complete these questions.

Scaffold Shopping list

- In what ways did your menu and snack plan differ from your usual weekly eating habits?
- Overall, were these differences healthier or less healthy? Why?
- Did you notice any change in the way you felt and acted by the end of the week? Explain some possible reasons for this change in attitude and behaviour.
- With your parents, compare the cost of the foods and drinks on your shopping list with what they normally spend on your school lunches and snacks for a week. Discuss possible reasons for any differences in cost.
- Propose some reasons why someone would not stick to a menu and snack plan such as this.

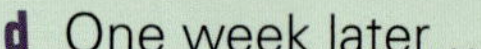

1 Consider the following statement: 'When young people are involved in family meal decision-making and preparation, they develop healthier eating habits for life.' Identify as many different stages and methods of family meal decision-making and food preparation as you can. Evaluate your own involvement in these stages and processes within your family. Could your involvement be increased? How could you propose this to your parents or caregivers?

2 Describe what the terms 'variety', 'balance' and 'moderation' mean to you and your diet, and how you will use these principles to guide your daily food choices. Predict some barriers in your day-to-day life that might limit your ability to follow these principles.

WHAT ARE MY NUTRITIONAL NEEDS?

You may have heard the phrase 'You are what you eat'. While you may not turn into a cabbage or a hot dog anytime soon, the foods that you eat do dictate how well you can function, whether that's out on the sporting field, concentrating in the classroom, leading a long and productive life or just getting through the day successfully. So what does it mean to make healthy choices regarding food and nutrition?

WHAT DO YOU NEED, AND WHAT HAPPENS WHEN YOU DON'T GET IT?

The five food groups promote the consumption of foods that provide your body with a wide range of nutrients. Remember, your body cannot store many of these nutrients, so you need to consume small amounts of each nutrient every day. Each of these nutrients plays an important role in the way your body functions.

Shutterstock.com/asife

Figure 2.10 Eating healthily is a choice.

Table 2.3 Nutrients in the five food groups

	Grain or cereal foods	Vegetables and legumes/beans	Fruit	Milk, yoghurt or cheese	Lean meat, poultry, eggs, nuts and seeds
Main nutrients	carbohydrate protein iron dietary fibre thiamine folate iodine	beta-carotene and other carotenoids vitamin C folate dietary fibre	vitamin C dietary fibre	calcium protein riboflavin vitamin B_{12}	protein iron zinc vitamin B_{12} (animal foods only) omega-3 fatty acids
Other important nutrients	energy magnesium zinc riboflavin niacin vitamin E	carbohydrate magnesium iron potassium	carbohydrate folate beta-carotene potassium	energy fat carbohydrate magnesium zinc potassium	dietary fibre (plant foods only) energy essential fatty acids niacin vitamin E (seeds, nuts)

Carbohydrates

Carbohydrates are the body's primary (and preferred) source of fuel. Simple carbohydrates (sugars) are used more quickly for energy than complex carbohydrates (starches), but they don't last as long.

Protein

Protein provides the building blocks for all the cells in the body. Bodies are constantly breaking down and making new tissues such as blood cells, muscle fibres, enzymes, hormones, skin and hair, so daily intake of protein is essential. Protein can also be used as an energy source in extreme circumstances.

Shutterstock.com/Jmiks

Figure 2.11 Pasta is a popular source of carbohydrates.

Vitamins and minerals

There are many different vitamins and minerals, and the functions they perform are many and varied. Generally, they support the body's biochemical processes and regulate metabolism. Examples of vitamins include A, C, niacin and folate. Minerals include calcium, iron and potassium.

Dietary fibre

Dietary fibre is technically classed as a carbohydrate, but because it can't be completely broken down, your body doesn't extract the same amount of energy from fibre as it does from other carbohydrates. The primary benefit of fibre is the effect it has on your digestive system. Fibre aids in digestion and helps prevent gastrointestinal problems.

Shutterstock.com/Africa Studio

Figure 2.12 While meat is a common source of protein, there are many other great options.

Shutterstock.com/Adisa

Figure 2.13 Vegetables and fruit should make up a large part of any healthy diet.

Shutterstock.com/Khudoliy

Figure 2.14 A range of sources of dietary fibre is critical to good health.

Fats

Fats are an essential part of a healthy diet, as some nutrients such as vitamins A, D, E and K are fat-soluble. This means they are best absorbed when consumed with fats, and are stored within the body's fatty tissue. Fat cells can be an insulating and protective barrier around internal organs. Fats contribute to the function of cell membranes and the immune system, and can also be a source of energy. There are 'good' and 'bad' fats. Bad fats are saturated fats and trans fatty acids, which are known to contribute to cardiovascular disease and should be avoided where possible. These types of fats are normally found in highly processed or manufactured foods, particularly deep-fried takeaway foods and commercially baked goods (biscuits, pastries, pies). Good fats are monounsaturated and polyunsaturated fats, including omega-3 and omega-6, and small amounts are essential in your diet.

Shutterstock.com/Barbara Dudzinska

Figure 2.15 Some sources of fats are healthier than others.

Shutterstock.com/joesayhello

Figure 2.16 Many discretionary foods are high in 'bad fats'.

9780170463096

Water

Shutterstock.com/Balazs Justin

Figure 2.17 Tap water is the best water!

Water is the most essential nutrient of all – you wouldn't survive more than a few days without it. Water is present in all of the cells in all organs of the body; it helps to regulate body temperature and transport nutrients, hormones and waste products around the body. You constantly use and lose water due to chemical processes inside your body, as well as when using the toilet, sweating and even exhaling. That water must be constantly replaced. There is varied advice on how much water to consume daily, but you should aim to drink around 2 litres per day, and more on days when you are very active or the weather is hot. If your urine is clear, light in colour and regular, and you aren't thirsty, you are probably drinking enough.

DOES MY WEIGHT MATTER?

Obesity

Making sure you are consuming an appropriate amount of energy to match your daily needs is critical for good health. Obesity is now the leading cause of premature death and illness in Australia, and if current trends continue, by 2030 75 per cent of Australian adults will be **overweight** or **obese**. Obesity is usually accompanied by many health disorders, such as cardiovascular disease, type 2 diabetes and some cancers.

overweight adults with a BMI from 25 to 30

obese adults with a BMI of 30 or higher

Healthy body weight can be estimated using three different methods: body mass index (BMI), waist measurement and waist-to-hip ratio (WHR).

Body mass index (BMI)

BMI is a method used to estimate total body fat. This helps to determine if your weight is within the normal range or if you are underweight or overweight. There are exceptions to this measure, such as elite athletes with large muscle mass, which can render a BMI figure inaccurate. BMI is calculated using the following formula: weight (kilograms) divided by height^2 (metres).

The chart shown in Figure 2.18 makes it easy to determine which weight range you are in without having to do any calculations. Just find your weight along the *y*-axis (left side) and your height on the *x*-axis (bottom) and trace the point where these two meet.

Waist measurement

This method is suitable for adults only. Using a tape measure, measure the width of an adult's body at their waist (about level with their belly button). The tape should be above their hip bones and below their rib bones. Compare the measurement with the information in Table 2.4.

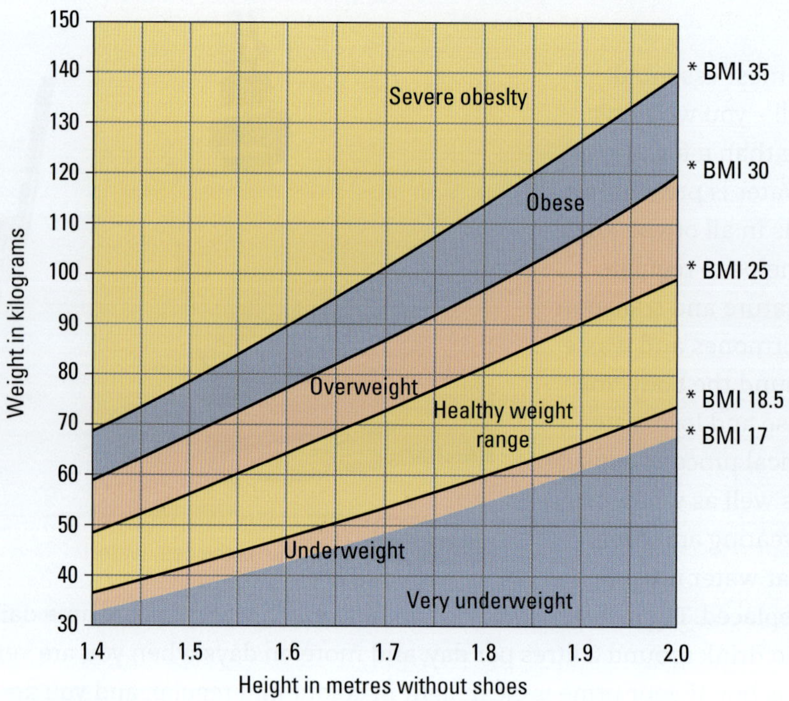

Figure 2.18 Body Mass Index (BMI) = Weight (kg)/Height^2 (m)

Table 2.4 What waist measurement means for health

Waist measurement	Weight-related health risk
Men less than 94 centimetres Women less than 80 centimetres	Low risk
Men 92–104 centimetres Women 80–88 centimetres	Increased risk, especially if their BMI is more than 25
Men more than 102 centimetres Women more than 88 centimetres	High risk

Waist-to-hip ratio (WHR)

This also applies only to adults. The waist-to-hip ratio requires a measurement of the width of an adult's body at their waist (this is the same as the waist measurement discussed above) and also of their hips. The width of the hips is measured by passing the tape measure around the body, level with the big bony parts of the hips. Waist-to-hip ratio is then calculated by dividing the waist measurement by the hip measurement.

A WHR of greater than 0.9 for men and 0.8 for women indicates an increased health risk.

THE AUSTRALIAN DIETARY GUIDELINES

The Australian Dietary Guidelines were created by the National Health and Medical Research Council, an Australian government body that develops guidelines on health-related matters for Australians. The Australian Dietary Guidelines provide information on the types and amounts of food required to promote health and wellbeing, reduce the risk of diet-related conditions and reduce chronic disease in the Australian population.

- **Guideline 1**: To achieve and maintain a healthy weight, be physically active and choose amounts of nutritious food and drinks to meet your energy needs.
- **Guideline 2**: Enjoy a wide variety of nutritious foods from the five main food groups every day, and drink plenty of water.
- **Guideline 3**: Limit intake of foods containing saturated fat, added salt, added sugars and alcohol.
- These three guidelines will be explored in further detail within this chapter.

ENERGY: INTAKE VERSUS EXPENDITURE

The energy you need to maintain a healthy, active lifestyle comes from the food and drink you consume. The more active you are every day, the more energy you need. In countries like Australia, where food is plentiful and relatively cheap, it can be quite easy to consume more than you need. This is a major cause of the increase in obesity and health conditions such as cardiovascular disease, diabetes and many types of cancer. These types of **lifestyle diseases** are the leading cause of death in Australia. Many people try **yo-yo dieting** programs to lose weight, but they can be expensive and don't work in the long term.

lifestyle disease a disease that potentially can be prevented by changes in diet, environment and lifestyle, such as heart disease, stroke, obesity and osteoporosis

yo-yo dieting the practice of repeatedly losing weight by dieting and subsequently regaining it Most overweight or obese people have an imbalance between their energy intake and energy expenditure. Getting this back into proper balance requires making long-term, sustainable changes to a person's entire lifestyle, not a drastic six-week weight-loss program.

Energy expenditure

There are three parts to energy expenditure:

1 **basal metabolic rate** (BMR)
2 the amount of energy used in the process of eating and digesting food
3 physical activity.

Of these three parts, the only one that you really have any control over is physical activity.

basal metabolic rate the energy used when at rest, in order to maintain life (breathing, brain function, heart function, etc.)

Then and now

Early in the 20th century, hard, physical work was common, especially around the home. People chopped firewood, scrubbed clothes by hand, grew their own vegetables and fruit, pumped or carried water by hand, and walked or rode horses or cycled everywhere. There were no 24-hour gyms, treadmills, personal trainers, aerobics classes or drastic weight-loss diets. All of these physical chores and duties were done without 'energy drinks' or 'energy snacks'. Very few people were overweight.

In the past 20 years, 'energy drinks' and 'energy snacks' have become popular, and the companies that make and sell these products have made billions of dollars worldwide. The advertising for these items promises that they will provide the 'boost of energy' needed to be able to work, play sports or party with friends.

At the same time, advances in technology have made life much easier, more relaxing and more convenient. As a result of all of the money, technology and time dedicated to making their lives physically easier, people are now spending time and money to attend gyms or invest in personal trainers in order to compensate for their more **sedentary** lifestyles.

sedentary a type of activity that does not use much energy, such as sitting

Alamy Stock Photo/imageBROKER

iStock.com/fatihhoca

Alamy Stock Photo/ClassicStock

iStock.com/Geber86

Figure 2.19 Sedentary lifestyles were uncommon in previous generations, but modern technology has enabled us to lead our lives with less physical labour.

Extra food

When more energy is eaten than used, the body tends to store the surplus energy in fat cells. This happens even if fat is not part of your diet, because your body can convert all forms of unused energy into fat. These fat cells become a back-up fuel tank.

This is why it is so important to limit discretionary foods (those that are high in energy, low in nutritional value) and make sure you are maintaining high levels of activity. Physical activity is not just playing sport, but also walking to school or helping with the gardening or housework!

FAST FACT

In Australia in 2018, 67 per cent of adults were obese or overweight. In 1980, 60 per cent of Australian adults had a healthy weight, and only 10 per cent were obese.

This process works both ways. If you expend more energy than you consume in a day, your body can use stored fat cells for energy, which may reduce your body weight. One kilogram of body fat contains 37 000 kilojoules of energy, which is the daily recommended energy intake for an average adult for four days!

calories a measure of the energy in food (a single calorie is the amount of energy needed to heat a gram of water by one degree Celsius). This measure is more common in other countries, but in Australia it may still come up in diets, books and everyday discussion

HOW MUCH ENERGY DO YOU NEED?

Energy is not a nutrient but a fuel necessary for your body to function, every minute of every day. Energy is measured in either **calories** (Cal) or kilojoules (kJ).

One kilojoule equals 0.24 calories, and 1 calorie equals 4.2 kilojoules.

It can be difficult to work out exactly how much energy a person needs, as age, sex, body size and shape, and daily activity levels all help determine the amount of energy required. Children and adolescents, whose bodies are growing rapidly, require extra energy. Some teenagers with healthy body weight can get into the habit of consuming a lot of sugar or fat without gaining body fat. This habit sometimes stays with them when (as adults) they are no longer growing and don't need the extra energy. When all that extra energy isn't being used to grow upwards, it can result in growth outwards!

Figure 2.20 Excess energy leads in only one direction!

Estimating daily energy requirements

There are several websites and apps that can be used to estimate how much energy you use, and need to consume, on a daily basis. Make sure you're using reliable sources and up-to-date information, and look for information in kilojoules instead of calories, as this will match the information you see on food and drink packaging. Refer to Worksheet 2.6 for some methods of estimating your daily energy requirements.

Worksheet 2.6

THE BEST SOURCES OF ENERGY

The total amount of energy you receive is the result of several factors:

- how often you eat (the number of snacks, meals and drinks other than water)
- how much you eat (the size of drinks, meals and drinks other than water)
- the types of foods you eat (foods with higher sugar or fat content will yield more energy).

Energy comes from most of the things you eat and drink, except water. The combination of carbohydrates (sugars and starches), protein and fat are your body's sources of energy.

Carbohydrates and protein give 16 to 17 kilojoules of energy per gram, whereas fat gives 37 kilojoules of energy per gram – more than double! Of these three sources of energy (carbohydrates, protein and fat), your body's preferred fuel supply is carbohydrates, particularly if you are performing physical activity.

But wherever your energy comes from, if the total energy intake is higher than your expenditure, the extra energy will be stored as body fat.

The recommended balance of energy sources is shown in Table 2.5.

Table 2.5 Daily energy allowance by source

Energy source	Percentage of daily energy allowance
Protein	15–25
Carbohydrates	45–65
Saturated fats	20–30

If you follow the variety, balance and moderation described earlier in this

Figure 2.21 Which of these images best represents the recommended balance of energy sources?

chapter, and consume the right amount from each of the five main food groups, you are probably already meeting this balance of energy sources!

Despite the messages from advertisements and other persuasive marketing campaigns, most people don't need to eat foods that contain large amounts of energy. Even elite athletes, who expend much more energy than the average person, can usually meet their energy needs from a regular healthy diet, without needing special energy drinks or snacks.

Worksheet 2.7

Ask yourself, 'What else am I providing my body with when I consume this energy-rich food?' Food or drink that contains a range of vitamins and minerals as well as energy is much healthier than food or drink with lots of energy but almost no other nutrients.

WHAT DO ATHLETES NEED?

glycogen the substance that is the main form of stored carbohydrates

Carbohydrates are the preferred fuel to provide the energy required for vigorous exercise, while protein and fat are used to provide energy to the muscles during rest and lower intensity exercise. The body can store carbohydrates in the form of **glycogen** in the muscles and liver, but this storage capacity is limited. If vigorous physical activity lasts longer than an hour, extra carbohydrates are needed. (For example, a football player may eat fruit during half-time of a game that lasts about 90 minutes.) Sometimes a small energy-rich snack should be eaten just prior to a game or competition, to 'top up' glycogen reserves. (This strategy should be used with a healthy regular diet, not instead of it.)

Immediately after vigorous sport, give your body the fuel and nutrients it needs to recover so you will be ready for more activity in the following days. Usually the best option is a small snack of carbohydrates and protein, such as tuna on crackers or muesli with yoghurt. This should not replace your regular healthy meals and snacks.

hydration combination with or absorption of water

Hydration is also vital for good physical activity. You only need to lose 2 per cent of your normal fluid stores to experience a decrease in your physical and mental effectiveness. On top of your daily needs, extra water is needed when exercising, particularly over a long period of activity or in hot conditions. Planning ahead is the key! Ensure your body is already hydrated before game day, and bring along a water bottle. Try to drink small amounts early and often during your activity. Replace any fluids you lose by drinking water after you have stopped playing or training. If you have lost weight while exercising, you must replace this fluid. If you have been replacing lost water properly, your body weight should return to its pre-exercise level by the following morning. Your urine should be regular and a light colour.

Sports drinks

You may be tempted to consume 'sports drinks', which have a range of additives such as sugar, salt, electrolytes or other minerals, and even protein or caffeine. These drinks are sometimes useful for athletes who are doing many hours of vigorous activity each week, but most people don't need them. They aren't that much different from cordial, or even soft drink.

iStock.com/apomares

Figure 2.22 Hydration is essential for optimal performance.

1. Summarise the three guidelines from the Australian Dietary Guidelines.
2. Compare the three Australian Dietary Guidelines with the Australian Guide to Healthy Eating. See if you can find at least three similarities, and discuss them with a classmate.
3. What is the purpose of the Australian Dietary Guidelines? Discuss reasons why they are developed and communicated to the Australian public.

1. Take a few moments to read the section headed 'Then and now' on page 77, and reflect on what our modern lifestyle means to you. Share your response with a friend or the class. Do you predict an ongoing trend?
2. Draw two more scales like Figure 2.23 and fill in the missing words in the sentences below. Your diagrams need to reflect the imbalance of energy intake and energy expenditure.

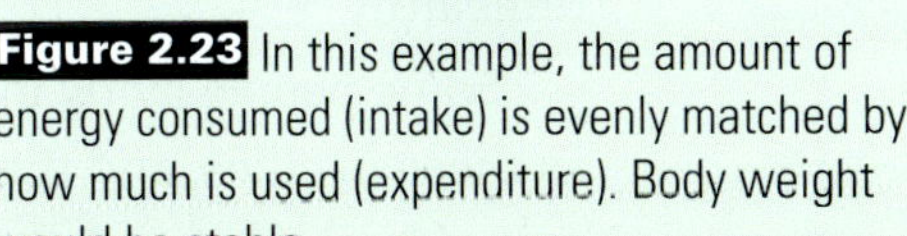

Figure 2.23 In this example, the amount of energy consumed (intake) is evenly matched by how much is used (expenditure). Body weight would be stable.

   a. In this example, the amount of energy consumed (intake) is less than the amount used (expenditure). Body weight would ____________.
   b. In this example, the amount of energy consumed (intake) is higher than the amount used (expenditure). Body weight would ____________.

1. Consider the following scenario: Next week is your school athletics carnival. You have been attending training and working hard on improving your results. Your Physical Education teacher thinks you might go close to being champion for your age group this year. Last year you were on track to achieve this, but with so many events to compete in and the weather being warm, your performances faded over the course of the day. You felt weak and light-headed, and had trouble focusing. By the end of the day you weren't even getting close to the personal bests you had achieved at training.

But that's not going to happen this year! To assist your performance, you want to be as prepared as possible. Apparently what you eat and drink the day before competing is important, so you are going to make sure to pack lunch, snacks and drinks for carnival day.

Develop a strategy for the days leading into the event and for event day itself. Consider the types and quantity of foods and drinks that will help you reach optimal performance levels. Construct a plan to represent your strategy in an organised format. Be sure to justify the different elements of your strategy.

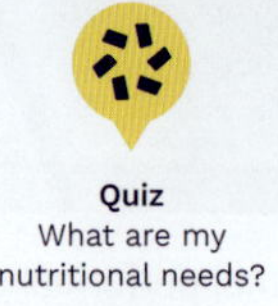

Quiz
What are my nutritional needs?

HOW IS MY DIET INFLUENCED?

There are many factors that influence your eating decisions and habits. It is important to know how to recognise and manage them, particularly if they are influencing your decisions and habits in a negative way.

Worksheet 2.8

FACTORS THAT INFLUENCE YOUR EATING HABITS

Each time you make a decision around food and eating, you make a justification, either consciously or subconsciously. An example could be, 'I don't think I will have any breakfast today because I ate so much last night before bed', or 'I'd better have a yoghurt and fruit snack after school today, as I already had some chocolate with my lunch.'

NUTRITION AND SOCIETY

Most Australians enjoy a high standard of food choices and nutritional health, along with easily accessible information about healthy diets and food choices. However, many Australians do not eat enough vegetables, fruit, wholegrain cereals and healthy dairy products. Many people eat large amounts of fast food, even though they know it is unhealthy.

Some people living in remote rural areas don't have the same access to healthy foods as those in cities, and this is often reflected in their overall health. The availability and cost of fresh fruit and vegetables can also be an issue for people living in remote locations.

Shutterstock.com/ArliftAtoz2205

Figure 2.24 Choices can be limited in some rural towns.

9780170463096

People on limited incomes may have trouble following nutritional guidelines because of the perception that the cheapest foods are often unhealthy choices. Areas with low **socio-economic status** have twice as many fast-food outlets as more affluent suburbs. However, studies show that diets high in fast food or pre-cooked meals are actually more expensive than those where meals are cooked at home using items purchased from supermarkets.

socio-economic status a measure of advantage or disadvantage of an individual or population group based on factors such as income, education and occupation

CASE STUDY ⇨ LOCATION, OBESITY AND COVID

Identify

The COVID-19 pandemic can rightly be described as the biggest health disruption in a generation. There has never been a time more challenging for food systems and supply chains often impacting access to food, not experienced previously. Our relationships with food, our feelings of connectedness, our exercise and wellness routines and our ability to maintain a healthy weight have all become much more difficult to achieve yet so much more critical to get right. If that wasn't enough, it's the long-term impacts, the financial burden on families, mental and physical health of our regional and remote communities, and the increasing demands on the entire health system that I fear is something we will be 'paying back' for decades.

Understand

Most importantly, it is our children who are doing it tough. The feeling of 'catch up' for them is real and starting to look never-ending. When you overlay COVID-19 impacts on a generation already threatened by obesity and what do we have? Catastrophic and life-long impacts on the health and wellbeing of our children, not seen in previous generations and presenting as one of the most lasting consequences for societies as a whole. Maintaining our health and wellbeing has become a wicked problem, more so than ever before. Can we recover? The answer is yes but only with good planning, commitment and deep understanding of prevention, our population and the importance of partnerships.

Two-thirds of all Queensland adults were already overweight or obese prior to COVID-19, with a self-reported 32% increase in obesity among Queensland adults between 2004 and 2020. New research from My Health for Life conducted in April 2021 showed almost half of Queenslanders surveyed reported to have gained weight, and 21% reported a gain of more than 5 kilograms.

According to the same research, lockdowns and restrictions have also led to changes in eating behaviours, with an increased consumption of processed and long-life foods, and reduced consumption of fruit and vegetables. Only one in three Queenslanders reported eating fruit and vegetables daily. Weekly fruit and vegetable consumption has also decreased, and in April 2021, 15% of Queenslanders reported they don't consume any vegetables and 18% don't consume any fruit, critical sources of energy and nutrients vital to health and wellbeing.

Food systems and supply chains have been affected by concerns of food shortages, and food insecurity has led to reduced access to healthy foods among the most vulnerable. Food security is achieved 'when all people, at all times, have physical and economic access to sufficient, safe and nutritious food to meet their dietary needs and food preferences for an active and healthy life'.

The most recent data (2011–13) in Australia shows that about one-third of people (31%) living in remote areas were food insecure, compared to 4% in the general Australian population. Central and North Queensland have scored 60% higher on the McKell Institute's Food Insecurity Index compared to inner Brisbane. Food insecurity has exacerbated pre-existing health disparities across population groups, with the most disadvantaged more likely to experience higher rates of obesity, illness, hospitalisation and premature death compared to other Queenslanders.

While we focus on vaccinations and navigating our way through the pandemic (and we should), we can't afford to lose sight of preventive health, in fact, it has become more important. According to research from the Centre for Disease Control, people living with overweight and obesity are at greater risk of severe COVID-19 complications. They are more likely to be admitted to a hospital for COVID-19 related pneumonia at a younger age, require mechanical assisted ventilation, and access intensive care units (ICUs) than healthy weight patients. Individuals with obesity are also twice as likely to die from COVID-19 than individuals with a healthy weight. Disadvantaged population groups are even more at risk.

Obesity prevention has always been important for population health and wellbeing; but, the dual impact of COVID-19 on obesity, and obesity on COVID-19, impacts our short-term and long-term health even more.

Health and Wellbeing Queensland was established by the Queensland Government in 2019, before the pandemic. The remit was about reducing chronic disease driving an equity lens with the purpose of reducing the pressure that chronic disease has on our health systems. It was known that preventative health was the only 'real' long-term solution to combatting the rate and risk of chronic disease. How insightful and how RIGHT and how timely was that decision?

Source: 'Obesity and COVID-19: It's time to double down' by Dr Robyn Littlewood, Health + Wellbeing Queensland, 16 September 2021, https://hw.qld.gov.au/blog/obesity-and-covid-19-its-time-to-double-down/

Discuss

1 The author of the article asserts that where people live within Queensland influences the rates of obesity, illness and death. Discern two statements that demonstrate this assertion.
2 Describe the relationship between obesity and COVID-19, and the role they play in exacerbating negative health impacts.
3 What is your position on this issue? Do you agree with the notion that, depending on where people live and their level of income, they may not have any control over their food choices and subsequent level of body weight? Justify your response.

I DON'T ALWAYS MAKE GOOD CHOICES BECAUSE...

While Australians generally have access to an excellent range of healthy foods and plenty of exposure to sound nutritional advice, there are millions of people around the world who do not.

Still, many Australians make bad decisions when it comes to their diet. They ignore advice about healthy eating, and they choose unhealthy foods and drinks over healthier options. Why?

Everyone is tempted occasionally by the prospect of a treat, even though it may not be a healthy choice. It's important to remember that there can be a place for high-sugar and high-fat treats in your diet, just not many of them and not all the time. Learning to eat food in moderation will help you with your choices.

Newspix/Gary Merrin

Figure 2.25 The amount of choice can be overwhelming.

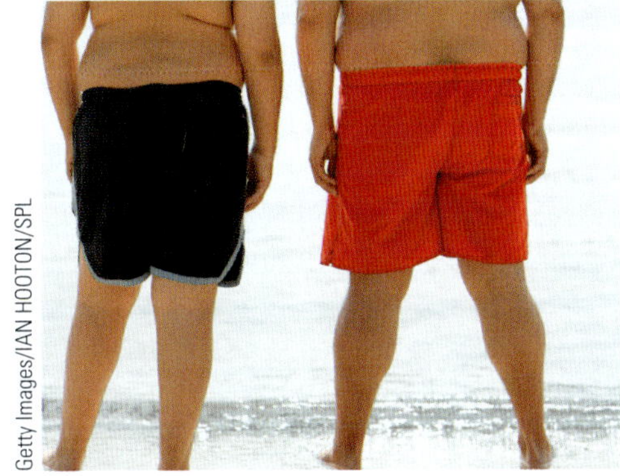

Getty Images/IAN HOOTON/SPL

Figure 2.26 Childhood habits can be hard to break.

Shutterstock.com/Roman Samokhin

Figure 2.27 Sweet treats can be enjoyed in moderation!

FACE TO FACE Choices, choices ...

Identify some possible reasons for not always making healthy dietary choices. For most people, there will be several reasons. Share your answers with your class and see how many you can come up with. Try working in small groups to survey or interview people in your school or local community.

Worksheet 2.9

WELLBEING CHECK IN

TRYING NEW FOODS

MoodMission

This activity has been developed in collaboration with Dr David Bakker of MoodMission

Identify

Sometimes it's hard to try new things. We might even find ourselves sticking to old things because it feels too risky to try something new. But what if the new thing is actually really good?

Understand

Eating familiar foods can be comforting, especially when compared to the risks of trying something new. Some people find themselves refusing to try new foods because they're afraid they won't like them. But trying new things and being able to cope with things that we don't like is an important emotional skill. This process can actually build resilience, especially when something bad happens and we realise that it's not actually the end of the world – it's just a bad taste that passes quickly.

Practise

1 Think of a food you've never tried before. It doesn't have to be an obscure delicacy, it can just be something you've seen other people eat but you haven't had the chance to try yet. You might like to talk to a friend about what they eat to come up with ideas.
2 Now make a plan to try the food. Will you need to talk to an adult, or can you get it from the shop yourself?
3 Once you've tried it, rate how much you enjoyed it on the scales below.

How tasty did you find the food?

Absolutely disgusting ← → Absolutely delicious

Would you try this food again?

Definitely not ← → Definitely

Reflect

1 Did you find it scary to try the new food?

If you did, what do you think you were scared of? Do you think you were right to be scared, or did you feel more fear than you needed to?

If you didn't find it scary, what do you think you learned from trying the new food? Was it just learning that you did or didn't like the food, or did you learn something else about how you deal with feelings?

1 Table 2.6 lists the factors that have been known to influence decisions people make about what, when, how much and how often they eat. Copy the table and rank each of the influences from 1 to 16, with '1' being the most influential to your decisions and habits about eating and drinking, and '16' the least influential. This may take you a few attempts. Then compare your results with the rest of the class.

Quiz
How is my diet influenced?

Table 2.6 Knowing your influences

	My rank	Influence
Most influential		Taste
		Hunger cravings
		Cravings for a specific food
		Convenience – easy to prepare or eat
		Food availability – if food is there, eat it
		Parental influence
		Peer influence
		Health benefits
		Mood
		Body image
		Habit
		Cost
		Time considerations
		Media and advertising
		Cultural or lifestyle decisions such as being vegetarian, or observing religious or cultural practices
Least influential		Certain social settings – at a sporting event or going to the movies
My top three influences		**Class top three influences**
My bottom three influences		**Class bottom three influences**

1 Compare your rankings with your classmates: are there any trends?
2 Share your top three influences and bottom three influences with the class.
3 Devise a class tally, either on the whiteboard or in a spreadsheet.
4 You should now be able to see which influences from the list have the most and least impact on the class's decisions and habits about eating and drinking. What are some similarities and differences between your classmates' influencing factors and your own?
5 Do your top three influences tend to make you eat healthily or unhealthily? Give an example.
6 Now that you have an understanding of influencing factors, how can you use this knowledge? Plan a way to maximise the times when you are influenced to eat healthily and minimise the times you are influenced to eat unhealthily.

1 Conduct online research into the following statement: 'A healthy diet is cheaper than junk food.' Access at least three articles that appear to be credible.
   a Summarise the key points by each author and compare these across the three articles, noting similarities and differences.
   b Evaluate how applicable you believe this statement is to you or your family. Justify your perspective from a financial and health basis.

HOW CAN I UNDERSTAND FOOD LABELLING?

What do you already know about food packaging? As a general rule, the less packaging a food has, the better that food is for your body and the environment. Although some foods need packaging to retain their freshness, packaging is usually included for marketing purposes, featuring visually appealing designs and claims about taste, health and value for money.

WHAT ARE YOU PUTTING INTO YOUR BODY?

Some of the information on food and drink packaging is reliable, but most of it is not closely regulated. Food manufacturers carefully choose the words on their packaging to encourage you to buy their products. The following are some of the techniques used.

- Stating obvious facts that make a food item sound healthier. For example, the label on a bottle of olive oil might state that it's 'cholesterol free'. But in fact, *all* plant-based foods have zero cholesterol, including canola, peanut and sunflower oil. While the statement is true, it is misleading.
- Making big, bold but unfinished claims such as '25% less added salt' but leaving the consumer to find the small print underpinning these claims.
- Making health-related claims using vague words such as 'support' or 'promote' instead of 'prevent' or 'protect'. An example is 'supports the immune system'. Using vague words means the statement doesn't have to be factually true.

- Using the terms 'light' or 'lite', which don't necessarily refer to the amount of fat, sugar or energy. They may refer to the taste, texture or colour of the food.
- Using terms such as 'natural', 'diet' and 'homemade', which all imply a higher level of nutrition. There are no regulations on how these words can be used and what they represent, so it is best to ignore them.
- Stating that a product has 'no added sugar' doesn't mean it is low in sugar – many products with fruit as ingredients will have high levels of sugar even without any more being added.
- Labelling something '90% fat-free' – this sounds quite healthy, but it actually means that the product is 10 per cent fat, which is a large amount!
- Along with some other states, the Queensland Government has introduced a kilojoule menu labeling scheme for chain restaurants, 'Fast Choices'. The requirements mean that fast food restaurants have to display the kilojoule content of their food and drinks so people can make healthier choices when eating fast food.

Weblink
Fast Choices: Kilojoule menu labelling

INVESTIGATION

MYTHBUSTING

Purpose

You will determine the prevalence of misleading statements on food packaging by analysing the observations of food products. The list of packaging techniques provided above is a **secondary source**. This investigation will develop a **primary source** of marketing myths promoted by food brands to increase sales.

Materials

- A digital camera or smartphone, if visiting a supermarket personally
- A computer for searching online catalogues and for data analysis

secondary source typically the interpretation or evaluation of primary sources. Secondary sources often describe or explain primary sources.

primary source a first-hand account of an event by someone who experienced it directly and is considered authoritative. Primary sources report on discoveries and represent original thinking.

Method

1. As a class, allocate a food product group to each pair of students, e.g. breakfast cereals, yoghurt, lunch/snack bars, bread, cooking oils, etc.
2. As a pair, either visit a supermarket personally or access a supermarket catalogue online.
3. Record how many different products exist for your food product group (look for different brands/manufacturers). Tally this as a figure, but also take photos in the supermarket aisle or screenshots from online catalogues.
4. Record how many occurrences you observe of each of the different marketing techniques listed above. Some products will use more than one technique, and you may come across some other techniques not listed above, so be prepared to record these too.
5. Tally your findings and include your accompanying photos and screenshots to support your observations.
6. Determine the prevalence of misleading statements on food packages by calculating the ratio of products that contain these types of statements compared to those that do not.

Weblink
For some ideas, watch 'Healthy Labelling' by the ABC's The Checkout.

Discussion

1 How prevalent were the misleading techniques used on food packages compared to products that choose not to?
2 Which particular misleading statements were most common for your food product group? Can you propose some reasons for this?
3 Share your findings with your class and compare results. What were the similarities and differences in results for the various food product groups? Why would these exist? Consider having a competition among your class – who can find the product with the most outrageous or nonsensical health claim?

WHAT'S ON THE PACKAGE?

The really useful information is often written in small print on the back or side of the packaging. The following are the most important items to check:

- 'use by' date – make sure your food is fresh
- nutrition information panel – this is required for all foods with packaging and is your main source of reliable information
- ingredients – listed in order of most to least, by weight
- storage or preparation instructions – to make sure your food stays fresh and is prepared safely
- allergy advice – for those with allergies, even if the list of ingredients doesn't mention the item you are allergic to
- daily intake guide – shows the percentage of the recommended daily intake of energy or nutrients you would receive by consuming one serve of this food or drink.

In recent years we have seen the introduction of the Health Star Rating system (see Figure 2.31), which was designed to provide consumers with an easier way to compare similar food and drink products and make healthier choices. People who don't take the time to read the nutrition information panel now get an at-a-glance overall rating of the healthiness of a product.

Figure 2.28 Daily intake guide for energy per serve

Nutrition Information		
Servings per package – 16 Serving size – 30 g (2/3 cup)		
	Per serve	**Per 100 g**
Energy	**432 kJ**	**1441 kJ**
Protein	2.8 g	9.3 g
Fat		
Total	0.4 g	1.2 g
Saturated	0.1 g	0.3 g
Carbohydrate		
Total	18.9 g	62.9 g
Sugars	3.5 g	11.8 g
Fibre	6.4 g	21.2 g
Sodium	65 mg	215 mg
Ingredients: Cereals (76%) (wheat, oatbran, barley), psyllium husk (11%), sugar, rice, malt extract, honey, salt, vitamins.		

Figure 2.29 Nutrition information

Australian Government
National Health and Medical Research Council
Department of Health and Ageing

www.eatforhealth.gov.au

HOW TO UNDERSTAND FOOD LABELS

What to look for...

Don't rely on health claims on labels as your guide. Instead learn a few simple label reading tips to choose healthy foods and drinks, for yourself. You can also use the label to help you lose weight by limiting foods that are high in energy per serve.

Nutrition Information

Servings per package – 16
Serving size – 30g (2/3 cup)

	Per serve	Per 100g
Energy	**432kJ**	**1441kJ**
Protein	2.8g	9.3g
Fat		
Total	0.4g	1.2g
Saturated	0.1g	0.3g
Carbohydrate		
Total	18.9g	62.9g
Sugars	3.5g	11.8g
Fibre	6.4g	21.2g
Sodium	65mg	215mg

Ingredients: Cereals (76%) (wheat, oatbran, barley), psyllium husk (11%, sugar, rice, malt extract, honey, salt, vitamins.

Total Fat ▶
Generally choose foods with less than **10g per 100g**.
For milk, yogurt and icecream, choose less than **2g per 100g**.
For cheese, choose less than **15g per 100g**.

Saturated Fat ▶
Aim for the lowest, per 100g. **Less than 3g per 100g is best.**

Other names for ingredients high in saturated fat: Animal fat/oil, beef fat, butter, chocolate, milk solids, coconut, coconut oil/milk/cream, copha, cream, ghee, dripping, lard, suet, palm oil, sour cream, vegetable shortening.

Fibre ▶
Not all labels include fibre. Choose breads and cereals with **3g or more per serve**

◀ 100g Column and Serving Size
If comparing nutrients in similar food products **use the per 100g column**. If calculating how much of a nutrient, or how many kilojoules you will actually eat, use the per serve column. But check whether your portion size is the same as the serve size.

Energy
Check how many kJ per serve to decide how much is a serve of a 'discretionary' food, which has 600kJ per serve.

Sugars
Avoiding sugar completely is not necessary, but try to avoid larger amounts of added sugars. If sugar content per 100g is more than 15g, check that sugar (or alternative names for added sugar) is not ◀ listed high on the ingredient list.

Other names for added sugar: Dextrose, fructose, glucose, golden syrup, honey, maple syrup, sucrose, malt, maltose, lactose, brown sugar, caster sugar, raw sugar.

◀ Sodium (Salt)
Choose lower sodium options among similar foods. **Food with less than 400mg per 100g are good, and less than 120mg per 100g is best.**

Other names for high salt ingredients: Baking powder, celery salt, garlic salt, meat/yeast extract, monosodium glutamate, (MSG), onion salt, rock salt, sea salt, sodium, sodium ascorbate, sodium bicarbonate, sodium nitrate/nitrite, stock cubes, vegetable salt.

Ingredients ▲
Listed from greatest to smallest by weight. Use this to check the first three ingredients for items high in saturated fat, sodium (salt) or added sugar.

Source: National Health and Medical Research Council

Figure 2.30 Tips for reading a nutrition information panel

Using the Health Star Ratings

Make healthier choices by using the health stars to compare similar packaged foods.

Eating healthier food helps maintain a healthier you.

Health Star Ratings are applied using a strict calculation.

Health stars are on the front of many packaged foods.

Health stars are simple to use: the more the stars the healthier.

Fresh is best, but when buying packaged food use the health stars.

Health stars can provide information about key nutrients.

To find out more, visit:
healthstarrating.gov.au

The Health Star Rating | A joint Australian, State and Territory governments initiative in partnership with industry, public health and consumer groups.

The more stars, the healthier.

Figure 2.31 A quick guide to using the Health Star Ratings

Shutterstock.com/LADO

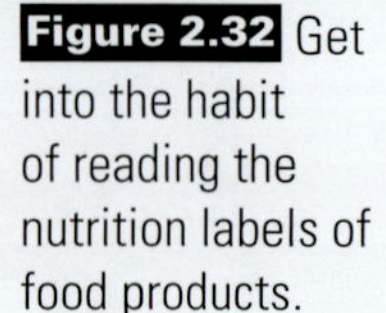
Figure 2.32 Get into the habit of reading the nutrition labels of food products.

Quiz
How can I understand food labelling?

REVIEW

1. Propose five useful tips for reading labels, and describe how you would put them into practice on a daily basis.
2. Find the nutrition information label on an item of food from your home, cut it out and bring it to your next lesson. Hand it to your teacher, who will spread the labels out on a table at the front of the room. As a class, see if you can arrange the labels in order from least healthy to most healthy. You don't need to know what foods they come from, just what those foods contain. There are lots of things to consider, such as the ingredients and nutrients and their varying amounts, so be prepared to discuss and justify your input. It is not always easy to agree on exactly what is healthy!

REFLECT

Look closely at Figure 2.30 and then complete the following.

1. Of all of the labels your class brought in from home, how many would you classify as coming from 'healthy foods'? Why?
2. Was there any particular aspect of a label(s) that caused a lot of debate or disagreement? Why?
3. Explain what was the most difficult part of putting all of the labels in order.
4. Discuss how you will you use nutritional information on labels to guide your food choices in the future.

EXTEND

Weblink
Health Star Rating system

1. Typically when a new public health policy or initiative is introduced, we have a great opportunity to review its impact. How were things working before its introduction? What is different since its introduction?
   Let's provide our own critique of the Health Star Rating system.
   a. Do you think the introduction of the Health Star Rating system has been effective?
   b. Is it something you use when deciding what food or drink product to buy? Justify why or why not?
   c. Do you tend to buy healthier products when you use the system?
   d. Could you propose some suggestions for improvements?
   e. How would your family members and friends answer these questions? Design a survey to gather responses.
   f. How do experts and academics evaluate the Health Star Rating system? Conduct web searches to ascertain their findings. Look for discussion about the design of the system, its implementation and suggestions for improvements. It is always good practice to find multiple articles from different authors. Compare their findings and develop a summary.

9780170463096

WHAT IS THE IMPACT OF ADVERTISING ON MY DIET?

Advertisements for food and drink products are everywhere; you see them several times a day. In fact, sometimes you see so many advertisements that you don't even notice them. But your brain takes in a lot of information about brands and products without you really thinking about it, and the people responsible for marketing foods know this. How much food and drink advertising do you think you are actually exposed to?

CASE STUDY ⇨ JUNK FOOD ADVERTISING

Identify

The Queensland Government has taken steps to limit the opportunities for junk food companies to target their advertising at children.

Understand

The Queensland Government will ban junk food promotions at government-owned sites in a bid to crack down on poor diets and childhood obesity.

Health Minister Steven Miles said the unhealthy marketing would be phased out at more than 2,000 outdoor advertising spaces, including bus stops, train stations and road corridors.

'Junk food advertisers target kids, we know that, and obesity in childhood is a leading indicator of obesity in adulthood', Mr Miles said.

'This is about doing what we can to protect our kids from the kind of marketing that leads them to make unhealthy choices.'

Mr Miles said the ban would affect leased spaces owned by the State Government.

'Obesity is a real challenge for our community, for our hospitals and the health services, but also for the individuals who are suffering – this is really just a decision about the Government leading by example and saying that we will use our spaces to advertise healthier options', he said.

Obesity Policy Coalition executive manager Jane Martin said the restrictions were a win for young people and their health.

'This is exactly the sort of action that's recommended by agencies around the world, like the World Health Organisation, to protect children from the influence of junk food marketing', Ms Martin said.

'Young people use public transport, they're exposed to this sort of marketing, it's wallpaper in their lives.'

Reproduced with permission from Cancer Council Western Australia

Figure 2.33 Junk food advertising directed at children has been banned on government-owned sites in several Australian jurisdictions, including Queensland, the ACT and Victoria. Other states are currently lobbying for this policy change too.

'This is a really important step to protect children when they're on their way to school, going to meet up with friends [where] it's very difficult for parents or anyone else to intervene.'

Ms Martin said Queensland would join the ACT in providing the lead for other jurisdictions considering similar reform.

She said other areas of concern, including digital and television marketing, were in the remit of the Federal Government.

Source: Reproduced by permission of the Australian Broadcasting Corporation – Library Sales. Lily Nothling and Kate McKenna © 2019 ABC

Discuss

Search online for another article that supports the main contention in this article.

1 Summarise the primary viewpoint in these articles.
2 Identify the health issues that were prevalent before the introduction of the new advertising ban.
3 Propose some other strategies that could be used to minimise the influence of junk food advertisers on children

UP AND MOVING Jingle battles

Pair up with a classmate and take turns singing a line of a jingle (catchy song) from an advertisement for a food or drink product. Whoever can't come up with another jingle on their turn loses that battle, and then pairs up with a new opponent who also lost their first battle. Winners pair up with other winners, and keep facing off until we have a Jingle Battle champion (the last remaining undefeated student). If needed, the game can be made easier by saying catchphrases as well as jingles.

Worksheet 2.10

goodwill a good relationship, a friendly attitude

Weblink
Damaging diet advice by social media influencers

WHO CAN YOU TRUST?

Food marketing and manufacturing companies exist to make profit, and they achieve this by encouraging you to believe whatever is necessary to make you buy their products. At times, certain brands will fund or sponsor community events or junior sports programs. Most companies do this as a way to build up **goodwill** among their potential markets.

It gets even trickier when food companies or food industry bodies fund research into the benefits of their products. You will not be surprised to know that the results of this type of research are usually quite favourable for the company paying for the research. An ever-increasing source of misinformation can be celebrity social media influencers. These people use their public profile and online popularity as a basis for providing seemingly 'expert' nutritional advice without any formal qualifications, and usually in an attempt to generate income for themselves. Doctors are asking consumers to re-think the messages and nutritional advice promoted online, which could be placing people's health at risk.

Images courtesy of The Advertising Archives

Figure 2.34 Three brands known for clever use of marketing strategies

Advertising works, and it works very well. Food and drink manufacturers spend millions of dollars each year on marketing and advertising because they know it generates sales. They wouldn't do it otherwise.

Some advertising messages are directly relevant to what the product provides, such as an energy drink advertisement describing how energetic you will feel after drinking the product. Other advertising campaigns just give the vague, general perception that if you use the advertised product, you will be cooler, more popular, more athletic, smarter or experience a better lifestyle. They can convey this idea without explicitly saying or writing these things; in fact, some advertisements have no spoken or written words at all!

Certain brands also use the power of 'nagging', or 'pester power'. These advertisements are aimed at children and rely on the child nagging their parents to buy them the product. Toys, giveaways and links to movies or superheroes are common examples of this strategy. Remember, most food and drink advertisements are for items you don't really need, so companies need to convince you otherwise!

AAP Image/Tony McDonough

Figure 2.35 Some brands pay a lot of money to have their image very visible.

Even the COVID-19 global pandemic has been utilised as a marketing opportunity. During periods of lockdown and social distancing, fast food corporations released marketing campaigns exploiting people's boredom and isolation as a marketing tactic.

Weblink
Junk food advertising capitalising on lockdown

INVESTIGATION

TALLY IT UP

Purpose

To analyse the factors that act as enablers or barriers to healthy eating among youth.

Method

There are three parts to this activity. It may take a week to complete them.

1 For a 24-hour period, make a note every time you see or hear an advertisement for a food or drink product. Note whether it was for a 'discretionary food' (low nutrient, high fat or sugar snack or drink, fast food or junk food). After 24 hours, add up how many food or drink advertisements you saw and how many of those ads were for discretionary foods.

2 Watch an hour of commercial free-to-air TV, during a show aimed at children or teenagers (these will usually be aired just as you get home from school). Record how many advertisements for food or drink you see, and note how many of the advertised foods would be classified as 'discretionary'.

Watch another hour of commercial free-to-air TV, when the show is not just for children or teenagers – the news might be a good example. Don't pick a sport program (see part 3). Record how many advertisements for food or drink you see, and note how many of the advertised foods would be classified as 'discretionary'.

3 Watch your favourite sport, either on TV or by going to the game live. Again, record how many advertisements for food or drink you see, and note how many of the advertised foods would be classified as 'discretionary'.

You will now have three tables of data that you can analyse during your next Health lesson. Here are three examples:

Part 1: Daily exposure	
Number of food or drink advertisements	Number of discretionary food or drink advertisements

Part 2: TV shows		
	Number of food or drink advertisements	Number of discretionary food or drink advertisements
TV show 1: ___________		
TV show 2: ___________		

Part 3: Sport		
	Number of food or drink advertisements	Number of discretionary food or drink advertisements
Sport: _______________		

9780170463096

Discussion

1 Create a graph(s) that will represent your data in a clear and logical way. You may choose to do this manually on paper or by using spreadsheet software. Think carefully about the type and format of your graph(s).
2 Identify and discuss two trends you have noticed in your data.
3 Why do you think these trends exist? Is it a deliberate strategy or just a coincidence?
4 Considering the increases in obesity and lifestyle diseases in Australia, do you think current advertising strategies are appropriate?
5 If not, what do you think can be done about this?
6 Were you surprised by any of the results of this activity? Give reasons for your answer.

1 Describe three different advertising strategies used to sell food and drink.
2 Describe the concept of 'goodwill' towards a company.

1 Reflect back on an occasion when you have been enticed into buying something, not necessarily because you really wanted or needed it, but because of the effectiveness of the advertising. What was the product, and what was the advertising strategy?
2 If faced with a similar situation in the future, what techniques could you use to help you resist temptation?
3 Examine at the image in Figure 2.36. Why do you think this particular company would use an image like this to try and sell more products? Who are they targeting and how are they doing it?
4 Using the internet or magazines and newspapers, gather a range of images that you can use to create a mural or poster. Your images need to represent as wide a range of advertising strategies as possible, from a range of different brands and products.

Alamy Stock Photo/Jeff Morgan 16

Figure 2.36 Well, if it's good enough for Santa Claus …

1 The marketing of discretionary food products to children is self-regulated by the food and beverage industry, which is usually represented by lobby groups such as the Australian Food and Grocery Council. They utilise initiatives that aim to 'reduce advertising and marketing to children for food and drinks that are not healthier choices' and 'only advertising healthier choices to children'. (Source: Australian Food and Grocery Council)

Let's put this to the test.

a Can you find examples of discretionary food and drink products that appear to be marketed at children? Gather evidence via photos, screenshots, or downloads from the web.

b What techniques are used to target children?

c Make a determination on how fair it is for companies to specifically target children when advertising discretionary products.

d Is it reasonable for the food and beverage industry to self-regulate their advertising techniques, or should the Australian government be regulating this more closely? Justify your response.

Quiz
What is the impact of advertising on my diet?

HOW CAN I EAT SUSTAINABLY?

Most Australians have access to a wide range of nutritious foods and the means to buy them. While Australia as a whole produces more food than its population needs, growing food puts strain on the environment.

As the world's population continues to grow, producing enough food to feed everyone will become increasingly difficult. There are limits as to how much raw food can be produced using the current resources, particularly nutrient-rich soil and water. There are a number of ways to make **sustainable** food choices; each of these choices makes a small contribution to helping conserve the natural environment.

sustainable in the context of food choices, refers to the ability to maintain or improve nutritional and lifestyle standards without exhausting natural resources or causing severe ecological damage

123RF.com/maxfx

Figure 2.37 Ultimately, our natural environment is the source of everything we consume.

Alamy Stock Photo/MBI

Figure 2.38 Beef production is a significant contributor to Australia's economy.

Weblink
Understanding bush foods First Nations Peoples harvested

FAST FACT

First Nations People have successfully lived off the land for more than 40 000 years. To do so requires a very intimate knowledge of how to sustainably harvest food from a range of plant sources without depleting the natural resources for ongoing use.

9780170463096

ORGANIC FOOD

Organic food is grown without using human-made chemicals such as fertilisers or pesticides, and it is free of **genetic modification** (GM). Organic foods can be plant based (fruit or vegetables) or animal products, including meat, eggs and honey.

FACE TO FACE Organic food

Organic foods have grown in popularity in the past decade. Discuss with a classmate why you think this might be the case. Consider the following questions:

1 What do people see as the benefits of using organic products, and how do they develop this viewpoint?
2 Are there any drawbacks to the increase in popularity of organic products?
3 Do members of your household buy organic products? If so, why? Compare your answer to your classmate's and discuss why your family chooses to use more (or less) organic products.

genetic modification using scientific methods to change the characteristics of an organism's DNA, often to make it resistant to disease or to produce larger crops

Some people choose organic foods for ethical reasons, because animals raised on organic farms are usually treated more humanely. Often organic farming relies on more traditional and sustainable farming practices, such as conserving water, using renewable resources, rotating crops, and natural recycling of nutrients.

UP AND MOVING Kitchen inspection!

Have a look through your fridge and pantry at home, and see if you can find an organically produced food item. Take a photo or bring the label into school (check with your parent or caregiver first). How many different items can your class collect in time for your next Health lesson?

MINIMAL TRAVEL, MINIMAL PRODUCTION

There are several other ways to minimise environmental impact when making choices about food and drinks:

- buy food locally, by using farmers' markets and small local stores
- grow your own food: at home, at school or in a community garden

Shutterstock.com/Ben Wehrman

Figure 2.39 Farmers' markets are a great way to buy fresh food from your local area.

Shutterstock.com/yevgeniy11

Figure 2.40 Wasted food ends up in landfill.

- choose foods that are less processed. The more effort that goes into manufacturing food, the greater the impact on the environment.
- read the labels and try to buy foods that were manufactured or grown in Australia
- look for foods with minimal packaging
- try to minimise overeating and food wastage. Consuming more food than you need is bad for your health and puts extra stress on the environment. Food wastage in Australia is huge. About 3 million tonnes of food per year (worth about $5 billion dollars) ends up going to landfill.
- compost fresh food scraps (such as fruit and vegetables) at home to look after your own garden better and to reduce landfill. Composting food scraps produces fewer greenhouse gases than when they are added to landfill.
- be mindful of where your seafood is sourced. Some species of fish are at risk of being overfished. There are smartphone apps available to help when your family is shopping for seafood.

Weblink
The Australian Marine Conservation Society has a handy online tool to help you choose seafood wisely.

It is difficult to buy local and/or organically grown foods when those foods aren't in season. For instance, strawberries don't grow in Tasmania during winter. So either you go without strawberries until closer to summer, or you buy strawberries that have been transported from somewhere else in Australia (or even overseas), which means you are no longer buying local. Some crops can be grown locally out of their normal season with the use of chemicals, but that means they are no longer organic.

Tap water versus bottled water

Australia has one of the cleanest and safest drinking water supplies in the world. The water that comes out of the tap is as good as any bottled water. Despite claims about the special 'origins' and 'purity' of bottled water, it's no purer than tap water.

9780170463096

Figure 2.41 Strawberries are a seasonal crop; buying them out of season means that they are not locally produced.

FACE TO FACE Bottled water in Australia

Conduct online research into the use of bottled water in Australia, then discuss the following questions with a classmate. You can use the Bottle Water vs Tap Water link provided online in Nelson MindTap as a starting point.

1 Why do some people choose to buy bottled water instead of drinking tap water?
2 What are the benefits of drinking tap water instead of bottled water?
3 Why is it important for us to reduce the amount of bottled water being purchased?
4 Propose some strategies for how we can reduce the amount of bottled water being purchased across the country.

Weblink
Water'ABC Behind the News, 'Bottled vs Tap Water'

Ecological footprint

An ecological footprint is a measure of the impact people have on the environment. It factors in the impact of food production, but also of timber and material production, and the space required for infrastructure and handling wastes. In 2016 the global ecological footprint was 22.6 billion hectares. However, there were only 12.2 billion hectares available, meaning that as a global community we use almost 50 per cent more natural resources than can be regrown or replenished. Australians currently use 6.6 global hectares per person, but the average for every person on the planet would need to be 1.63 for the long-term viability of our ecosystems. If people in every other country consumed resources at the rate Australians do, it would take three 'Earths' to support them.

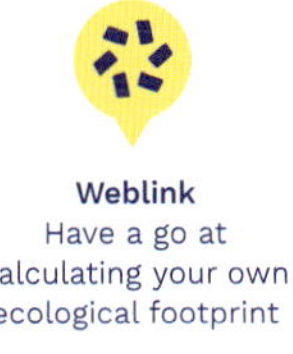

Weblink
Have a go at calculating your own ecological footprint

Local projects

The best way to tackle large environmental problems is to start in your own backyard. As a class, see if you can get one of the following local projects up and running in your school.

1. Start a school vegetable garden

   Many Australian schools are starting to include vegetable gardens within their school grounds. There are many uses and benefits of vegetable gardens. Does your school have one? How could you go about getting one started? Perhaps try investigating other schools that have one and find out how they started theirs.

2. Canteen audit

   How sustainable is your school canteen? Develop a checklist of things you'd like to see taking place in your canteen. Things to focus on include the types of meals and snacks, where the food comes from, how it is grown and processed, how much is wasted and recycled, and where the food scraps go. Offer some strategies for improvement.

3. Nude Food Day

   World Nude Food Day is on 16 October each year, but you can choose to promote this cause whenever you like. It is an event to encourage healthy, nutritious lunches that are environmentally friendly, using only fresh foods and eliminating all unnecessary wrapping and packaging.

Weblink
Nude Food Day

Quiz
How can I eat sustainably?

1. List three sustainability initiatives mentioned in this chapter, and find three more on the web. Make sure they are related to food. Write a short description of each initiative.
2. Does your school currently utilise any food-related sustainability initiatives? Outline how you could introduce one to your school.

1. Propose what you believe to be the primary environmental issues facing Australians and the way we use food.
2. Hypothesise if these issues would be the same in other parts of the world. How would you answer question 1 if you lived in a developing nation?

1. Can you take on the challenge? The 'Love Food Hate Waste' program invites you and your family to save money and help our environment by reducing your food waste. Use the weblink provided and follow the steps to monitor how much food you and your family throws away at home in one week.

Weblink
Track your food footprint

9780170463096

CHAPTER 2 REVIEW

1. List the five main food groups in order of most recommended serves per day to least.
2. Explain what discretionary foods are.
3. What is Australia's nutritional model?
4. What is the difference between energy intake and energy expenditure?
5. What are the key principles of a healthy diet? (Hint: they start with V, B and M.)
6. State what the three food sources of fuel for the body are.
7. A recovery snack after exercise should include which two nutrients?
8. What function does fat play in the body?
9. What is the difference between a serving size and a portion size?
10. What is the difference between 'overweight' and 'obese'?
11. What is your ecological footprint?
12. List four different types of information that can be found on the nutritional information panel on food labels.
13. What percentage of Australian adults are overweight or obese?
14. Summarise in what order the ingredients on a food product are listed on its packaging.
15. List six influences on eating or food choices.
16. List five tips or tricks for choosing healthy food options.
17. Describe two advantages of having a compost system in your backyard.
18. Define 'sedentary'.
19. Why was the Health Star Rating system introduced in Australia?
20. List three ways to consider the environment when making decisions around food and drink products and dietary habits.

HEALTH BENEFITS OF PHYSICAL ACTIVITY

IN THIS CHAPTER

You will learn about the importance of staying fit and active throughout your life and how to make exercise and physical activity a part of your everyday life.

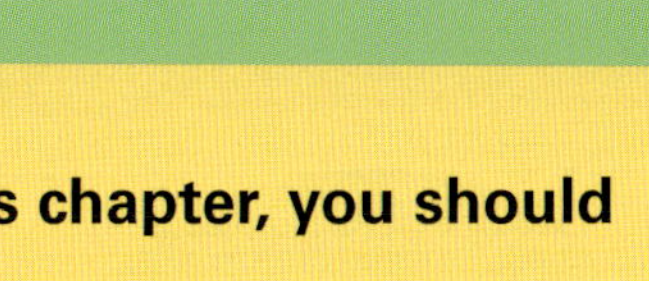

Shutterstock.com/Nicetoseeya

By the end of this chapter, you should be able to:

- define and categorise types of physical activity
- identify opportunities for including activity in your daily routine
- recall national physical activity, sedentary behaviour and sleep recommendations for various age groups
- determine your typical amount of weekly physical activity as well as sedentary behaviour
- understand the health benefits of physical activity including physical, social, emotional, spiritual and cognitive/mental health
- identify the factors that influence your participation in physical activity and/or sedentary behaviour.

HOW DOES PHYSICAL ACTIVITY PARTICIPATION IN AUSTRALIA COMPARE TO OTHER COUNTRIES? 133

WHAT IS PHYSICAL ACTIVITY?

Quiz
Pre-chapter

Before you start, take the pre-chapter quiz to find out how much you already know.

physical activity movement of large muscle groups that requires the use of energy

Worksheet 3.1

Video
Case study: How can we make exercise and physical activity part of our every day life?

Everyone knows that participation in regular **physical activity** results in improved personal and community health. But what does physical activity actually mean? Is it walking the dog or going for a run? Is it hanging out at the skate park or training for netball? How often do you need to be active? Being active on most days – five or more days a week – is considered to be regular physical activity. There are many ways to be active; you don't necessarily need to join a gym or participate in competitive sport.

FACE TO FACE Physical activity

With a classmate, brainstorm all the things you can think of that you would consider to be physical activity.

Alamy Stock Photo/Myrleen Pearson

Figure 3.1 What activities count as physical activity?

TYPES OF PHYSICAL ACTIVITY

All the physical activities you listed in the brainstorm activity can be sorted into two categories:

1. incidental physical activity
2. structured or planned physical activity.

Incidental physical activity

Incidental physical activity is unplanned activity you do during the day, usually in the process of doing something else. For example, if you catch the bus to school, you may have to walk to the bus stop. This walking is considered incidental physical activity. Incidental physical activity can happen in many different ways: walking up and down the stairs in your home or walking around the shops are just two ways to be active without even realising it. Other types of incidental activity are shown in Table 3.1.

Table 3.1 Types of incidental physical activity

Type of incidental physical activity	Description	Examples
Household tasks and gardening	Completing household tasks that use energy	Vacuuming the floor, mowing the lawn, cleaning the bathroom
Active transport	Travelling to a place using own means and efforts	Riding a bicycle to school/ work, riding a scooter to the shops
Occupational activity	Work-related tasks that result in **energy expenditure**	Landscape gardening, riding a bicycle to deliver mail or food deliveries
Play	Non-structured, informal, fun activities	Playing at the park, playing tag, climbing on playground equipment

energy expenditure the amount of energy used to complete an activity, measured in kilojoules

Structured physical activity

Structured physical activity is a type of activity that is planned. Exercise, recreational and leisure activities and organised sport are all types of structured physical activity. There are many examples of structured physical activity: playing sport, going to the gym, doing a spin or dance class, skateboarding and bushwalking are just a few!

DOMAINS OF PHYSICAL ACTIVITY

There are many types of physical activity, and it is not difficult to fit these into your daily routine. Think about all the times you could be active throughout the day. There are many places where this can happen: at home, at school, in the workplace and while getting yourself from one place to another. The places where you are active are called domains. There are four domains in which all activities take place:

- leisure-time activity domain
- household/gardening domain
- occupational domain
- active transport domain.

Newspix/Ben Swinnerton

Figure 3.2 Is this activity incidental or structured?

Leisure-time activity

This refers to what you do in your spare time. You choose the activity you would like to do and you do it for your own enjoyment.

Household/gardening activity

This activity refers to the things you do around the house and garden. Vacuuming, sweeping the floor, hanging out the washing, weeding the garden and raking the leaves are all examples of physical activity in the household and gardening domain.

Occupational activity

sedentary a type of activity that does not use much energy, such as sitting

Occupational activity is done as part of your job or work. Some jobs are highly **sedentary** and involve sitting at a desk all day, while others involve different amounts of activity. Labourers, cleaners, physical education teachers and tradespeople are examples of people who are likely to be very active as a result of their jobs.

Active transport activity

This is physical activity undertaken to reach a destination. Walking, cycling, scootering and skating are all forms of active transport, where the transport is 'self-propelled/powered'.

Worksheet 3.2

Walking to the bus stop is also a type of active transport. What other forms of active transport can you think of? Apart from the physical benefits for the person being active, what benefits are there for the community?

CASE STUDY: PHYSICAL ACTIVITY IMPROVES STUDENTS' ATTAINMENT

Identify

Students who take part in physical exercises like star jumps or running on the spot during school lessons do better in tests than peers who stick to sedentary learning, according to a UCL-led study.

Understand

The meta-analysis of 42 studies around the world, published in *British Journal of Sports Medicine*, aimed to assess the benefits of incorporating physical activity in academic lessons. This approach has been adopted by schools seeking to increase activity levels among students without reducing academic teaching time.

Typical activities include using movement to signify whether a fact is true or false, or jumping on the spot a certain number of times to answer a maths question.

The study concluded that incorporating physical activity had a large, significant effect on educational outcomes during the lesson, assessed through tests or by observing pupils' attention to a given task, and a smaller effect on overall educational outcomes, as well as increasing the students' overall levels of physical activity.

Lead author Dr. Emma Norris (UCL Centre for Behaviour Change, UCL Psychology & Language Sciences) said: 'Physical activity is good for children's health, and the biggest contributor of sedentary time in children's lives is the seven or eight hours a day they spend in classrooms.

'Our study shows physically active lessons are a useful addition to the curriculum. They can create a memorable learning experience, helping children to learn more effectively.'

Co-author Dr Tommy van Steen (Leiden University, The Netherlands), added: 'These improvements in physical activity levels and educational outcomes are the result of quite basic physical exercises. Teachers can easily incorporate these physical active lessons in the existing curriculum to improve the learning experience of students.'

In one of the 42 studies analysed, eight- and nine-year-olds simulated travelling the world by running on the spot in between answering questions relating to different countries. The research team, also led by Dr Norris at UCL, concluded the children were more active and more focused on the task than peers in a control group, following teachers' instructions more closely.

In another study in the Netherlands, primary school children who took part in physically active lessons three times a week over two years made significantly better progress in spelling and mathematics than their peers – equating to four months of extra learning gains.

Source: 'Physical activity in lessons improves students' attainment', University College London, 16 October 2019, https://www.ucl.ac.uk/

Discuss

1 Given the findings of this study, do you think students are more likely to learn better after lunch break or before lunch break?
2 Why do you think this?
3 What assumption did you make about student activity during lunch in your answer?
4 Why do you think students who are engaged in physical activity show higher levels of problem-solving, memory and overall academic achievement?

INVESTIGATION  THE LOOK STUDY

Purpose

The LOOK study is a longitudinal project investigating the effect of physical activity on the health and development of young Australians. Your investigation should increase your understanding and appreciation of how participation in regular physical activity improves the quality of life of Australians across multiple areas.

Method

1 Access detailed information about this project at look.org.au.
2 Use the information presented on the website to understand how the project was structured and is still being undertaken as a longitudinal study. Read the key discussion points outlined below prior to commencing your research/investigation, and take summary notes as you navigate through the website to obtain important key findings.

Weblink
LOOK

Discussion

1 One of the current main findings of the LOOK project is: 'In brief, strong evidence has emerged of a negative impact on the health and well-being of children of the 21st century when they are not afforded opportunities for regular well-designed physical education provided by specialist teachers; when they are insufficiently active; and when they do not participate in any form of organized sport'

'Early research findings' of the LOOK lifestyle study, Research Institute for Sport and Exercise (UCRISE)

Complete the following table to summarise the negative impacts on the following health areas:

Physical health	Social health	Emotional health	Cognitive/mental health

2 Are boys more physically active than girls? Briefly discuss the findings, and outline possible reasons for any change in levels of physical activity that boys and girls experience while at school.
3 Highly active people can also experience high levels of sedentary behaviour.
   a How is this possible?
   b Provide three simple strategies to reduce the amount of sedentary behaviour 12–17 year olds engage in.

Video
Australians' physical activity, sedentary behaviour and sleep: How many hours a day do you spending exercising, sitting down and sleeping? What changes could you make to your daily routine to be healthier? Watch the video and join the discussion!

NATIONAL PHYSICAL ACTIVITY, SEDENTARY BEHAVIOUR AND SLEEP RECOMMENDATIONS

To help understand how much activity is necessary for health, and how often it should occur, the Australian Government created Australia's Physical Activity and Sedentary Behaviour Guidelines for adults, and the Australian 24-hour Movement Guidelines for Children and Young People (5–17 years) for children and teenagers. These guidelines outline the minimum levels of physical activity people should do in order to gain health benefits and suggest ways to be more active in everyday life. The guidelines for children, teens and adults are shown in Table 3.2.

Table 3.2 National physical activity, sedentary behaviour and sleep recommendations

Age group	Physical activity guidelines	Sedentary behaviour guidelines
5–17 years	Accumulating 60 minutes or more of moderate to vigorous physical activity per day involving mainly aerobic activities Several hours of a variety of light physical activities Activities that are vigorous, as well as those that strengthen muscle and bone should be incorporated at least 3 days per week. To achieve greater health benefits, replace sedentary time with additional moderate to vigorous physical activity, while preserving sufficient sleep.	Break up long periods of sitting as often as possible. Limit sedentary recreational screen time to no more than 2 hours per day. When using screen-based electronic media, positive social interactions and experiences are encouraged.
18–64 years	Doing any physical activity is better than doing none. If you currently do no physical activity, start by doing some, and gradually build up to the recommended amount. Be active on most, preferably all, days every week. Accumulate 150 to 300 minutes (2½ to 5 hours) of moderate intensity physical activity or 75 to 150 minutes (1¼ to 2½ hours) of vigorous intensity physical activity, or an equivalent combination of both moderate and vigorous activities, each week. Do muscle strengthening activities on at least 2 days each week.	Minimise the amount of time spent in prolonged sitting. Break up long periods of sitting as often as possible.
SLEEP (Newly added recommendations – 2019) An uninterrupted 9 to 11 hours of sleep per night for those aged 5–13 years and 8 to 10 hours per night for those aged 14–17 years. Have consistent bed and wake-up times.		

FAST FACT

1 Inactivity is a risk factor for lots of **chronic** conditions. People who do not meet physical recommendation guidelines are at greater risk of cardiovascular disease, type 2 diabetes and **osteoporosis**.

2 2.5 per cent of the total **disease burden** is attributed to physical inactivity.

3 10–20 per cent of the disease burden from diabetes, bowel cancer, uterine cancer, **dementia**, breast cancer, coronary heart disease and high blood **cholesterol** is attributed to physical inactivity.

chronic a condition that lasts a long time or frequently reoccurs

osteoporosis a disease where bones lose their density and become fragile and brittle, leading to a higher risk of broken bones

disease burden the impact of a health problem as measured by financial cost, mortality or morbidity

dementia a disorder that affects the functioning of the brain and interferes with a person's ability to live a normal life

cholesterol a fatty substance produced naturally by the body and found in blood

dimensions the parts or features that make up a situation, problem or thing

DIMENSIONS OF PHYSICAL ACTIVITY

The three **dimensions** of physical activity in the national physical activity guidelines are known as the FIT formula:

F = frequency: how often people need to be active (days per week)

I = intensity: how hard the activity needs to be

T = time: how long the activity should be done for.

Frequency

The frequency of physical activity required for children and youth is very straightforward: you need to be active every day!

Intensity

Intensity is a little more difficult to understand. Intensity is a measure of how hard the activity is, or how much energy is needed to do it (energy expenditure). Intensity can be classified as light, moderate (medium) or vigorous (hard), depending on how much energy is used. Some examples of light, moderate and vigorous activity are shown in Table 3.3.

Worksheet 3.3

Table 3.3 Examples of light, moderate and vigorous activity

Light intensity	Moderate intensity	Vigorous intensity
Will not leave you out of breath. Example: washing the dishes.	Will leave you feeling warm and slightly out of breath. Example: bike riding.	Will make you 'huff and puff' and raise your heart rate. Example: playing soccer.

Left to right: Shutterstock.com/Denizo71; 123RF.com/Jacek Chabraszewski; iStock.com/leisadavis

VO_2 maximum the maximum amount of oxygen that can be taken up, transported and utilised, and measured in ml/kg/min in a laboratory

rate of perceived exertion a measure of how hard you think your body is working (subjective measure)

There are a number of ways to work out the intensity of an activity. Some, such as heart rate monitoring and **VO_2 maximum**, require some equipment and calculations, but others are simple to use. One simple method is the talk test. If you can continue to talk comfortably while being active, then you are working at light or moderate intensity. There are also a number of devices and smartphone apps that can be used to calculate how hard you are working during exercise.

Another method that can be used is **rate of perceived exertion**. Using a scale, you estimate how hard you think you are working, based on how you feel. Table 3.4 shows a modified version of the Borg rating of perceived exertion (RPE) scale.

Table 3.4 Modified rating of perceived exertion (RPE)

	This is easy! No effort at all. **RPE = 1**		I am slightly breathless, but I can still answer you if you talk to me. **RPE = 6**
	I can do this all day. **RPE = 2**		I am puffing quite a bit now and even though I can talk to you, I don't want to. **RPE = 7**
	I'm breathing a bit faster but still doing it easy. **RPE = 3**		I can't keep this pace up for long! I can just answer you, but I don't want to. **RPE = 8**
	Starting to warm up now. I'm sweating a bit, but I can still chat while I exercise. **RPE = 4**		This is extremely hard and I can't do much more. **RPE = 9**
	This is getting a little bit harder, but I am comfortable and happy to chat. **RPE = 5**		I am working as hard I can and can't go on any longer. **RPE = 10**

UP AND MOVING Get up and get active!

Use the modified rating of perceived exertion (RPE) in Table 3.4 to determine how hard you are working when performing each of the following activities:

- walking two laps of the oval
- doing 10 burpees
- sprinting 100 metres
- climbing up a set of stairs as quickly as you can.

Record your RPE for each activity. Compare your results with those of others in your class, looking for similarities and differences. Discuss reasons why the RPE for the same activity might be different for different individuals.

Time

The amount of time you need to spend is simple: you need to be active for at least an hour a day. This hour can be made up of a number of shorter periods of time, as long as they add up to 60 minutes.

Type

The type of activity determines the type of benefits that are gained from being active. Type (T) can be added to the FIT formula to make FITT:

F = frequency

I = intensity

T = time

T = type.

Earlier in this chapter you looked at different types of physical activity and how to classify them. The type of activity you do is usually chosen for a purpose. Household activity is done to complete chores, active transport is done to reach a destination, occupational activity is part of a person's job and leisure-time activity is for enjoyment. The type of activity can also be chosen for a certain purpose. For example, you might want to improve your fitness, strength or sporting ability. This will determine the activities you choose. The physical activity pyramid categorises physical activity based on the FITT formula of frequency, intensity, time and type. When including physical activity in your daily routine, lifetime physical activities (the bottom layer of the pyramid) should be the physical activity you do most.

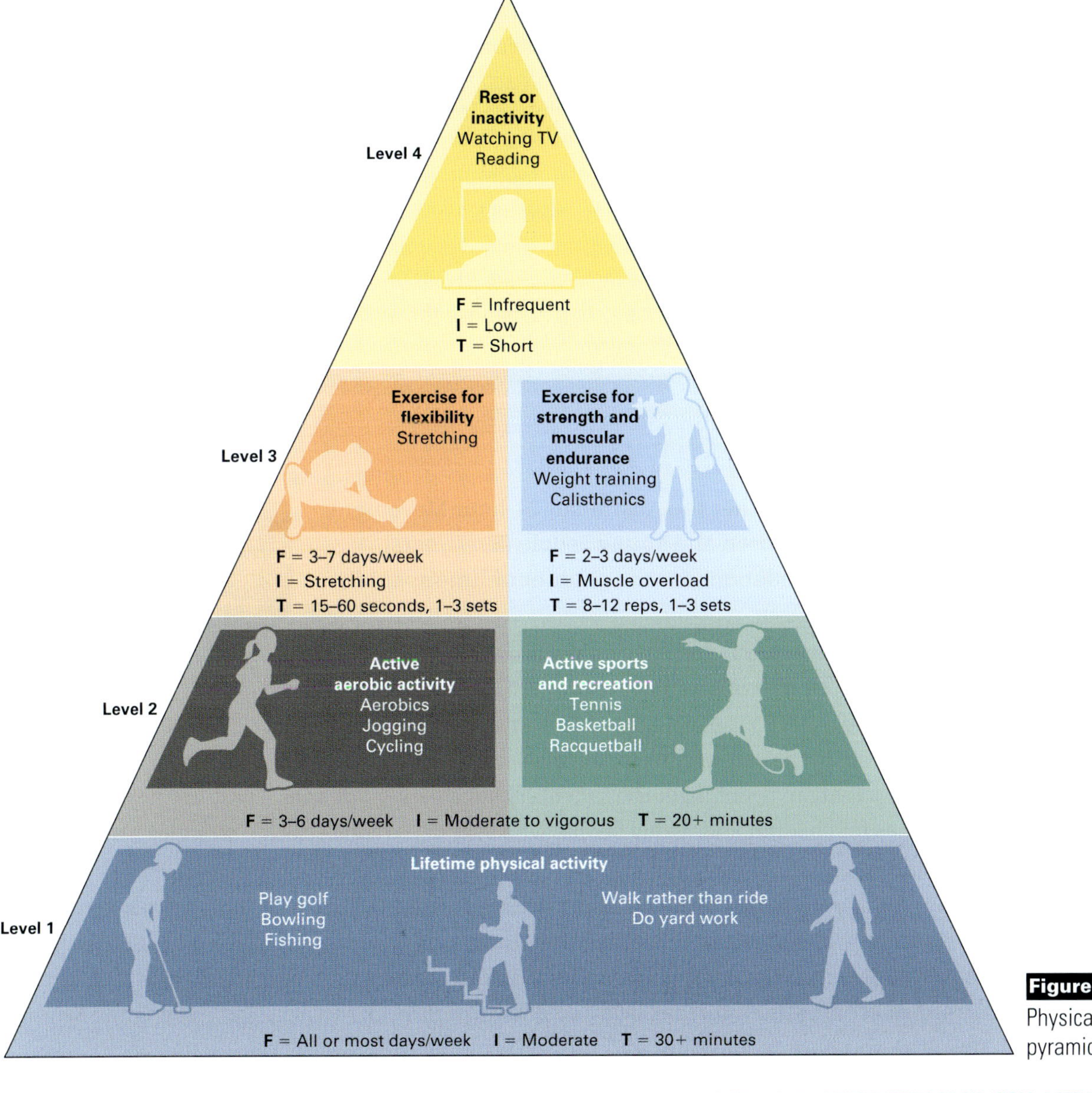

Figure 3.3 Physical activity pyramid

INVESTIGATION

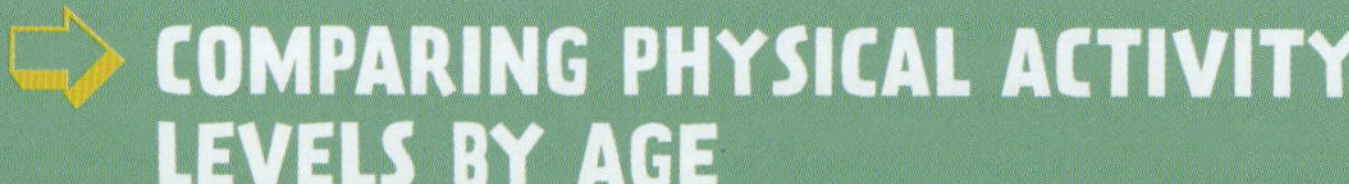

COMPARING PHYSICAL ACTIVITY LEVELS BY AGE

Purpose

To compare and contrast physical activity and sedentary behaviour patterns of older Australians and young people.

Method

1. Ask an adult, perhaps a parent or a grandparent, to list all of the physical activity they did for one weekday and one weekend day in the past week. Remind them to consider each of the domains (leisure-time activity, household and gardening, occupational, and active transport).
2. Ask them to also write down how much time they spent doing each activity and the intensity (light, moderate or vigorous).
3. Record the physical activity you did on one weekday and one weekend day last week.

Discussion

1. Make a table to present the data you have collected. Include the frequency, intensity, time and type (FITT) of activity done for your older Australian, as well as your own data.
2. Did you and the adult both complete activities in each level of the pyramid? Were there any levels in which you didn't do any activity? If so, suggest reasons why both you and the adult didn't perform any activity in those levels.
3. Why do you think the Level 1 activities are at the bottom of the pyramid?
4. Did you and the adult you asked meet the guidelines for physical activity and sedentary behaviour? Use the information you collected to justify your answer.
5. Research the guidelines for older adults (65+ years) and children 0–5 years. Compare them to the guidelines for adults and your age group. What are the similarities and differences? Present your information in a Venn diagram.

SEDENTARY BEHAVIOUR

Weblink
Australia's Physical Activity and Sedentary Behaviour Guidelines

You will also notice that the Australian 24-hour Movement Guidelines for Children and Young People (5–17 years) makes recommendations for the amount of time spent in sedentary behaviour.

FAST FACT
Over half of all golfers are aged 55+!

Sedentary behaviour involves sitting or staying in one place for long periods of time, where only small amounts of energy are used. You spend a lot of time sitting throughout the day. Think about all the times you sit: in the car, in front of the television or in class. Even people who meet the physical activity may not meet the sedentary behaviour guidelines. There are positive health outcomes for children and young people who spend less time sitting – so sit less, move more!

Minimising the amount of sedentary behaviour in your day is important for your health and **wellbeing**. There are many different ways to increase the physical activity in your daily routines and to reduce the amount of time you are sedentary. Some small changes you could make include using the stairs rather than the lift or escalator; getting dropped off for school further away so you have to walk further; and putting the remote away so you have to stand up and move to change the channel on the television.

wellbeing an overall feeling of wellness that combines physical, mental, social, emotional, cognitive and spiritual health

SLEEP

Sleep recommendations were added to the guidelines in 2019 after health authorities confirmed the widespread health benefits associated with having consistent sleep and wake-up times, in addition to quality sleep.

In summary, these recommendations require:

- an uninterrupted 9–11 hours of sleep per night for those aged 5–13 years, and 8–10 hours per night for those aged 14–17 years
- consistent bed and wake-up times.

hypokinetic disease diseases that are caused by inactivity

FACE TO FACE Debate

As a class, debate the statement 'Advances in technology are making people lazy'. Research your point of view and find examples to support or argue against the statement. Include all labour-saving devices (for example, cars, dishwashers, remote controls) in your interpretation of 'technology' in this task.

Quiz
What is physical activity?

1 The physical activity 'guidelines' are recommended to enable people to experience good health. Explain why regular physical activity is recommended over five to seven days, rather than just accumulating the same amount on the weekend.
2 Everyone engages in sedentary behaviour, but excessive sedentary behaviour has been linked to **hypokinetic diseases**. Identify an acceptable amount of sedentary behaviour.
3 Discuss how you would know if you were participating in an activity at the recommended moderate intensity.

1 Using all of the activities you brainstormed (see page 106), create a diagram to show your understanding of how physical activity can be categorised. Your diagram must describe the physical activity as incidental or structured. Each activity type can then be further classified into its domain.
2 Construct a 'road map' of your sitting time for a typical school day (Monday–Friday). Include the periods of time when you are active and those when you are sedentary. Add up the total time you spent sitting for the day. You can model your road map on the diagram below, or simply enter your data into a table.

Time	Amount of time sitting
7–7.30 a.m.	30 mins eating breakfast while watching TV
8.40–9.35 a.m.	55 mins sitting in Maths class

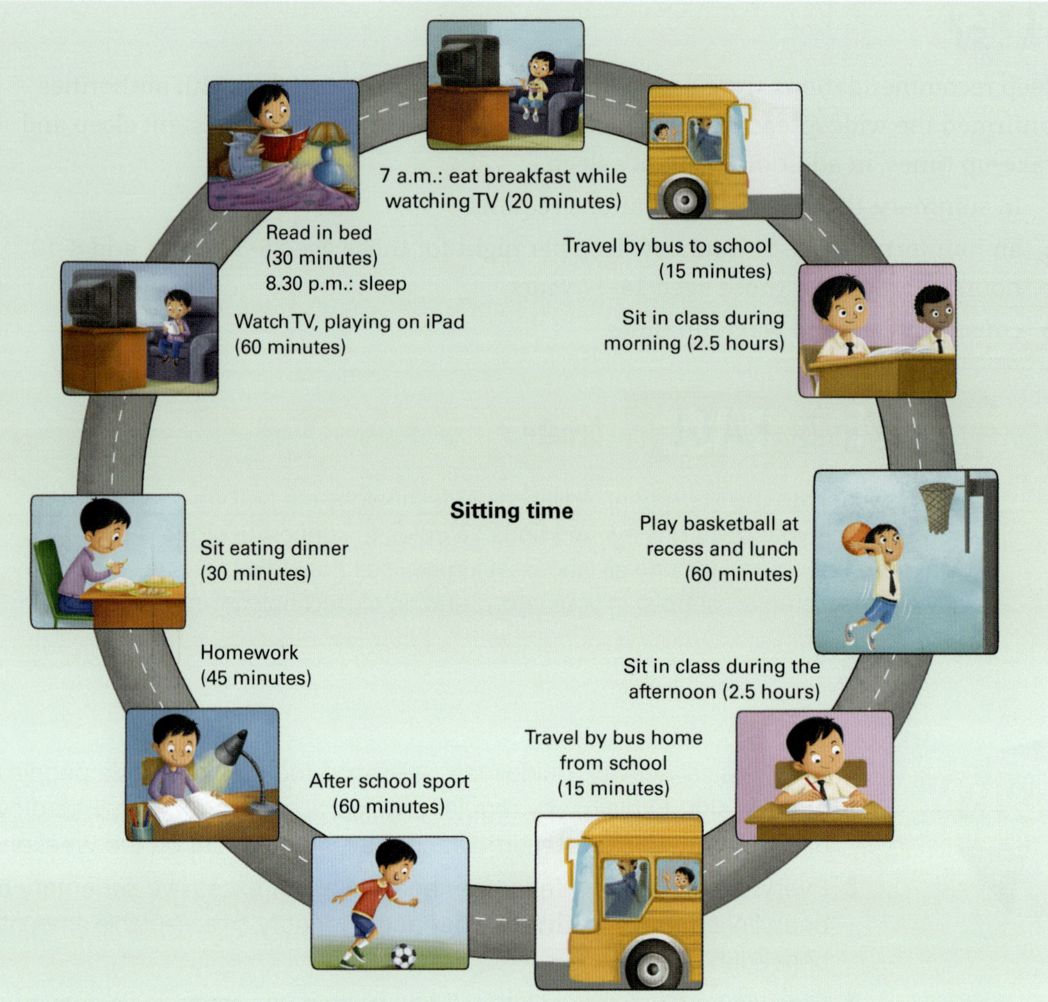

3 Recall a time when you did not have a good night sleep, either because you had a late night and got less than eight hours of sleep, or because you couldn't sleep comfortably due to illness, heat, etc. Briefly discuss how you performed the following day at school, on the sporting field or just generally around the house.

1 Referring to the physical activity pyramid (page 113), read each of the cases below. Create a table using the information given for each person into the dimensions (frequency, intensity, time and type) provided by the Australian 24-hour Movement Guidelines for Children and Young People.

- **Case 1:** Emma, aged 12, walks 10 minutes to and from school each day with her friends. She also does athletics and trains for an hour and a half twice a week. She competes on Saturdays for three hours.
- **Case 2:** Brianna, aged 13, catches the bus to and from school each day. Every lunchtime, she plays basketball with her friends and is often hot and sweaty by the time she has to go back to class. At home, Brianna has to walk the dog for half an hour every day, and they usually end up playing fetch and running around chasing a ball together.
- **Case 3:** Steven, aged 11, is driven to school by his mum, who also picks him up. Steven loves to play on his iPad and spends most afternoons after school playing computer games. He plays football on the weekend in winter; the games go for 90 minutes.

Refer to the guidelines to work out if each person is meeting the recommendations.

WHAT ARE THE BENEFITS OF BEING PHYSICALLY ACTIVE?

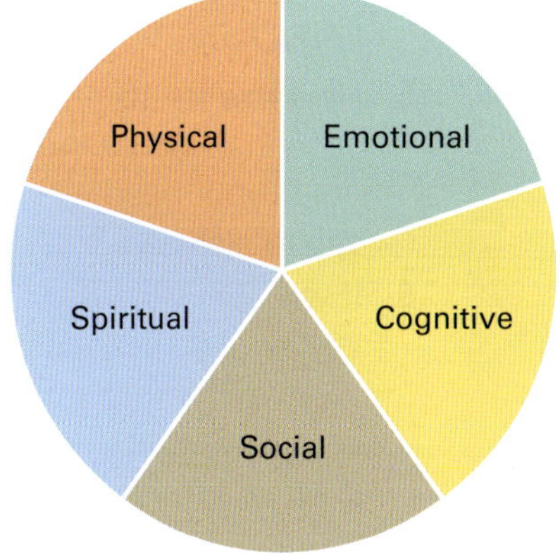

Figure 3.4 Dimensions of health and wellbeing

Physical activity has many physical, social, emotional and **cognitive** benefits. These benefits affect you as a person and the community in which you live. When considering the benefits of being physically active, words like 'fit' and 'healthy' are often used interchangeably, but while they are related, they each refer to different aspects of overall wellbeing.

Wellbeing is how a person feels about life and how effectively they can function. A person who is 'well' is satisfied at school or work, is spiritually fulfilled (can find peace and purpose in life), enjoys leisure time, is physically fit, is socially involved (has friends) and has a positive emotional outlook (is mostly happy).

Physical, emotional, cognitive, social and **spiritual health** are all parts of wellness. The World Health Organization defines health as 'a state of complete physical, mental and social wellbeing and not merely the absence of disease' (Constitution, World Health Organization, © 2020 WHO). Fitness can be defined as 'the ability to carry out daily tasks with vigour and alertness, without undue fatigue, and ample energy to enjoy leisure time pursuits and meet unforeseen emergencies' (Clarke HH. Basic understanding of physical fitness. Physical fitness research digests series 1971;1:2.).

cognitive relating to your ability to think, learn and remember

spiritual health having a sense of belonging and connectedness

Worksheet 3.4

WELLBEING CHECK IN

EXERCISE CAN MAKE YOUR BRAIN FEEL BETTER

This activity has been developed in collaboration with Dr David Bakker of MoodMission

Identify

Exercise isn't just good for physical fitness. It has positive effects all over the body, including the brain.

Understand

There are many reasons why exercise makes us feel good, and two of these are particularly significant. Firstly, exercise can promote the release of feel-good chemicals in the body, called endorphins. Secondly, it can help us feel like we've achieved something, which can boost our confidence and optimism. You don't have to run a marathon to get these benefits. A really quick exercise can be a step in the right direction. Planking is a way of exercising muscles in your arms, torso and legs, all without moving.

Practise

1. Find a flat spot where you can lie down.
2. Begin in the plank position, with your forearms and toes on the floor.
3. Keep your body rigid, with a straight line from your ears to your toes and no sagging or bending.

4 Relax your head and look at the floor.
5 Hold this for 10 seconds – count these out in your head while you're planking.
6 If you can, try going to 20, 30 or 60 seconds. You may have to work your way up to this with practice.

Shutterstock.com/G-Stock Studio

Reflect

How do you feel after doing this short exercise? Do your stomach and core ache in sort of a good way? Do you think if you did this over and over, maybe every day, you'd get better at it?

PHYSICAL BENEFITS OF REGULAR PHYSICAL ACTIVITY

Fitness is specific to each person. Your daily tasks might not involve using much energy at all, so you might not require the same level of fitness as an athlete who runs 10–15 kilometres every day! The many parts to fitness are called the components of fitness. Health-related fitness components are those that are important to your health and reduce the risk of disease. Skill-related components are those that are important to your performance in motor skills and sport. The physical benefits of participating in regular physical activity can be health-related and/or skill-related. Table 3.5 lists some of the components of fitness.

Table 3.5 Health- and skill-related fitness components

Health-related fitness	Skill-related fitness
Aerobic capacity	Balance
Anaerobic capacity	Reaction time
Body composition	Coordination
Muscular strength	Agility
Muscular endurance	Speed
Flexibility	Muscular power

The benefit of physical activity depends on the frequency, intensity and type of activity. If you look at the physical activity pyramid again (page 113), you will see that the frequency, intensity, time and type of activity are specific to the outcome of the activity. For example, the exercise frequency, intensity and time you need in order to achieve flexibility is different from that required to improve your strength. Exercise programs and activities to improve different components of fitness are examined in more detail in Chapter 9 (pages 359–67).

Benefits for the heart

cardiorespiratory system the functioning of the heart and lungs

Regular physical activity has many benefits for the heart and **cardiorespiratory system**. It can increase the size and strength of the heart, meaning it can pump more blood with each beat and doesn't have to work as hard to deliver the blood around the body. Regular physical activity also helps keep blood vessels healthy and free of blockages, which improves the circulation of blood around the body. Having healthy **cardiovascular** and **circulatory systems** means less chance of heart disease, high blood pressure (**hypertension**), high cholesterol, heart attacks and heart failure.

cardiovascular system relating to the heart and circulatory systems

circulatory system the system that circulates blood around the body, including the heart, blood and blood vessels

hypertension abnormally high pressure of blood in the blood vessels

FAST FACT

Did you know there are two types of cholesterol or fatty substances in your blood? One is good and has protective properties, but the other is bad if there is too much of it in your blood, because it can block your blood vessels so the blood can't flow easily. Regular physical activity helps to reduce the amount of the 'bad' cholesterol, the low-density lipids (or LDLs), and increase the 'good' cholesterol, the high-density lipids (or HDLs).

Worksheet 3.5

Weblink View the series of illustrations showing how a build-up of cholesterol can lead to a heart attack on the American Heart Association website.

Benefits for the bones

Strong bones are important because they provide the framework of the body and, much like the frame of a house, they hold you up! Physical activity in childhood can make bones stronger and also helps to stop bones becoming weaker as people get older. **Weight-bearing exercise** has also been shown to increase **bone density**, reduce the risk of osteoporosis and help manage joint-related diseases such as **arthritis**.

Figure 3.5 LDL vs HDL cholesterol

Benefits for the muscles

Regular physical activity can increase the strength, endurance and flexibility of muscles, **ligaments** and **tendons**. Increased strength allows muscles to work more efficiently, reduces the risk of muscular and joint injury and, as you get older, allows you to stay more mobile and independent.

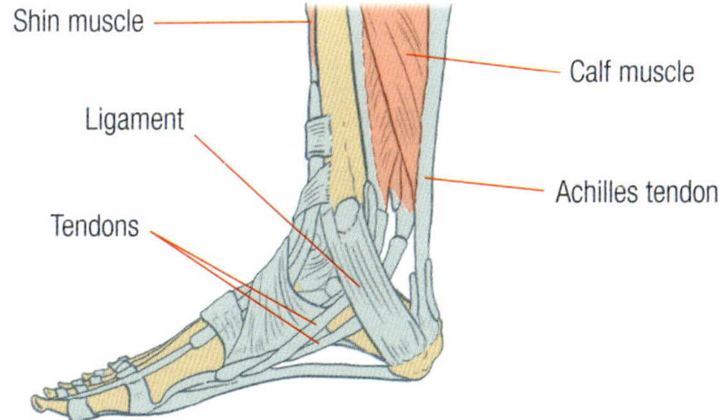

Figure 3.6 Ligaments and tendons of the ankle

FAST FACT

Ligaments join bone to bone and tendons join muscle to bone.

weight-bearing exercise exercise that requires the bones to support the weight of the body

bone density the amount of calcium and other minerals in bones, which is used to indicate how strong they are

arthritis a disease that causes joints to become stiff, swollen and painful

ligament a strong structure that connects bones to other bones, such as at the knee, ankle and shoulder

tendon a flexible but inelastic cord-like tissue that connects muscles to bones

There are many other physical benefits of regular physical activity. Some of these are summarised in Table 3.6.

Worksheet 3.6

Table 3.6 Physical benefits of regular physical activity

Healthy heart	
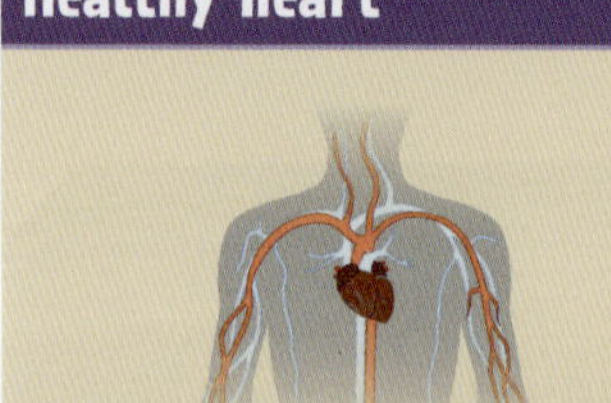	Less risk of high blood pressure Less risk of cardiovascular disease (CVD) Lower 'bad' cholesterol
Healthy bones	
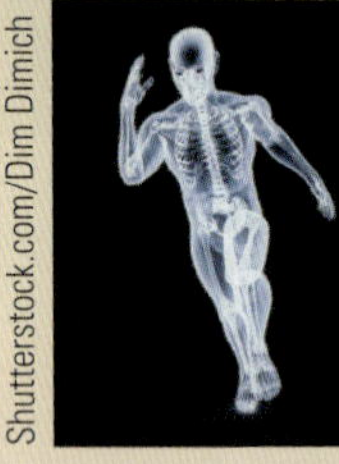 Shutterstock.com/Dim Dimich	Stronger bones Decreased loss of bone density Stronger and healthier bones and joints May help arthritis sufferers
Healthy muscles	
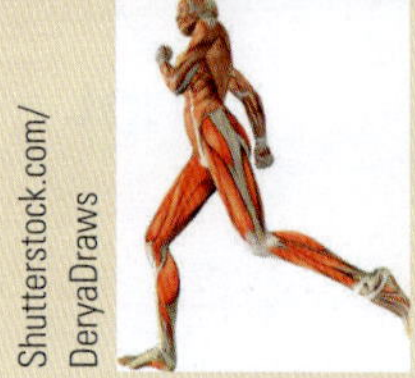 Shutterstock.com/DeryaDraws	More strength Bigger muscles Greater flexibility Less chance of muscle injury
Healthy body	
 Shutterstock.com/michaeljung	Less body fat and risk of obesity Less risk of type 2 diabetes Less risk of bowel and breast cancer and other cancers Longer life Delayed physical effects of ageing Increased energy levels (takes longer to get tired) Improved immune system (don't catch colds often) Improved posture

Weblink
Watch the YouTube video to discover more benefits of exercise on the brain and body.

INVESTIGATION  HOW CAN I REDUCE MY RISK OF CARDIOVASCULAR DISEASE?

Purpose

To discover some of the statistics about cardiovascular disease (CVD) and become more aware of how to reduce the risk of cardiovascular disease.

Method

1 Go to the Heart Foundation website and click on 'Data and statistics'.

Weblink
Heart Foundation

Discussion

Use the statistics to answer the following questions.

1 How many people does CVD affect each year?
2 What is the leading cause of death in Australia?
3 What are 'modifiable' risk factors?
4 How does physical activity contribute to the prevention of CVD?
5 Investigate the preventative health practices available in your community for CVD.
6 Using the information you have found, design a health promotion campaign for these preventative health practices. Present your campaign in one of the following formats:
   - TV commercial
   - radio commercial
   - print media campaign (poster, brochure, newspaper, magazine, etc.)
   - web-based campaign.

Remember, your target audience is people your age.

OTHER BENEFITS

Being physically active is not only good for physical health, it can also have social, emotional, cognitive and spiritual health benefits.

Social health

Social health is your ability to make and keep healthy relationships with the people around you, such as family, friends and teachers.

You are socially healthy when you have a network of friends and family you can rely on for support and to share life experiences. It is important to talk to people every day about what is going on in your life, including both the good and the bad!

nutrition food that is needed for health and growth

Social health is just as important to overall wellbeing as physical health. Poor social health can be as bad for your health as poor **nutrition** or not exercising. Physical activity can improve social health through interactions with others while playing team sports and in recreational settings. Being active with someone else also has the added benefit of making physical activity more enjoyable.

There are many ways to improve your social health through physical activity. Some examples are volunteering at the local Auskick program, offering to walk an elderly neighbour's dog or getting involved in a community garden. The benefits are more

iStock.com/kali9

Figure 3.7 Why is it important to have a range of people to talk to, including adults as well as friends?

than just being physically active: you get to meet new people, help others and contribute to the community. These benefits help to improve your social health.

Five tips for improving your social health

1 Develop healthy relationships with yourself and others.
2 Avoid unhealthy and destructive relationships.
3 Find out where to get support when you need it.
4 Involve yourself in your community.
5 Help other people.

It is important to have people in your life who you can trust and who will provide help and support when you need it. There will be people you might talk to every day and share both little and really important things with. These might be your parents, or perhaps a close friend. You will only interact with some other people, such as school counsellors, when you need them.

FAST FACT

1 Activities with the greatest proportion of female participation include Pilates (90 per cent), netball (89 per cent) and dancing (89 per cent).

2 Activities with a greater proportion of men participating include cricket (88 per cent), AFL (84 per cent) and golf (81 per cent).

Emotional health

Emotional health is the ability to recognise, understand and effectively manage your emotions and to use this knowledge when thinking, feeling and acting. Regular physical activity can reduce stress, anxiety positive effect on self-esteem and sleep habits. How you think and feel can play a big part in how you behave.

Figure 3.8 You can tell someone, or sometimes everyone, how you feel with one little picture.

Life as a teenager is not always happy, and there will be times when you have to face situations that are difficult to deal with emotionally. These situations might include the breakdown of a friendship, starting at a new school, being bullied by classmates or even the death of a family member. On top of the expectations of just being a teenager, these experiences can be very hard. The ability to thrive despite these challenges depends on your **resilience**. Being resilient means being able to adapt well to difficulty, trauma or tragedy – anything that causes stress – and manage feelings of anxiety and uncertainty.

resilience the ability to cope with and recover from difficult situations

FAST FACT

Which of the two statements below do you think is true?

- Successful people are more likely to be happy.
- Happy people are more likely to be successful.

Research has found that 'chronically happy people are in general more successful across many life domains than less happy people and their happiness is mostly because of their positive emotions ... When people feel happy, they tend to feel confident, optimistic and energetic and others find them likeable and sociable'.

Adapted from 'The benefits of frequent positive affect: does happiness lead to success?', Lyubomirsky, King, Diener, and The Gallup Organization, Psychological Bulletin of the American Psychological Association, Vol. 131, No. 6, November 2005

9780170463096

FAST FACT

Geelong Grammar School in Victoria has been collaborating with Professor Martin Seligman and his team from the University of Pennsylvania. They have developed 'Positive Education' – a whole-school approach to teaching and learning from early learning to Year 12. The Handbury Centre for Wellbeing develops strategies to help students deal successfully with modern living, allowing them to feel confident, resilient and optimistic.

Cognitive health

Cognition is the ability to think, learn and remember. It is the basis for how you reason, judge, concentrate, plan and organise. Good cognitive health, like physical health, is very important.

Regular physical activity improves cognitive health. Many studies have shown that the structures in the brain increase in size and the brain performs better with regular physical activity. This boosts memory and learning, improves decision-making and allows you to think more clearly and learn more effectively. That means that by increasing physical activity levels, you can also do better in your school work!

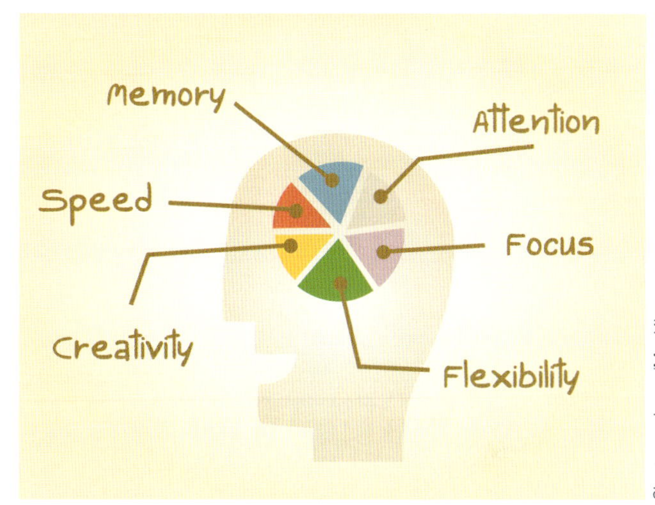

Figure 3.9 Cognitive skills

Memory game

What to do

1. Play a game of memory. You can use a deck of cards or make your own with 10 pairs of words or pictures.
2. Place the cards face down on the table and take turns locating the pairs.
3. Record how many turns you took to find all of the pairs.
4. Now, go outside and do five 50-metre sprints.
5. Repeat steps 1 to 3.

Discussion

1. Did the number of turns required to locate all of the pairs stay the same, increase or decrease after your sprinting?
2. Based on your results, write a statement that shows the relationship between exercise and short-term memory.
3. Define cognitive health.
4. What are the benefits of regular physical activity for cognitive health?
5. Design another experiment like this one to test the effect of physical activity on cognitive health.

Regular physical activity doesn't just improve cognitive function. It can also help older adults maintain their brain function so they continue to lead healthy, meaningful, independent lives. There are many benefits, to both the individual and the community as a whole, of reducing age-related decline in physical, social, emotional and cognitive health. Older adults who are healthy are likely to have fewer illnesses, be more mobile, have greater independence and have a lower risk of dementia.

FAST FACT

Regular physical exercise, such as a brisk 30-minute walk three times a week, can increase brain power and maintain good brain function in older adults.

Shutterstock.com/Tom Wang

Figure 3.10 Keeping active as you get older not only benefits the body, but also the brain.

WELLBEING CHECK IN

HOW TO RUN WITHOUT RUNNING

MoodMission

This activity has been developed in collaboration with Dr David Bakker of MoodMission

Identify

Want to go for a run without going for a run?

Understand

It doesn't take much to get the body's endorphins flowing and you feeling good. Anything that gets your blood pumping can do the trick, such as running on the spot. Make sure you don't hurt yourself with any of these exercises though, and always keep good form – don't move your body in ways that could cause injury.

Practise

1. Run on the spot and lift your knees as high as possible.
2. Time yourself for 30–60 seconds.
3. Do you feel your heart pumping a bit harder? If not, go for another 30–60 seconds.

Reflect

How do you feel after doing this short exercise? You might feel a bit sore, hot and sweaty, but you also might feel a little more engaged. You will certainly feel different to when you started. This is helpful to notice, as this is another way exercise can help us feel better when we're feeling down or anxious – it can make us change states.

Shutterstock.com/puhhha

9780170463096

Spiritual health

Spiritual health refers to a person's sense of belonging, meaning and purpose in life. Spiritual health is a personal and unique part of you that allows you to make sense of your world. People who are physically active are more likely to make connections with others and with nature and to be more aware of their spirituality. For some people, spirituality may involve their religious beliefs and faith; for others it may be a sense of inner peace. Many cultures believe the mind, body and spirit are all connected. When these elements are all healthy, a person will experience overall wellness.

iStock.com/FatCamera

Figure 3.11 Meditation and mindfulness is a great technique for connecting mind, body and spirit.

MINDFUL WALKING

This activity has been developed in collaboration with Dr David Bakker of MoodMission

Identify

Going for a walk isn't just good for your physical health, it can also be a good opportunity to see the world around you.

Understand

We know that mindfulness is good for mental health. Mindfulness is often practised through meditation, but it actually just means paying attention to the present moment with openness, curiosity, and without judgement. This means we can do almost anything mindfully, not just meditate. Walking can be a great way of exercising mindfully, as you can pay attention to all the sights, sounds, and sensations that you may not normally notice.

Shutterstock.com/Sacho films

Figure 3.12 Walking can be a great way of exercising mindfully.

Practise

The next time you walk anywhere, keep mindfulness in mind. You don't have to plan a special walk; it could just be the next time you walk from one class to the next, or to your front door. Make sure you don't have any big distractions when you walk. That means going alone and leaving your headphones out.

While walking, try to pay special attention to your surroundings. What colour and patterns are on the ground? If you're outside, look at the trees and count the branches. Notice their shapes. Notice how many colours you can see in each plant, and compare these with neighbouring plants.

If you get distracted by something, first notice that you've become distracted, put the distraction to one side and then refocus on your surroundings. You might have lots of distracting thoughts about yourself, other people, the past, the future, etc. That's okay – in fact, it's totally normal. Refocusing on your surroundings will help you let go of any anxieties or negativity.

Reflect

What did you notice on your walk? Did you discover something that you hadn't noticed before? Were you able to be present and not distracted by your own thoughts? It's OK if you were distracted, but the more you practise mindfulness, the easier it can become to be present and fully enjoy experiences.

CASE STUDY

CLOSING THE GAP REFRESH

Identify

There are a number of health and wellbeing initiatives designed to promote better health outcomes for Aboriginal and Torres Strait Islander Peoples. The Australian Indigenous Health*InfoNet* website contains a wealth of information about Indigenous (also known as First Nations Peoples in this book) health issues.

Understand

In 2019, all levels of Australian government and a Coalition of Aboriginal and Torres Strait Islander Peak Organisations signed a formal agreement to work in genuine partnership to reduce disadvantage among First Nations Australians. This initiative has been called 'Closing the Gap Refresh' and aims to eliminate inequality between Indigenous and non-Indigenous Australians with respect to life expectancy, child mortality, education and employment outcomes, with a special focus on shared decision-making. It comes after more than a decade of mostly unsuccessful attempts by governments to meet specific 'Closing the Gap' targets.

Between 2008 and 2018, four out of seven 'Closing the Gap' targets were **not met**. They were:

- the target to halve the gap in child mortality rates by 2018
- the target to halve the gap in reading and numeracy by 2018
- the target to close the gap in school attendance by 2018
- the target to halve the gap in employment by 2018.

Two of the three continuing targets are **on track**. They are:

- the target to ensure 95 per cent of all Aboriginal and Torres Strait Islander four year-olds are enrolled in early childhood education by 2025, with 86.4 per cent of children enrolled in 2018
- the target to halve the gap in Year 12 attainment, or equivalent, for Indigenous Australians aged 20–24 by 2020.

The final target, to close the gap in life expectancy by 2031, is not on track.

Every year, the Prime Minister releases a Closing the Gap report to Parliament that details the progress on these targets. Those involved in Closing the Gap Refresh believe that, going forward, effective programs and services need to be designed, developed and implemented in partnership with First Nations Australians.

Weblink
Health*InfoNet*

Explore the other health and wellbeing programs on the Health*InfoNet* website and then answer the following questions.

Source: Australian Indigenous Health*InfoNet*

Discuss

1 Describe what the Closing the Gap Refresh program aims to do.
2 Explain how this initiative is tackling chronic disease risk factors in the Aboriginal and Torres Strait Islander communities.
3 Conduct some research to find a program running in your state that's aimed at improving the health and wellbeing of Aboriginal and Torres Strait Islander Peoples.
4 Summarise the main aims of the program.
5 Discuss why it is important that people who share similar cultural values are involved in the development of health promotion activities for their own culture.
6 Investigate reasons why a sense of connection to country/place is important for sustaining health and wellbeing for Aboriginal and Torres Strait Islander Peoples.

Worksheet 3.7

Worksheet 3.8

FAST FACT

Indigenous Australians are more likely to participate in basketball and netball than the overall Australian population.

1 Referring to Table 3.6, state the likely physical health benefits of the activities undertaken by these teenagers:

- Michael, aged 13, loves to run. He goes running with his dad three to four times a week for 30 to 40 minutes and he is usually really puffed at the end of his run.
- Holly, aged 12, doesn't really like sport but she loves to walk her dog, Jett. Holly walks Jett every day for 30 minutes and enjoys being outside in the fresh air.
- Zach, aged 14, does gymnastics twice a week. He does lots of strength-based activities where he has to lift or support his own body weight.

2 Explain why it is important to have 'healthy bones' and what contribution regular weight-bearing activity plays in achieving this.

3 Participation in regular physical activity promotes cognitive development and growth. In your own words, what does having good cognitive health mean?

4 Create a diagram like Figure 3.13. Write your name in the innermost circle, then in the next circle write the names of the people who are most important to you. In the outer circle, write the names of the people who are in your life, but who you might not interact with as often.

From these people, list who you would turn to in each of the following scenarios:

- Something great has happened and you want to share it.
- You are having trouble at school and need someone to talk to.
- You have had an argument with your parents and you feel like they don't understand you.
- You are excited and nervous about starting at a new dance school.

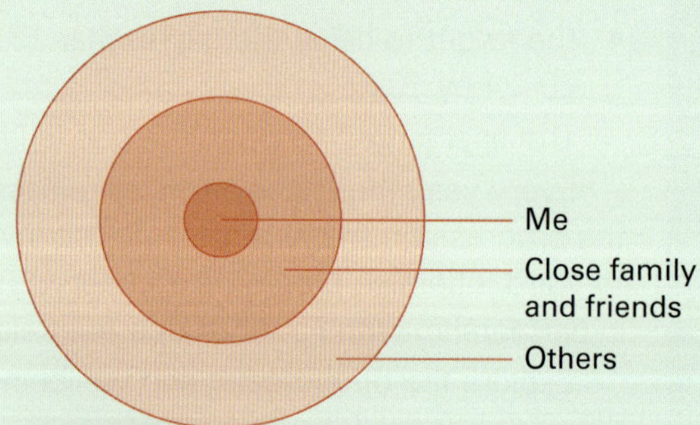

Figure 3.13 Who do you interact with most?

1 When someone feels stressed, why might people suggest that they go for a run, a swim or a gym workout? How do exercise and physical activity decrease stress?

2 Recall and briefly describe two occasions when participating in physical activity has improved your social health. This might be when you played sport for the school, walked the dog on the weekend or participated in a fitness class.

1 You have researched the modifiable risk factors that contribute to cardiovascular disease. These are different at different life stages.

a Prepare a list that highlights the top three risk factors for cardiovascular disease for people who are 20–30 years of age and for those who are 50–60 years of age.

b Why do you believe the factors might be different?

2 Positive psychology and a growth mindset are interrelated concepts that can be promoted by participation in regular physical activity and can be learnt and built upon by everyone. Conduct online research to identify at least three similarities that exist between these two principles.

3 Create a checklist for yourself listing conditions that should be followed each night to ensure good quality and quantity of sleep. You can include the following three prompts in your checklist, but add at least another three suggestions:
- Do not eat any foods after 8 p.m.
- Leave mobile phone and tablet on shelf in dining room recharging for next day
- Ensure room cannot be affected by external 'noise' – close door, ensure blinds block out all light, etc.

WHAT INFLUENCES PARTICIPATION IN PHYSICAL ACTIVITY?

Quiz
What are the benefits of being physically active?

Worksheet 3.9

The health benefits of being physically active are very clear. It is important to participate in regular physical activity and to reduce or limit sedentary activities. There are many opportunities to be active, and many choices to make. It is important to understand how and why you choose which physical activities to participate in.

SOCIAL INFLUENCES

The people who you are active with or who encourage and support you to be active are called social influences. Often physical activity is done with other people – walking with a friend, going for a bike ride or a day at the beach with family, playing team sports with mates. Friends and family are two of the biggest influences on your physical activity behaviours, but a dog can also encourage you to be active. Did you know that dog owners are more likely to be active than non-dog owners?

Shutterstock.com/michaeljung

Figure 3.14 Why do you think dog owners are more active than non-dog owners?

FAST FACT

Female participation in AFL has increased 154 per cent from 2017 to 2018. This change is attributed to the introduction of the AFLW.

Parents can be good role models for physical activity. By being active themselves, they are setting a good example and sending positive messages about the importance of lifelong physical activity. They can also support your participation in physical activity in many ways, including driving you to training and games, paying for uniforms and club fees, buying equipment or doing physical activity with you.

CULTURAL INFLUENCES

Australia is a country with many cultures and long-established traditions. First Nations Peoples have a rich and wonderful way of life that includes physical activity as part of its tradition. Dance is an important and unique part of the traditional ceremonies and is among the knowledge and stories that are passed down from one generation to the next. Chapters 8 and 10 have more information about some of the traditional games and dance of the First Nations Peoples.

All other Australians are migrants or descendants of migrants, meaning that there are many cultural influences on participation in physical activity. More than 25 per cent of all Australians were born overseas, and these people have all brought their own cultural influences with them.

stereotype an overly simple generalisation about a group of people or things

There are many **stereotypes** about Australians. One is that Australians love leisure time and watching or playing sport. Sport has played an important part throughout Australia's history. Have you heard of Phar Lap (a racehorse), Sir Donald Bradman (cricketer) or Dawn Fraser (swimmer)? What about the traditions of the Ashes, the Davis Cup and the Melbourne Cup? These all contribute to the way Australians are regarded today.

Alamy Stock Photo/robertharding

Figure 3.15 Dance is an important part of storytelling in First Nations Peoples' cultures.

CASE STUDY  SOCCER STEREOTYPES

Identify

Football is at the heart of many African-Australian communities and increasingly young women are joining the sport. But it has taken hard work, on and off the pitch.

Understand

Every year for the last five years, the African Nations Cup has been a highlight of Adelaide's sporting calendar … Until now though, there's been a lack of women and girls participating. [Coordinator Arsene Iribuka] says it's partly because of traditional cultural views on women participating in sport. 'A lot of families out there still see that sport is not a place for women, especially football,' he said. 'And for the culture that we come from, that is definitely a challenge in terms of progressing the women's game.'

But as more young women join the sport, that mindset is starting to change. Elizabeth Taban, an 18-year-old from Adelaide's northern suburbs, is among those lobbying for greater participation. She's been playing the sport since 2013.

'I do really like the athletic side of it and the winning side of it,' she said. 'But I [also] really enjoy the culture side of football.' Ms Taban is from South Sudan and says many of her teammates have faced similar barriers, including having to juggle training with responsibilities at home.

'As a woman, because I'm in the house, I'm expected to do cleaning, to do cooking as well. So sometimes I'd have to make sure I quickly do my homework, quickly cook and then go to training,' she said.

Football South Australia/Adam Butler

Figure 3.16 Women from African-Australian communities are breaking stereotypes to participate in soccer.

Ms Taban is hoping to play in a women's only tournament when the African Nations Cup returns in October. She says it has taken hard work on and off the pitch to be allowed to participate. 'The biggest challenge is probably the support,' she said. 'We don't get enough support from the guys' teams, and we don't get enough support from the community, so it's hard.'

A women's exhibition game was introduced two years ago. That turned into four games last year. 'We are making progress,' said Mr Iribuka. 'One of the reasons is because we recognise that the women's game is something that we've neglected for a long time.' He says the potential among young female players is impressive …

Wendy Carter, general manager of football operations at the Football Federation of South Australia, says … having women and girls involved could lead to greater participation at more elite levels, like the men before them. 'It actually gives the exposure to the girls … they are becoming a very strong part of the football community.'

For Ms Taban, the chance to play at competition level is emblematic of a wider battle. 'I'm not so much trying to change the fact that we have the responsibilities, but just to make sure that the community's a little bit more open to girls following their dreams and following their passions.'

Source: 'Adelaide's African-Australian women are fighting barriers stopping them playing football', by Rhiannon Elston, SBS News, 30 August 2019 © SBS.

Discuss

1 List at least two stereotypes that appear in the case study.
2 Why should local, state and federal agencies support programs such as the one outlined in this case study?
3 Briefly discuss what is meant by the term 'gateway to mainstream sport' as stated by Wendy Carter.
4 How does participating in team sports such as soccer build a sense of community?

1 Use the following table to list the physical activities that you do and why you do them, those that you would like to try and reasons why you may not have the opportunity to give them a go.

What I do	Why I do this activity	What I would like to do	Why I may not be able to do these activities
e.g. Play tennis	My friends are in my team	Go snow skiing	Too far to go, too expensive

2 Enablers are factors that positively influence people to do something. List three enablers for physical activity. (Hint: use your table from question 1 to help you!)
3 Briefly discuss how parents might be poor role models when it comes to encouraging their children to meet the recommended physical activity guidelines.

1 Discuss at least two ways your parents have provided you with the opportunities to participate in physical activity. Would your participation have been the same without their input?
   a Think of a stereotype that is associated with people participating in physical activity or sport and discuss how this stereotype might have a negative influence on participation.
   b Discuss instances where gender has provided a barrier to people taking up or participating in sport. How might these barriers be overcome?

1 Some overseas cultures place a great emphasis on academic development at the expense of being able to use leisure time for physical activity. Assume you are on a debating team and need to argue that leisure time is more beneficial for teenagers than doing homework and study. Write three paragraphs, one for each key point, to highlight why physical activity is vital to include in everyone's daily/weekly routine – especially up to the minimum recommendations for Australians, as discussed earlier in this chapter.

Quiz
What influences participation in physical activity?

HOW DOES PHYSICAL ACTIVITY PARTICIPATION IN AUSTRALIA COMPARE TO OTHER COUNTRIES?

As discussed earlier in the chapter, the Australian Physical Activity and Sedentary Behaviour Guidelines and the Australian 24-hour Movement Guidelines for Children and Young People (5–17 years) lay out recommendations for physical activity levels of different age groups.

CHILDREN AND ADOLESCENTS

The most recent data comes from a 2011–2012 ABS study on national nutrition and physical activity. This study found that the majority of children and young people were not meeting the guidelines for physical activity.

FAST FACT

1. In the most recent analysis, only 17 per cent of Australian children aged 2–5 met both guidelines.
2. Six out of 10 children aged 2–5 met the physical activity guideline.
3. Only 25 per cent of children aged 2–5 met the guideline for sedentary behaviour.

Table 3.7 Summary of Australia's recommended physical activity levels

	Ages 2–5	Ages 5–17	Ages 18–64	Ages 65 and over
Physical activity	At least 180 minutes per day with at least 60 minutes of energetic play	Several hours of light activities with at least 60 minutes of moderate to vigorous activity per day	Be active on most, preferably all days with at least 150 minutes of moderate to vigorous activity per week	Be active on most, preferably all days with at least 30 minutes of moderate activity per day
Sedentary or screen-based activity	Should not be restrained for more than 60 minutes at a time No more than 60 minutes of sedentary screen time per day	No more than 120 minutes of screen use Break up long periods of sitting	Minimise and break up prolonged periods of sitting	Be as active as possible
Strength	N/A	Vigorous and muscle strengthening activities three times a week	Muscle strengthening activities two times a week	Incorporate muscle strengthening activities

Source: Australian Institute of Health and Welfare 2019. Insufficient physical activity. Web report. Canberra: AIHW. Licensed under Creative Commons 3.0.

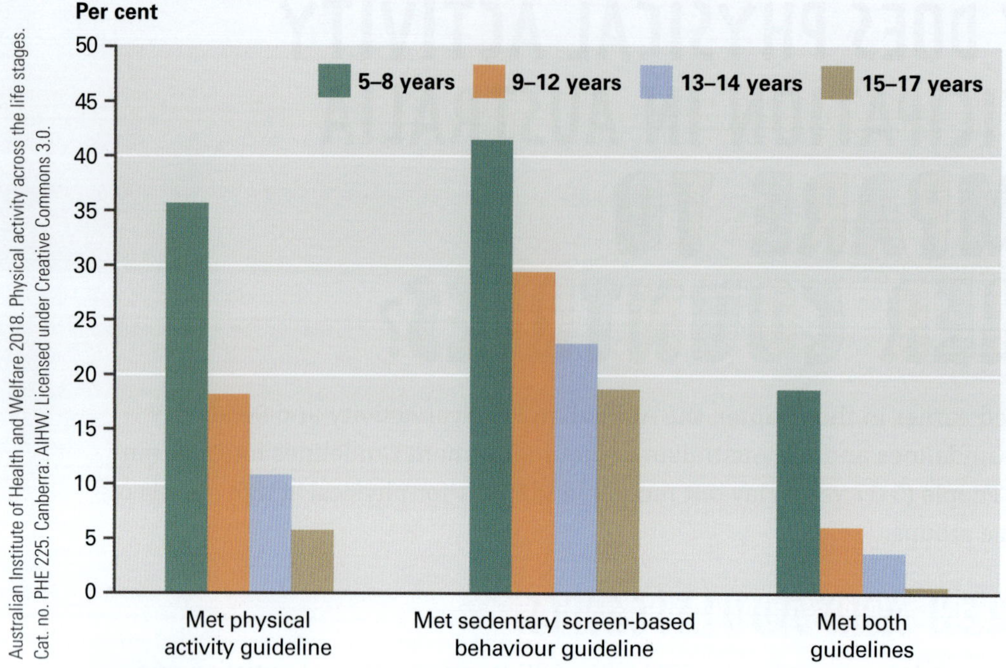

Figure 3.17 Percentage of children and young people meeting the recommended guidelines.

Research shows that the percentage of those meeting the guidelines decreases as age increases (Figure 3.17). In addition, there are significant differences between Indigenous youth and non-Indigenous youth, with substantially more Indigenous Australians meeting physical activity guidelines from ages 5–17 (Figure 3.18 and Figure 3.19).

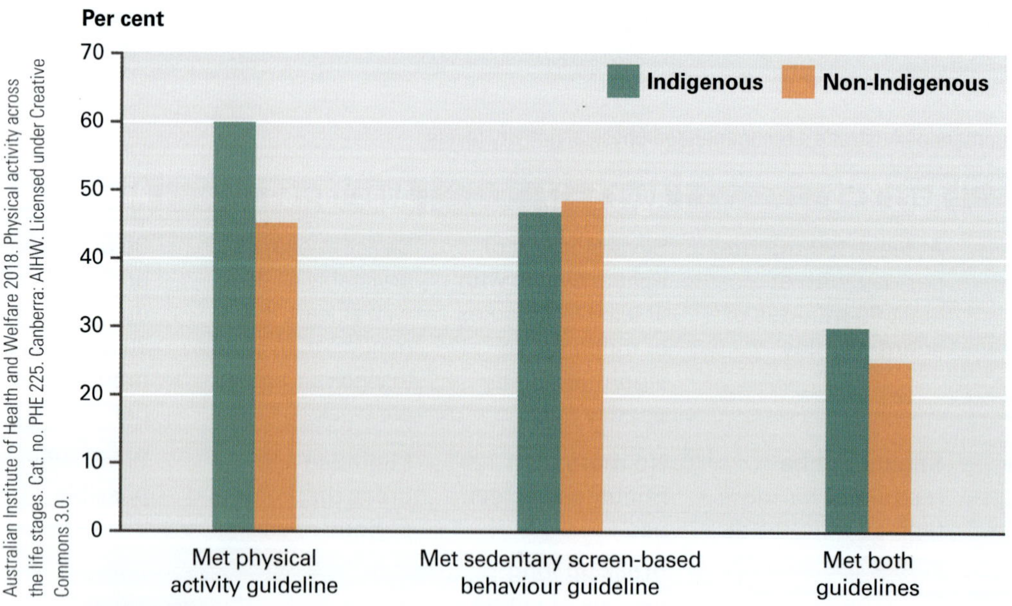

Figure 3.18 Comparing percentage of Indigenous vs non-Indigenous children aged 5–12 meeting the guidelines.

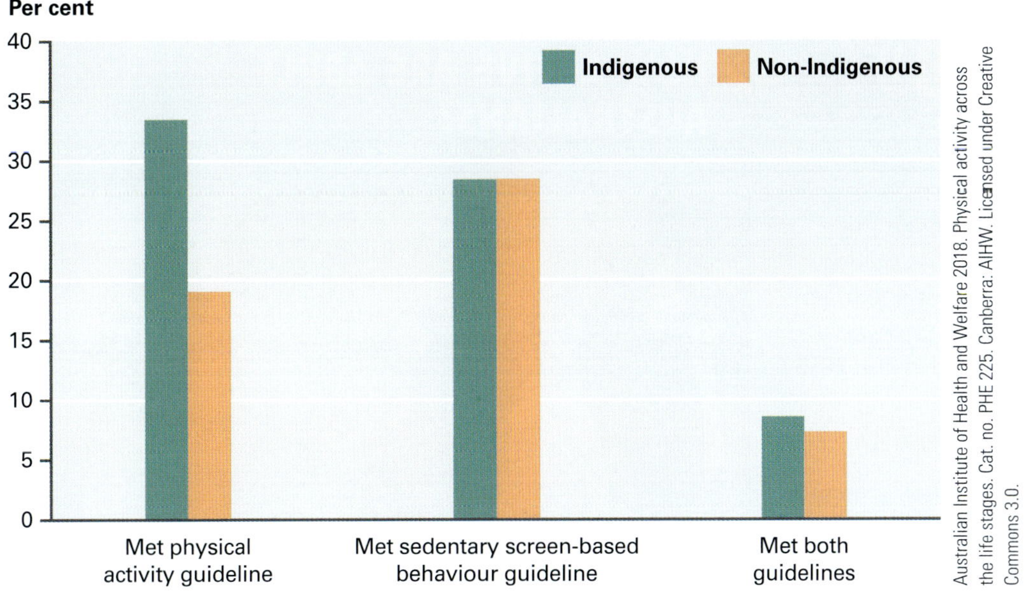

Australian Institute of Health and Welfare 2018. Physical activity across the life stages. Cat. no. PHE 225. Canberra: AIHW. Licensed under Creative Commons 3.0.

Figure 3.19 Comparing percentage of Indigenous vs non-Indigenous young people aged 13–17 meeting the guidelines.

Adult participation in physical activity

FAST FACTS

1. 71.1% of adults aged 65+ years engaged in some form of exercise in the last week, only 26.1% of older adults engaged in 30 minutes or more of exercise on 5+ days in the last week.
2. Only 15.0% of adults met both the physical activity and muscle strengthening guidelines.

Australian Bureau of Statistics 2017, Physical activity – 2017–18 financial year', https://www.abs.gov.au/statistics/health/health-conditions-and-risks/physical-activity/2017-18, accessed 03 June 2022. Based on Australian Bureau of Statistics data.

In 2017–18, 24.9 per cent of 18–64 year olds did strength or toning activities on two or more days in the last week as per guidelines. A higher proportion of men than women did two or more days of strength and toning activities 26.6 per cent compared with 23.3 per cent.

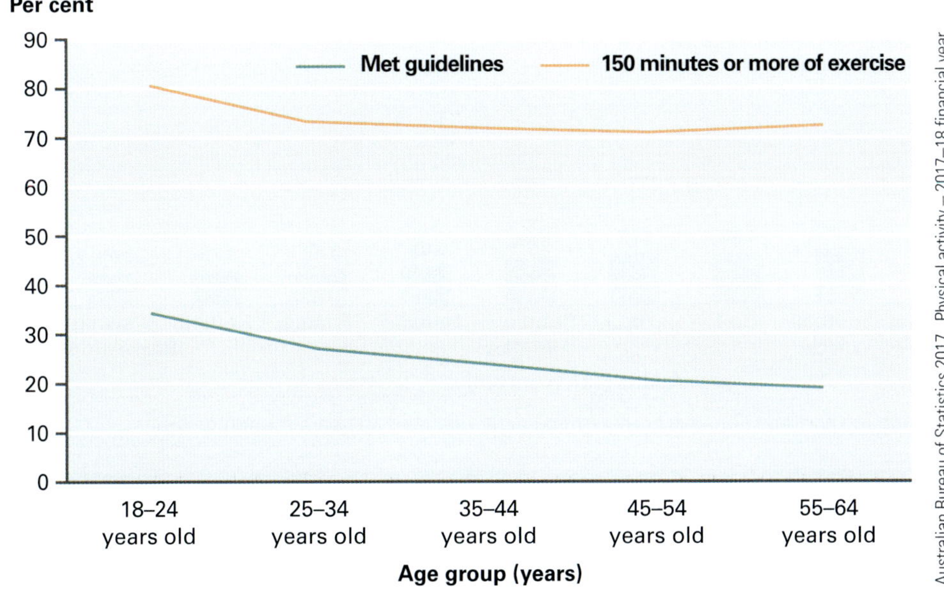

Australian Bureau of Statistics 2017, Physical activity – 2017–18 financial year, https://www.abs.gov.au/statistics/health/health-conditions-and-risks/physical-activity/2017-18, accessed 03 June 2022. Based on Australian Bureau of Statistics data.

Figure 3.20 Persons aged 18–64 years – whether met guidelines or undertook 150 minutes or more of exercise, 2020–21

Weblink
Read the article discussing Sport Australia's position on physical literacy.

CASE STUDY ⇨ HOW ACTIVE ARE ADULT AUSTRALIANS?

Identify

Sport Australia has released the latest annual data from its AusPlay survey, Australia's largest and most comprehensive sport and physical activity survey launched in late-2015. This data has revealed walking is Australia's top physical activity.

Shutterstock.com/Syda Productions

Figure 3.21 Walking is the number one activity for Australians.

Understand

AusPlay provides national, state and territory data on almost 400 different participation sports and activities in Australia and who is participating in them. Recreational activities such as fitness/gym and swimming top the list with walking popular with both males and females. AusPlay data revealed increases in the overall number of Australians participating in sport and physical activity since 2015, however COVID caused this to drop significantly in 2020.

[Sport Australia CEO Kate]Palmer says it's a step in the right direction, but solving Australia's inactivity crisis is far more complex and requires generational change.

'The positive news in this data is that it shows Australians are making the effort to get moving because they are becoming more aware of the importance of sport and physical activity to their health and wellbeing,' Palmer says.

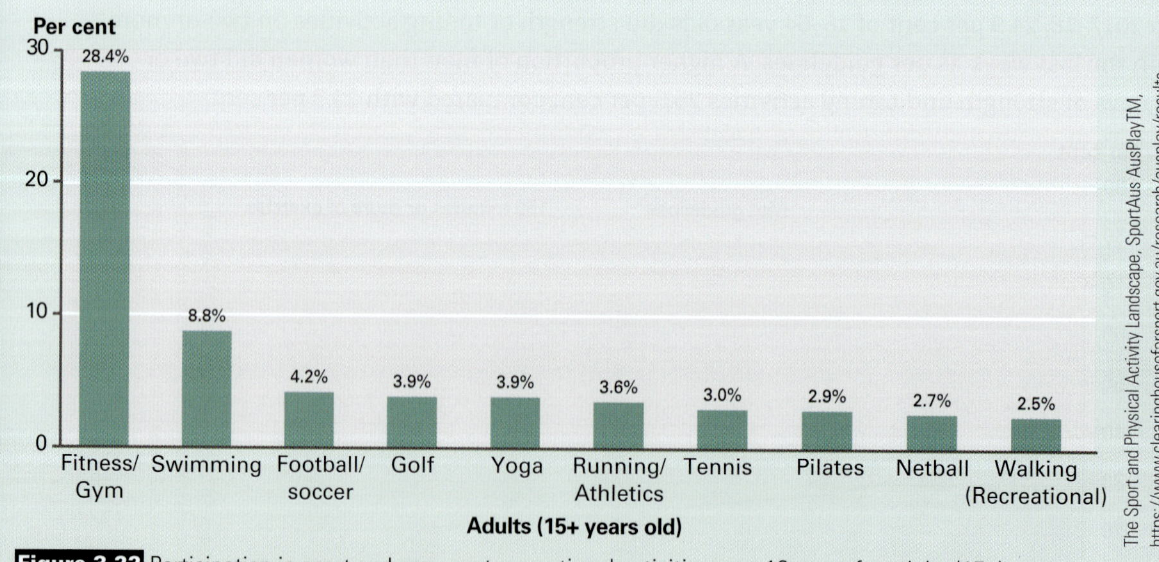

The Sport and Physical Activity Landscape, SportAus AusPlayTM, https://www.clearinghouseforsport.gov.au/research/ausplay/results

Figure 3.22 Participation in sport and non-sport recreational activities over 10 years for adults (15+)

9780170463096

'It's a small step in the right direction, but we're still falling a long way behind when it comes to meeting recommended physical activity guidelines. For example, research tells us only 19 per cent of children meet the recommended one hour of physical activity a day.

'Our general lifestyles are becoming more sedentary than ever before because of things such as technological advances, so that makes it critically important to find dedicated time for sport and physical activity in our lives.

'We need to move more and our lives depend on it. It is estimated physical inactivity now contributes to the deaths of 16 000 Australians every year. That's shocking, it's almost 14 times the national road toll …

'It's interesting to see how activities we participate in evolve as we age. Swimming is a key skill in our formative years, while team sports are popular around the early teens for social as well as physical development. Fitness and gym becomes a key motivation from late-teens onwards, while walking is the number one activity from 35 onwards. The message here is there's a sport or physical activity to keep you moving throughout your entire life.'

Source: 'Australia's top 20 sports and physical activities revealed', 30 April 2019, SportAus, https://www.sportaus.gov.au/media-centre/news/australias_top_20_sports_and_physical_activities_revealed

Men

Activity	Participants	Participation Rate
Fitness/Gym	2,498,483	24.6%
Swimming	790,112	7.8%
Golf	714,080	7.0%
Football/soccer	607,904	6.0%
Running/Athletics	392,317	3.9%
Cricket	357,912	3.5%
Basketball	357,144	3.5%
Tennis	356,408	3.5%
Australian football	333,833	3.3%
Walking (Recreational)	198,796	2.0%

Women

Activity	Participants	Participation Rate
Fitness/Gym	3,304,248	31.6%
Swimming	1,053,104	10.1%
Yoga	689,706	6.6%
Pilates	565,131	5.4%
Netball	490,341	4.7%
Running/Athletics	387,768	3.7%
Walking (Recreational)	382,034	3.7%
Tennis	290,382	2.8%
Dancing (recreational)	226,588	2.2%
Football/soccer	207,851	2.0%

Source: The Sport and Physical Activity Landscape, SportAus AusPlayTM, https://www.clearinghouseforsport.gov.au/research/ausplay/results

Figure 3.23 Top 10 sports and physical activities associated with adults (15+), by gender

Review

1 Kate Palmer has said it will take a 'generational change' for more Australians to meet the physical activity guidelines. What does this mean?

2 Physical inactivity contributes to more than 20 000 deaths per annum in Australia, and more and more researchers are linking this to hypokinetic diseases. List three types of hypokinetic diseases and how increased sedentary behaviour might contribute to these.

3 Analyse Figure 3.23 and propose why walking, running, cycling and fitness/gym have all seen significant increases over the last 10 years.

4 Analyse Figure 3.23 and explain how stereotypes and other sociocultural factors might influence participation in certain sports/activities.

INVESTIGATION

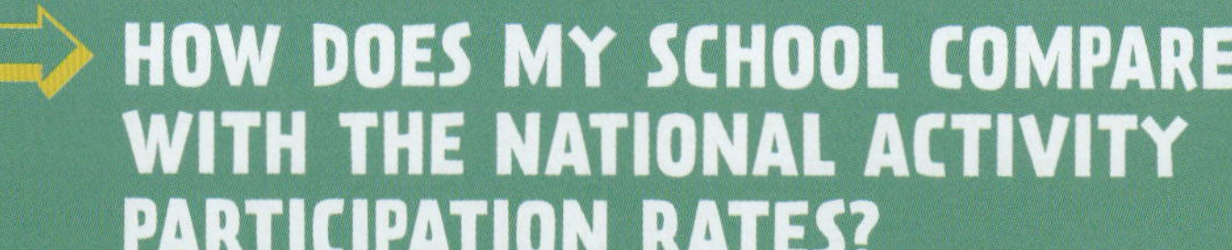

Purpose

To collect data on the most popular activities among students aged 9–18 and compare this with the most popular activities among adults aged 25–64 in our school.

Method

1 Design a simple survey to collect information on which activities people associated with your school (classmates, other students and teachers) engage in, both in and out of school. Hint: You might suggest 10 activities and ask respondents to rank them in order from highest participation to lowest participation.
2 Include an 'other' option to allow respondents to enter activities that you might not have listed.

Materials

The easiest way to conduct a survey and collect data is to use one of the many digital versions available, such as SurveyMonkey. Most of these are free to use as long as the number of questions is limited and the types of questions and responses are basic.

Discussion

1 Summarise your findings and highlight similarities and differences between the two age groups. Make sure you present your results in a visual format, such as a graph.
2 Suggest possible reasons why the types of activities people participate in changes as they get older. Make sure you make specific reference to the activities in your discussion.
3 It has been found that sedentary behaviour increases as we age. Suggest two ways that sedentary behaviours can be reduced for:
   a 12–17-year-olds
   b 18–64-year-olds
   c 65+year-olds.

THE ENVIRONMENT AND PHYSICAL ACTIVITY

Worksheet 3.10
Worksheet 3.11
Worksheet 3.12

Research has found that communities that have lots of outdoor spaces, such as parks and playgrounds, walking and bike tracks and recreational programs, have higher rates of physical activity than those with fewer recreational facilities. People who live near a park that has well-lit walking tracks are more likely to use this facility to be active.

CASE STUDY ⇨ THE IMPACT OF LOCKDOWN

Identify

The COVID-19 pandemic presented many challenges and hurdles to sport and physical activity participation, especially through lockdowns. All across Australia, social distancing requirements had significant negative impacts on participation in community sport, while at the same time seeing increases in recreational and non-sport activities that allowed social distancing to occur with a degree of safety, e.g. walking/jogging, bike riding, bush walking, horse riding, exercise/gym workouts at home, etc.

The following results from an AUSPLAY survey clearly show the impact COVID lockdowns had.

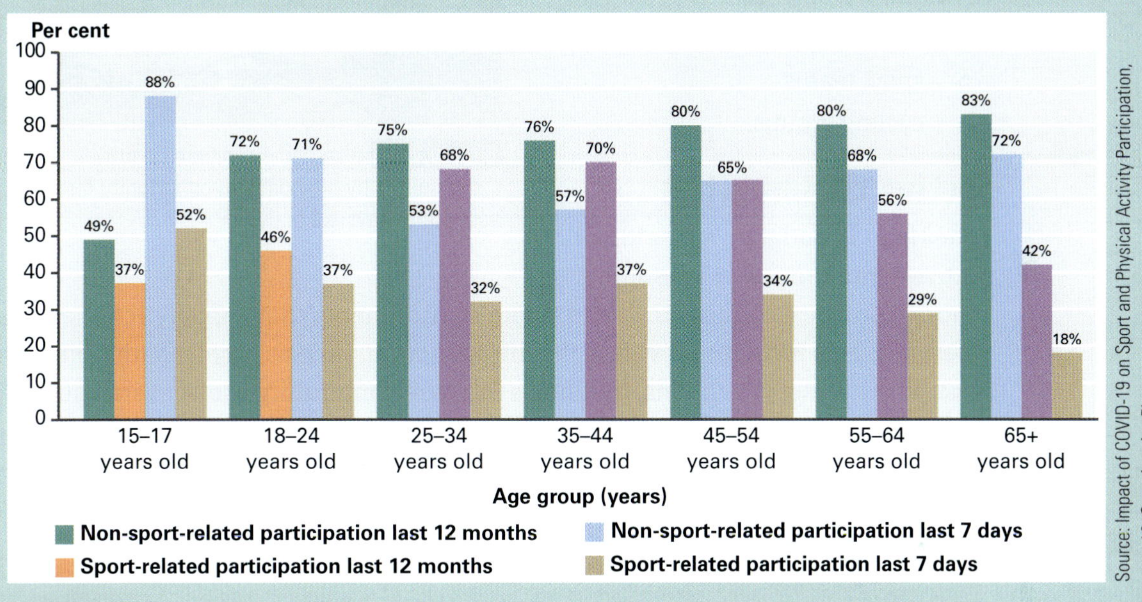

Figure 3.24 Sport and non-sport related activities during COVID 2020 lockdown for adults 18+

Discuss

1 Which age group was least affected when considering non-sport participation during COVID? Briefly explain why you believe this may have been the case.
2 Hypothesise why participation in non-sport related activities increases as we age.
3 Propose why people living close to parks, beaches, walking tracks, etc might have had lower levels of physical activity decreases during COVID than people living in more population dense communities.

The great outdoors

Being outdoors and active in a natural environment is great for health, but the availability of parks and green open spaces in communities can affect the amount of physical activity that people do. Primary school–aged children are more active when the school playground has more 'green' features such as trees and gardens. Being active outdoors has been shown to reduce stress, blood pressure, heart rate and muscle tension and to improve mood. Another benefit of being outdoors is the ability to 'unplug' from electronic devices, although more and more devices are becoming portable, so it's a good idea to leave them at home!

infrastructure basic facilities and services needed for the functioning of a community or society, such as roads, transportation, power lines and schools

Weblink
Heart Foundation: healthy, active communities

Local community

Local communities provide **infrastructure** and opportunities for residents to participate in physical activity in their local area. These may include:

- aquatic facilities
- sports grounds
- walking tracks
- bike paths
- skate parks.

There are also many opportunities to be involved in physical activity within the community, such as:

- walking groups
- sport teams
- cycling groups.

iStock.com/monkeybusinessimages

Figure 3.25 Bike paths promote healthy communities by providing an opportunity for accessible physical activity.

Active communities

The Heart Foundation's 'Healthy, active communities' is a resource package for local councils to use to create environments that support active and healthy lifestyles. This includes providing the infrastructure, such as bicycle paths, bike lanes and public transport options, that helps increase community use of sustainable and physically active modes of transport, as well as ensuring that every member of the local community has access to outdoor and indoor physical activity facilities.

Worksheet 3.13

FACE TO FACE Bike riding

Bike riding has many benefits for your health and the community. In pairs, discuss ways that your community would benefit if more people used cycling as a form of active transport.

INVESTIGATION ⇨ HOW CAN MY LOCAL COMMUNITY INFLUENCE HEALTHIER TRAVEL BEHAVIOUR?

Purpose

To research countries that have high uptake of bicycle programs and compare them to local circumstances in our own communities.

3

Method

There are various reports that focus on user-friendly bicycle programs around the world. Select any of the following sources of information to provide you with more knowledge about the characteristics of effective bike programs:

- Lonely Planet
- Wired
- National Geographic
- MentalFloss.

You could also choose to do your own research.

Discussion

1. There are a number of ways people can change their behaviour to reduce car trips in favour of walking, cycling, carpooling and public transport. Select one of the initiatives from the list and explain how it may influence someone to change their travel behaviour.
   - Active School Travel Program
   - Bike Buses
   - Bike Ed program
   - Buy/Borrow a Bike scheme, such as Melbourne Bike Share
   - Community-based cycling proficiency training
   - Cycle events, such as Ride to Work day
   - Bicycle recycling schemes
   - Bikes on buses schemes
   - Walk to School
2. What are some of the common factors among cities ranked highly in terms of 'effective bicycle programs'?
3. In addition to these factors, what would it take in your local municipality for more people to be encouraged to use bikes?
4. Design a proposal to your local council detailing three ways to influence healthier travel behaviour in your local community.

INCREASE PHYSICAL ACTIVITY, DECREASE SEDENTARY TIME

This chapter has shown how important physical activity is for overall health and wellbeing. Initiatives such as Australia's Physical Activity and Sedentary Behaviour Guidelines and the Australian 24-hour Movement Guidelines for Children and Young People (5–17 years) provide information about how much physical activity people of various age groups should be doing each day. Including physical activity in your daily routine is easier than most people think. Ideas include riding or walking to school, or getting off the bus or train one stop earlier; using the stairs at the shopping centre rather than the escalator; or being active at lunchtime. Increasing physical activity results in improvements in health and overall wellbeing.

1. Refer to Figures 3.17 and 3.20 and state the general trends observed regarding the proportion of people meeting the physical activity and sedentary guidelines with age.
2. Discuss any differences that may exist between Indigenous and non-Indigenous populations when it comes to meeting the physical activity and sedentary behaviour guidelines.
3. Considering the most popular activities in Australia, summarise at least two trends that exist.

1. Muscle strengthening activities are recommended for various age groups in order to experience health benefits. But not everyone can access a gym to work out. Propose four different exercises people your age might engage in to achieve muscle strengthening without necessarily going to a gym.
2. Create a list of three activities that contribute to your sedentary behaviour – remember, sleep is not considered sedentary behaviour! How could you modify these three activities to reduce the amount of time they contribute to your sedentary behaviour? Suggest one modification for each activity.
3. You have been approached by the local council to design an 'ideal' outdoor space for your community. The design needs to be aimed at increasing physical activity levels among all age groups within the community. You must consider access to the space, safety and aesthetics (the way it looks), as well as the types of facilities you will include.

   Present your design as an annotated poster or multimedia presentation.

1. Many new housing estates require the developers to include a proportion of land that is dedicated to both passive and active leisure. Research a recent development and provide specific examples of each of these options.
2. Devise a policy or an environmental idea for your school that could be implemented to increase the amount of physical activity students are involved in. You could score 'bonus points' by also including a couple of strategies to reduce sedentary behaviours within schools.

Quiz
How does physical activity participation in Australia compare to other countries?

9780170463096

CHAPTER 3 REVIEW

1. Physical activity can be classified in two ways: incidental or structured. Describe the difference between the two.
2. In our busy lifestyles, how can we ensure that we get a sufficient amount of physical activity in our day, despite only having a few hours after school or work that might be 'free'?
3. If people meet Australia's physical activity, sedentary behaviour and sleep guidelines for each age group, list three benefits they are likely to achieve.
4. Which acronym can be used to remember the dimensions to determine how often, how hard and how long to perform physical activity? Discuss their purpose.
5. What is the purpose of the physical activity pyramid?
6. Sedentary behaviour does not include periods of sleep, but what other behaviours does it include? Identify three examples.
7. Describe two small changes that can be made to your daily routine to minimise sedentary behaviour.
8. How do the heart, bones, muscles and the whole body benefit from regular physical activity?
9. There are many health campaigns aimed at increasing the physical activity levels of Australians. List at least three.
10. Social health is an important benefit of participating in physical activity. Provide an example of how this can be promoted both in and out of COVID lockdowns.
11. There are many opportunities to be active in your community. The choices you make are influenced by social (friends and family), cultural (who you are) and environmental (spaces and facilities) factors. Briefly discuss how each of these influences promote participation in regular physical activity.
12. Most Australians don't participate in enough regular physical activity. List at least three potential problems this may cause at the individual, social and community levels.

MENTAL HEALTH AND WELLNESS

4

IN THIS CHAPTER

You will learn that keeping your mind healthy is as important as keeping your body healthy. As you develop and grow as a person, your identity is continually forming and is influenced by many factors. You need to be able to identify these and make decisions about how they will impact you. You have a role in maintaining your own mental health, and can also help out friends and family. To do this, you need to be aware of how and where to go to access reliable sources of help.

Shutterstock.com/Nicetoseeya

By the end of this chapter, you should be able to:

- define mental health and its impact on our overall health
- identify the impact body image and self-worth have on mental health
- identify strategies to remove the stigma attached to mental illness in the community
- understand resilience and the skills that support resilient behaviour
- express health concerns and identify support resources
- identify how empathy plays a role in understanding and helping mental health issues
- understand how diversity impacts on understandings about mental health
- understand emotions and the different ways that they can be expressed
- support friends and family who are going through a challenging time
- plan and use positive health practices, behaviours and resources in the community
- evaluate health information
- identify how protective factors can be implemented in schooling environments
- propose and implement strategies for building support networks and connecting to the environment to promote health and wellbeing in the community.

HOW CAN I ACCESS SUPPORT IN MY COMMUNITY? 171

HOW CAN I ANALYSE HEALTH INFORMATION? 180

WHAT IS MENTAL HEALTH?

Quiz
Pre-chapter

Before you start, take the pre-chapter quiz to find out how much you already know.

In Chapter 3 you learnt about physical health, social health, emotional health, cognitive health and spiritual health. A further aspect of your overall health and wellbeing is your **mental health**.

mental health
a state of wellbeing where a person is able to cope with stress and live productively in their community

The World Health Organization (WHO) defines mental health as 'a state of wellbeing in which an individual realises his or her own abilities, can cope with the normal stresses of life and is able to make a contribution to their community.' Positive mental health and wellbeing is a combination of both positive feelings and positive functioning.

Source: 'Mental health: strengthening our response', World Health Organization, 30 March 2018

FAST FACT
Approximately one in seven young people (4–17 years old) experience a mental health condition.

Video
Mental health in Australia: What do you think good mental health looks like? What supports could you access in your local community if you were struggling with your mental health? Watch the video and join the discussion.

FACE TO FACE Mental health word cloud

In pairs, brainstorm all the words that come to mind when you hear the words 'mental health'. Use your words to make your own word cloud, like those shown in Figure 4.1. These might be appropriate to hang up around the school.

Figure 4.1 Positive and negative attitudes towards mental health

When we think about 'physical health' we usually think about all the good things our bodies can do. Sometimes the words 'mental health' have negative connotations, and some people think it's a subject that shouldn't be talked about. Often this is because people don't understand what mental health is, and therefore don't discuss it openly. Some people are even scared to talk about it.

Revisit the word cloud activity above. Count how many words on your list are negative and how many are positive. There are both positive and negative aspects of mental health, but if we concentrate on making the words 'mental health' positive, more people will be willing to talk about it. If mental health becomes easier to talk about, just like we do with 'physical health', more people might feel confident seeking help if they need it, and won't feel embarrassed. Having a mental health issue is not something to feel embarrassed about, just as having a physical health issue isn't. The best thing to do is identify the issue and get help. Challenging ideas in society is hard to do, and it will take time, but the more people speak up about **mental health issues**, the more accepted they will become.

mental health issues issues that affect the way people feel, think and behave, the most common being depression and anxiety disorders

People with good mental health:

- accept the world as it is and are realistic
- like themselves and have positive self-image
- act freely; they can be themselves and make their own decisions
- can express their feelings and handle their emotions well
- have good relationships and communicate well with others
- take responsibility for their own actions and don't blame others
- work towards a future and try to achieve their goals.

People with good mental health are resilient. As discussed in Chapter 3, **resilience** is the ability to deal constructively with change or challenge. Resilient people are able to maintain or return to their social and emotional wellbeing when they encounter difficult events. Imagine a yo-yo that comes back up after you throw it. This capacity to bounce back is what resilience is all about. When something challenging happens to you, how quickly do you bounce back?

Alamy Stock Photo/Kristina Kokhanova

Figure 4.2 Resilience is the ability to bounce back.

resilience the ability to cope with and recover from difficult situations

You could think of mental health as a scale or **continuum**, with good mental health at one end and **mental illness** at the other.

continuum a type of scale where the two ends are very different from each other (sometimes opposite), and there are varying elements in between

Mental health issues such as low self-esteem, anxiety, depression and panic disorders would fall somewhere around the middle of the continuum, depending on how much they affect your life. People may not stay at the same place on the continuum for their whole life; everyone has times of good mental health and times when they may need help. Mental health issues that go on for a long period of time can develop into mental illness, which is why it is so important to get help early.

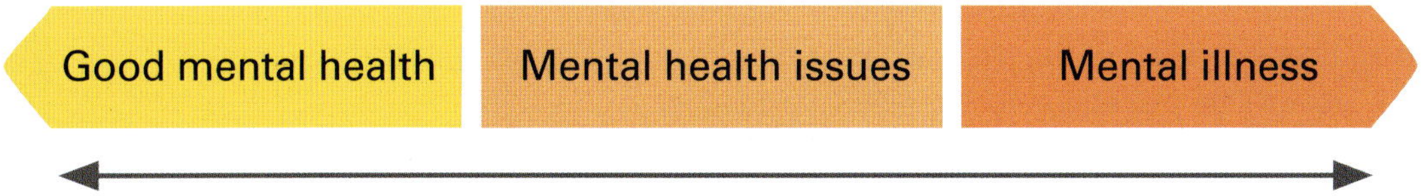

mental illness a diagnosable illness that has a deep effect on quality of life; examples include schizophrenia and bipolar disorder

Good mental health | Mental health issues | Mental illness

Figure 4.3 Continuum of mental health

CASE STUDY  MENTAL HEALTH ISSUES

Identify

Sometimes people might appear to have good mental health even though they are actually struggling.

Understand

Suzie gets up and goes to school every day, even though she doesn't feel like it. She has to drag herself out of bed to get herself organised each morning. She knows she is not pleasant to her family in the mornings, but it takes all her energy just to get dressed and eat some breakfast. She leaves with a hurried goodbye while she still has the energy. The walk to school is so draining and she desperately hopes she won't run into anyone she knows. As soon as she is at her locker with her friends, a different Suzie appears. She looks happy and is energised, laughing and having fun with her friends. When Suzie gets home after sport practice she goes straight to her room. She tells her parents she is studying, but she often falls asleep. She asks to have her dinner in her room as she has so much homework. Her parents don't like this but she has been so grumpy, their dinners usually end up in a fight with her angrily walking off anyway.

Suzie's teachers see a happy, confident and fully functioning student who excels at anything she puts her mind to. When her mum goes to talk to the Wellbeing Coordinator, she is truly shocked. The school says they will keep an eye on her, but that she is doing well at everything and they don't see a problem. Suzie's mum is so confused, she is not sure what to do.

Suzie doesn't know how long she can keep this charade up. School is getting harder and harder, her friends appear to be getting more distant, and she knows her family hate her at the moment. She's using all her energy at school trying to be 'normal'; she has nothing left when she gets home.

Discuss

1. Consider the 'People with good mental health' list on page 147 – does Suzie have good mental health?
2. Generalise and suggest why it might be hard to explain to Suzie that she has a mental health issue.
3. Propose a reason why it might be hard for others to see that Suzie is struggling with a mental health issue.
4. What might happen to Suzie if she doesn't get help?
5. Discuss how, as a friend, could you help Suzie.

INVESTIGATION  WORKING ON THE CONTINUUM

Purpose

To develop an understanding of the mental health continuum by using real-world examples. For this task, you will be placed into groups by your teacher.

Method

1 Your teacher will give each group a topic from this list to investigate. Good mental health; good self-esteem; feeling low; low self-esteem; anxiety; depression; panic disorders; eating disorders (this topic may be broken up depending on numbers); schizophrenia; psychosis.
2 Referring to the infographic provided, research your topic to produce your own infographic.
3 Your teacher will set up the mental health continuum in your classroom. When asked, place your infographic on the continuum (refer to Figure 4.4). Be ready to discuss your issue.

Weblink
What is an infographic?

Discussion

1 Summarise whether the class agreed on where each of the infographics should go on the continuum.
2 List the similarities between each topic.
3 **Explain** what the differences between each topic are.
4 Propose why you think people don't like to talk about the topics at the mental illness end of the continuum.

National Eating Disorders Collaboration (NEDC) at nedc.com.au

Figure 4.4 Example infographic on eating disorders

The other aspects of health – physical, social, emotional, cognitive and spiritual – can influence mental health. As discussed in Chapter 3, good physical health has a positive impact on the other aspects of health, including mental health and wellbeing. Similarly, positive mental health has a positive impact on the other aspects of health. Even if, for example, your physical health isn't all that good, it is still possible to have a positive outlook and see ways of improving your physical health. If your mental health is not good, you may not feel like socialising with your friends or family, which could lead to you feeling isolated or alone. This would affect your social health.

When all aspects of health are good, you are more likely to look after yourself and have a positive outlook, including being positive about yourself on the inside and out!

Worksheet 4.1

Knowing who you are and having a realistic body image are key aspects of having good mental health and wellness. People around the world in different countries and cultures deal with mental health issues in different ways. In Western countries it is generally thought that going to the doctor and then a mental health professional is the best way to help. We cannot generalise though, many people living in Western countries do not believe this is the best way. Other cultures use faith healers, religious or spiritual leaders, alternative medicine providers to help or some choose to manage on their own. Regardless of the way that suits you best, it is better to use preventative strategies all the time, and then if something does come up, get some help as soon as you can.

1 What is mental health?
2 Conduct an internet search to find out how many people aged 12–17 in your state experience mental health conditions.
3 **Identify** three personal characteristics people with good mental health have.

1 Think about Suzie's story. Why do you think she is hiding her mental health issue?
2 Discuss how diversity plays a role in accessing help.

1 Check out any mental health policies at your school. Do they cater for diversity? In a few sentences, list the ways it does this. If they don't, what things could/should they include?
2 Create a list of organisations in your community that help with mental health issues, locate them on a map. If you live in a rural/remote area where there aren't many, make a list of the organisations and their phone numbers that you can access for help.

WHAT IMPACTS MY MENTAL HEALTH?

identity the characteristics that make us who we are

There are many factors or influences that have an impact on your mental health, including **identity** and self-worth, body image and stress.

IDENTITY AND SELF-WORTH

Video
Identity and self-worth: What has already influenced your identity and what might influence your identity in the future? Why is healthy self-esteem not the same as arrogance? Watch the video and join the discussion.

WB Worksheet 4.2

Knowing who you are is a great step towards being resilient. If you know who you are, you know:

- what you like
- what you will and won't do
- where you want to go in life.

We often think about this as having good self-esteem. Good self-esteem helps you make good decisions because you are less likely to choose something that will have a negative impact on your life. For example,

Shutterstock.com/New Africa

Figure 4.5 I am me and I'm happy with that.

9780170463096

if you identify as being good at sport, you are less likely to stay up playing computer games all night before a big game, or to skip practice, because you know how important practice is to becoming a better player. Someone who is resilient usually has a strong sense of identity.

Stereotypes

Identity can be influenced by the stereotypes that exist around the world. Stereotypes based on gender, culture, religion, sexuality and ability can all have an impact on how people live their lives. The media, families, friends, school, church, community groups and/or workplaces often reinforce these stereotypes. The strength of your mental health can determine whether you feel like you need to fit a stereotype. If you have good mental health, you are more likely to feel comfortable being yourself, whether that fits a stereotype or not. A strong sense of identity will make it easier to challenge stereotypes, as you are less likely to worry about what other people think of you.

iStock.com/PIKSEL

iStock.com/hurricanehank

Figure 4.6 Look beyond the stereotype – what if you knew these two were straight A students? Would it change your opinion of them? Appearances can be deceiving.

FAST FACTS

In 2018–19, based on self-reported survey responses, an estimated 24% of Indigenous Australians reported having a diagnosed mental health or behavioural condition with a higher rate among females (25%) than males (23%). More than one in 10 individuals reported having diagnosed anxiety (16.5%) or depression (13.3%).

For many Aboriginal and Torres Strait Islander people (Indigenous Australians), good mental health is indicated by feeling a sense of belonging, having strong cultural identity, maintaining positive interpersonal relationships, and feeling that life has purpose and value (Dudgeon & Walker 2015; Dudgeon et al. 2014). Conversely, poor mental health can be affected by major stressors such as removal from family, incarceration, death of a close friend or family member, discrimination and unemployment, as well as stressors from everyday life (PM&C 2017; Gee et al. 2014).

The legacies of colonisation and the ongoing trauma experienced by Indigenous Australians also affect mental health. Dispossession from land, forced removal of Indigenous children from families, and institutionalised racism have enduring effects on social and emotional wellbeing (Dudgeon & Walker 2015).

FIRST NATIONS PEOPLES' IDENTITY

First Nations Peoples' identity is deeply historical and fundamental to the way that they live their lives. Everything in this world is interconnected, from the land, plants, animals, people, spiritual forces, water and space, and as such everyone is morally obligated to treat everything with respect, for example if we look after the land the land will look after us. We cannot ignore the impact we have on these. First Nations Peoples are given a totem at birth and are required to learn all the aspects of the totem, such as stories, dance, songs and ceremonies. Each clan or family, have responsibility for different parts, so the whole gets looked after by all.

A totem is a natural object that is given, as an inheritance to a family or clan, as an emblem of their responsibilities. They are responsible and required to care of everything that is in that environment, including sacred sites, ceremonies, flora and fauna.

FACE TO FACE Totems

Think about your family and it's history. **Identify** something that you would use as a totem.

» What object did you choose?

» What is the significance of the object? Why did you choose it?

» How does this represent historical members of your family?

» **Explain** how these totems affect identity. Think about personal, environmental and social factors.

» What other cultures/countries use meaningful expressions like the totems? For example a family coat of arms.

FAST FACT

"Aboriginal spirituality is defined as at the core of Aboriginal being, their very identity. It gives meaning to all aspects of life including relationships with one another and the environment. All objects are living and share the same soul and spirit as Aboriginals. There is a kinship with the environment. Aboriginal spirituality can be expressed visually, musically and ceremonially." (Grant, 2004)

Spirituality is considered to be an essential component of positive mental health, but will mean different things to different people. Spirituality is sometimes closely linked to religion but is often now considered to be a part of a person's value system and the way they think and treat the world around them, rather than a religious value.

Source: Grant, E.K., Unseen, *Unheard, Unspoken: Exploring the Relationship Between Aboriginal Spirituality & Community Development*. 2004, University of South Australia: Adelaide, pg. 8-9

BODY IMAGE

Weblink
Uncle Graham Paulson shares his insights into Aboriginal spirituality

Body image is the way you see yourself and what you believe others think of you. As you grow up you become more aware of what you look like and how you appear to others. Moving away from your family and forming your own identity is a normal process that everyone goes through. This can be hard at times, because there are many influences in your life. What your family or carers think might be very different from what your friends think; you need to decide what *you* think.

Sometimes your idea of how you look is quite different from reality. You might think your ears stick out more than they really do, or that you are fatter or thinner than you really are. Sometimes what you see and what others see are two different things. When what you see negatively affects the way you think about yourself, it can become a mental health issue.

Forming our body image

There are various things that influence your body image, including the media, your family, friends and other people, and transitions in your life, e.g. puberty.

The media

There is a great deal of pressure on both boys and girls to look a certain way. Boys are encouraged to look strong and athletic; girls are encouraged to look thin and pretty. Where do you think these ideals come from?

Every day we see advertisements, magazines, television programs, movies and books that show these 'ideal' body types. Messages seem to come from everywhere about having a certain 'look'. These messages imply that if you achieve this look, you will have more fun and more friends, be popular, wear cool clothes, get a partner, have money and be successful.

Alamy Stock Photo/Penny Tweedie

Figure 4.7 Aboriginal sisters Tessa and Jane enjoy painting their faces with their clan motives and totems.

FACE TO FACE Picture yourself

Do this activity in pairs. Your teacher will give each person two pieces of paper and pens or pencils for drawing. If you don't want to draw, you can use words to describe yourself and your partner instead.

1. On one piece of paper, draw a picture of yourself. Do not let your partner see the picture.
2. On the second sheet of paper, draw a picture of your partner. Do not let your partner see your drawing.
3. When both of you have finished, exchange the picture you drew of your partner. You will now have two drawings of yourself.
4. Look for similarities: what are they?

Figure 4.8 You may not see yourself the same way as others see you.

5. Look for differences: what are they? Why are they different?
6. How might the picture of you be different if it was drawn by another member of your class?

Can your body type really bring all of these things? Of course not, but these pressures are real, and can actually cause harm. All bodies are different. For some young people, it will be impossible to fit into a size 6 outfit, only a very small percentage of the population can! Too often, ideals about the perfect body are associated with weight, but in fact you could be severely underweight and still not meet that ideal. It is important to be realistic about what our bodies can and cannot be.

Identify

CASE STUDY

DANCE MONKEY SINGER TONES AND I OPENS UP ABOUT 'RELENTLESS BULLYING'

Source: news.com.au 30 November 2019

Dance Monkey hit maker Tones and I has posted an emotional message on social media, detailing the horrific bullying she's dealing with.

"People always say 'tones how does it feel, it's must feel great, what are you feeling, you must be over the moon'

It does and I don't want to take anything away from my well deserved achievements and to my fans I love you unconditionally but I have been hiding a big black hole for a while now and feel if I hide it like most artists do then how are we going to help the next generations of young artist to come.. truth is (and we have all seen it) with success comes judgement and opinions, this I was prepared for, it's normal (which is sickening) but the relentless bullying that follows every proud moment tears my mind in two. I make music, I have chosen to follow my passion in life and stick to it until it stuck to me. I am a very open honest, caring, good person and in the dark times of death threats and very harsh judgements from strangers I have never met, I have decided to push past it and show any artist that you can get through it and maintain your sense of self, even though I don't see an end in sight, this is how I will live my life now.

I am Toni Watson,

A female artist from Australia

I am going through the best and worst time of my life.

And today I am OK.

I love you and as always.. we are in this together ♥"

Getty Images/ Don Arnold

Figure 4.9 Tones and I – finding the strength to be herself even after severe bullying.

It comes after Tones and I, real name Toni Watson, gave a heartfelt speech at ARIAs, removing a handwritten note from her pocket and saying she doesn't think she's the 'most relatable female artist'.

'I'm not into make-up or dresses or typically girly things. But to me, those things don't really define what it is to be a female artist in this industry anymore', she said, to huge cheers and applause.

'It's being brave and courageous and true to yourself. No-one could have ever prepared me for the whole world judging me and comparing you to other artists. But what's most important is that you have to be a good person and care about others and carry yourself well', she continued.

'Thank you for Australia letting me know that I'm OK just the way I am.'

And while her Facebook post and speech garnered a swarm of support, it seems the singer, who just six months ago was busking in Byron Bay, has struggled with the negative side.

Concluding her post, Tones and I said she would push past it and 'show any artist that you can get through it and maintain your sense of self'.

4

Discuss

1 Discuss what Tones and I means by saying she is 'not the most relatable artist'.
2 Consider if how she looks and acts has an impact on her singing and performing ability.
3 Summarise the messages she is trying to get across.
4 Discuss what you see when you think of Tones and I.
5 Australia is a nation of diversity. Do you think that all young people see versions of themselves in the media? As well as weight, consider height, race, ethnicity, religion, gender, sexuality, ability, etc.
6 Do you have a celebrity you look up to? Why is that? Discuss the similarities and differences they have with you.

INVESTIGATION

STEREOTYPES IN THE MEDIA

Purpose

The media often portrays men and women in very similar ways – women as thin and beautiful and men as fit and strong. How often are we exposed to these stereotypes, and do we even notice them when we see them?

Materials

- A TV or a device to watch an hour of television with advertising
- A Word document or pen and paper to record your findings
- A list of stereotypes for both men and women or girls and boys. For example: Women make family dinners; Men cook the BBQ. Try to think of lots of different types of stereotypes – height, weight, occupation, who does the housework, what they do for fun, what they wear, how they treat people, their emotions, sports they choose, etc.

Method

1 Set aside one hour to watch a TV show/s featuring advertising. During this hour you are to record:
   a what show/s you watched and what time they were on
   b whether the characters in the show/s depicted stereotypes of males and females – you could put a tick next to your stereotype list every time a character is seen to conform to that stereotype
   c characters that do not fit the stereotypes
   d each advertisement you see and what they are selling. Do they fit your stereotypes? For example, is mum in the kitchen getting the children's lunches made? Or do they go against your stereotypes? For example, is dad at home doing the vacuuming and looking after the children?

Discussion

1 Consider the TV show you watched.
   a Who is the show/s targeted at? Who will mostly watch the show/s?
   b Did the characters mainly fit the stereotypes you identified?
   c Why do you think this is the case?

2 Consider the advertisements you watched.
   a What are they selling?
   b Who are they targeting? Who will mostly buy the product?
   c Did the characters mainly fit the stereotypes you identified?
   d Why do you think this is the case?
3 Consider both the show and the advertisements.
   a Were there any examples of characters who do not fit the stereotypes you identified?
   b Were these characters portrayed as positive or negative characters (i.e. hero/villain)?
   c Why are stereotypes so prevalent in TV shows and advertising?
   d What messages does this send to young people?
   e How can we stop this from happening and show individuality on TV?

Advertising

Worksheet 4.3

The media benefits from making us think we need to change ourselves. Advertisements are made to sell products. If they can convince you that you are lacking in some way, and that a certain product would improve you, you may be more inclined to buy the item. If you believe the advertising and buy the product, the advertisement has done its job. With so many similar products on offer, advertising needs to be inventive in order to appeal to buyers. Have you ever noticed that if you talk about something or search something online, you will then see advertising promoting that very thing on platforms such as Facebook, TikTok, Instagram, etc? Marketers are finding clever ways to target consumers as technology advances.

Family, friends and other people

Unfortunately, pressure to look a certain way sometimes comes from family, friends and peers, either directly or indirectly. For example, a friend or family member might make a comment about your body that makes you think you need to change something, even if that is not what they meant. Sometimes our parents or carers have their own body image issues, and make comments that affect their children. The messages they send may not even be verbal. For example, if you see an adult in your life regularly standing on the scales and making negative comments about putting on a kilogram or two, you may receive the message that putting on weight is a bad thing, or that we should pay very close attention to our weight. For young people, this can be dangerous, because you are going through a period in your life when you will put on weight because you are growing. It would be very

'Your jeans are getting a bit tight', Mum says.

'You will never be a supermodel, Sienna, you have such a big appetite', Dad said.

'I wonder when your muscles will start to grow, Ben', says Coach.

'Don't you think you should do something about that acne?', says your friend.

Figure 4.10 Comments from those around us can affect our body image.

9780170463096

harmful to interfere with this process! Look at the following images and think about how the comments would affect the way you behave or the decisions you make.

Transitions in life

Puberty can come at any time. Just when you begin to feel comfortable with your body and understand what it looks like and what it can do, all these enormous changes occur. You might start feeling differently about your body during this time. In Chapter 5 you will learn more about the physical social and emotional changes that happen during puberty.

THE IMPACT OF STRESS

Stress is a normal part of life. We often think of it as a bad thing, but it isn't always! Good stress (**eustress**) is when something good happens and you have a physical response to it, such as the sense of excitement you might feel if you won a prize and had to make an acceptance speech. Bad stress (**distress**) is when you have the same kind of physical response, but to something that isn't so positive, such as doing a final exam. The extent of the **stress response** will depend on how stressful the situation is. This 'fight or flight' response keeps you attentive, and is your body's way of keeping you safe. In some cases it can even save your life.

eustress stress experienced from a positive event

distress stress experienced from a negative event

stress response when the body senses danger, whether real or imagined, defences kick into high gear in a rapid, automatic process known as the 'fight or flight' response

What causes stress?

Stressors are the things that cause people stress. Figure 4.11 shows some examples of stressors that might happen in your adolescent years.

Some stressors can be easy to deal with and some can be really hard. The types of stressors you experience will differ at different times of your life. There will be things that stress you out consistently throughout your life – even as an adult – and others that will become less stressful over time. If you learn good coping strategies, the things that cause you stress won't have such a big impact.

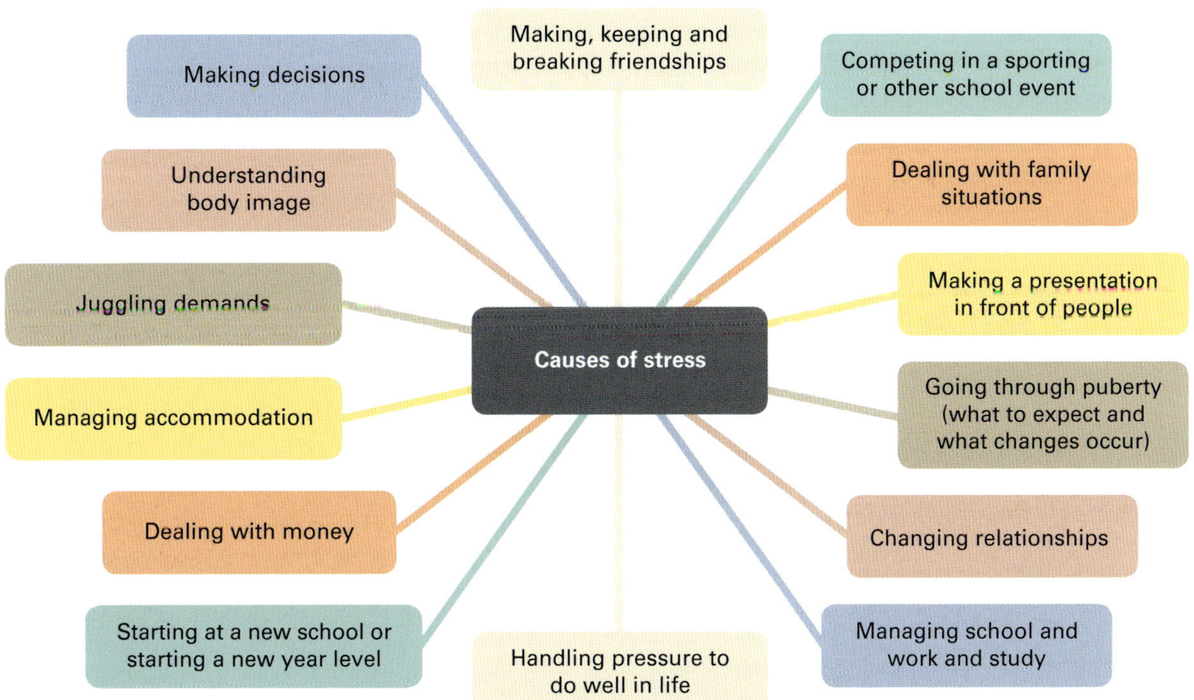

Figure 4.11 Causes of stress

INVESTIGATION

WHAT CAUSES STRESS?

Purpose

To discover what sorts of things cause people stress.

Materials

You will need a paper and pen or recording device to record people's responses.

Method

1 Working in a group, develop a set of questions you will ask a range of different people about what causes them stress, how they react to stress and what they do when they are stressed.
2 Make a list of people you could approach to interview. These could include parents/carers, family members, siblings, friends, family friends, coaches, teachers, the school crossing supervisor at the crossing, etc.
3 Interview those you have identified, making sure to ask for permission to record their responses.
4 Return to your group and compare the results. Tally the different types of things that cause people stress, different ways they react to stress and what they do when they are stressed.

Discussion

Come back to the class and discuss your findings; your teacher will lead this discussion. Record responses on the whiteboard. Can you add to this list? What other things cause you stress? Consider if there are good forms of stress and what your life would be like if it was 'stress-free'?

FAST FACT

In times of stress, your body tells your nervous system to make you more alert, and your hormone system gets your heart going. These work together to help you get through, and in some cases survive, a difficult situation. This 'fight or flight' response is instinctive. Remember this can happen and have the same effect whether it is eustress of distress.

WHAT'S MY BODY'S ALARM?

Identify

What does your body do when you get scared or angry? Knowing this might help you stay calm.

This activity has been developed in collaboration with Dr David Bakker of MoodMission

Understand

This chapter covers the fight or flight response, in which our body prepares us to deal with a threat by either running away or fighting. This can be thought of as the body's alarm system. When we can recognise that our body has detected a threat, we can decide on the best way to deal with it before the fight or flight response takes over.

Practise

1. Think about the last time you felt afraid, scared, frustrated or angry. There's a good chance your fight or flight response was being triggered to some degree at that time.
2. Using the diagram in Figure 4.12, identify the body changes that you noticed the most.

Reflect

The next time you notice one of these body changes occurring, remind yourself that it's okay, it's just your body responding to the way you're feeling. If you're safe but still feeling fight or flight, try not to run away or avoid the thing that's triggered the response. That way, you can show your body there's nothing to be afraid of.

Reactions to stress

Your body may have a physical reaction to stress. You may feel or have the symptoms shown in Figure 4.12.

If your body is under stress for too long, it can become difficult to function normally. You may start to experience physical, mental, emotional, social, cognitive or behavioural symptoms. Some examples of these are shown in Figure 4.13.

If you learn to recognise your body's responses to stress early, you can take steps to avoid negative changes.

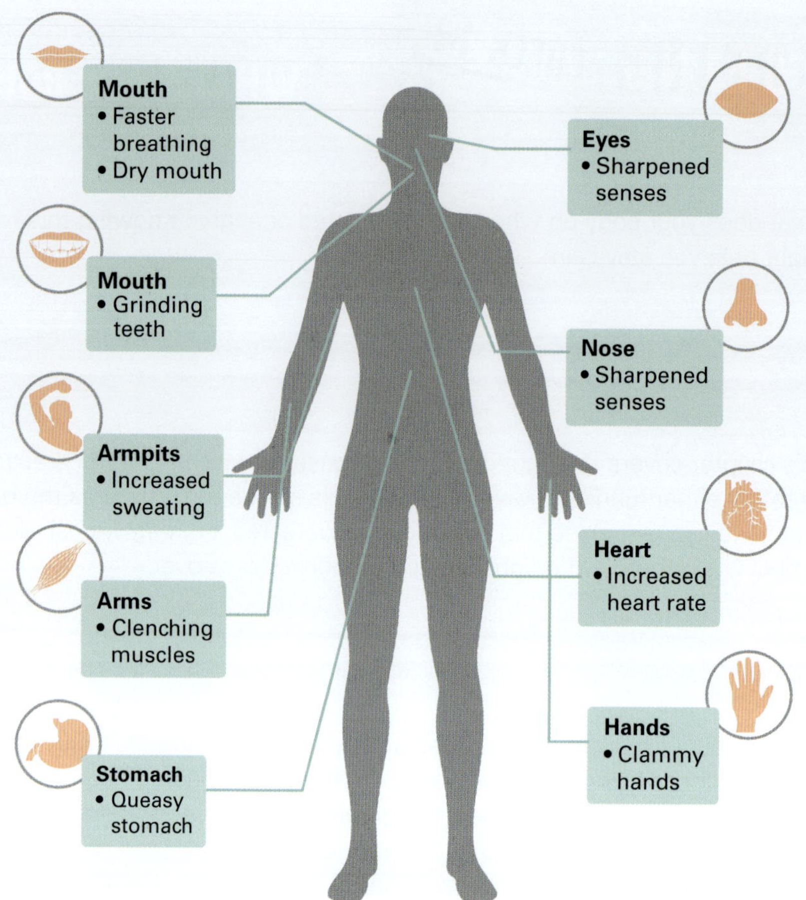

Figure 4.12 Physical responses to stress

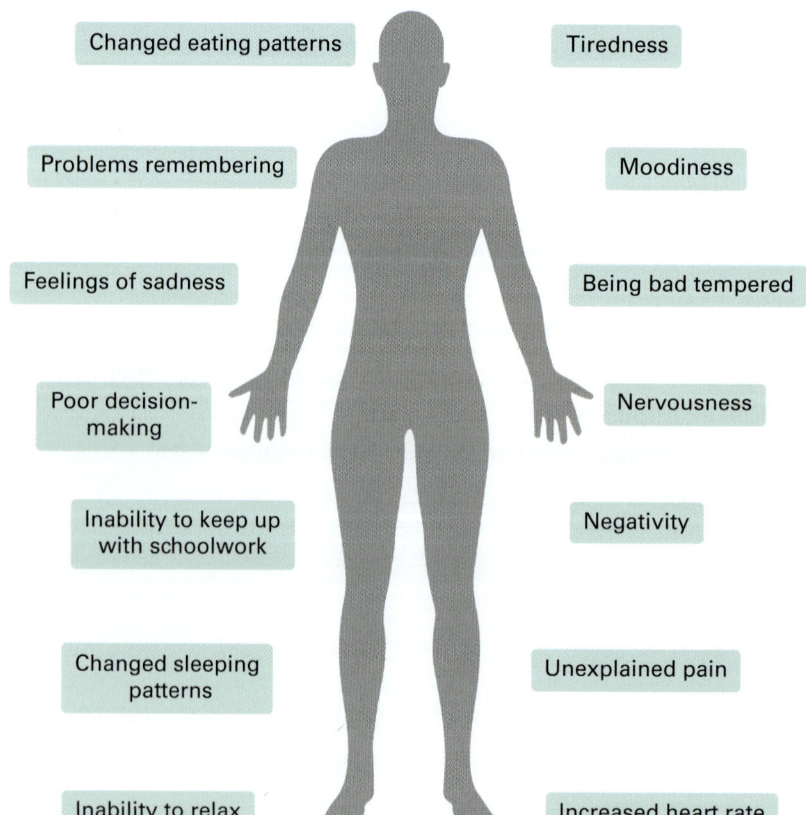

Figure 4.13 Symptoms that can occur after a prolonged period of continued stress

WayAhead - Mental Health Association NSW

Figure 4.14 Ten tips to stress less

Worksheet 4.4

FACE TO FACE Dadirri: deep listening

Dadirri is a form of deep meditation, or deep listening, a technique First Nations Peoples use as a method to contemplate. Watch the video on 'Dadirri: deep listening' before answering the following questions.

- What is the purpose of Dadirri?
- What is meant by 'enduring the flames'?
- What can you learn about First Nations Peoples from this clip?

Weblink Dadirri: Deep listening

WELLBEING CHECK IN ⇨ USING BREATH FOR STRESS AND ANXIETY

This activity has been developed in collaboration with Dr David Bakker of MoodMission

Identify

Sometimes when we're feeling bad, a big deep breath can help. Taking some time to breathe deeply is even better.

Understand

We tend to breathe faster and shallower when we're feeling upset or stressed. It's part of the *fight or flight* response (see *Reactions to stress* on **page 159**), where our body prepares us for dealing with a threat by running away from it or fighting it. However, there are many times when this response gets triggered by a false alarm and we don't have to run away or fight. In these moments, we can actually trick our bodies and brains into feeling calmer by breathing slower and deeper.

Practise

1 Find a quiet place where you can sit down. You can do deep breathing while standing, walking, or doing other things, but it might be easiest to practise now while sitting with nothing else to do.

2. Place your hand on your belly, just beneath your ribs. This is where you'll be breathing from – pushing your belly out against your hand when breathing in and letting your hand fall back towards you when breathing out.
3. Set a timer for a minute.
4. Take a slow, deep breath in for a count of 4.
5. Hold the breath for a count of 2.
6. Breathe out for a count of 4.
7. Hold for a count of 2 before breathing in again.
8. Continue this 4-2-4-2 counting while breathing in and out slowly.
9. After a minute, finish up and note how you're feeling.

Tip

Some people may feel a little dizzy, as their blood is getting more oxygen in it than normal. This is totally fine and safe, and you can change this by making the out breaths slightly longer.

Reflection

1. **Compare** how you felt before the breathing exercises to the way you felt after using them as a means of self-awareness and control.
2. When might you be able to do this deep breathing?

1. List five things that have an impact on your mental health.
2. Discuss how the media influences the way we think about our body image.
3. Discuss how the media's influence on our body image might contribute to stress.

1. Research what a gratitude diary is. Discuss how using this strategy for managing mental health would be effective.
2. Discuss the signs you might see in a friend if they are not managing their stress.
3. **Identify** the people at school who you could go and talk to if you didn't feel able to manage your stress by yourself.

Weblink
Explore the Black Dog Institute website and learn about how to manage stress to help you answer the following questions.

1. **Identify** how the Black Dog Institute got its name. What does the 'Black Dog' refer to?
2. How does the Black Dog Institute help individuals who are dealing with mental health issues or illnesses?
3. Investigate how the Black Dog Institute helps workplaces and schools improve their understanding of mental illness.
4. Consider why the Black Dog Institute are trying to reach out to as many people as possible with their programs. Propose how you could help 'spread their message'.

HOW CAN I BE RESILIENT?

As previously discussed, resilience is the ability to bounce back. This section will discuss some personal skills that can help build resilience.

MAKING THE RIGHT CHOICES

Sometimes people think they are dealing with their stress effectively, but some decisions you make can actually be harmful. As Chapter 1 discussed, some drugs that you think are harmless can in fact be quite harmful. How do you know whether the health information you get is correct? You can't make good decisions without getting the right information and weighing up the consequences. Chapter 5 also discusses decision-making.

FAST FACT

Some people smoke cigarettes to calm themselves down. But the drug contained in cigarettes, nicotine, actually speeds up the central nervous system! They are only feeling 'calm' because they are getting the drug they are addicted to.

COMMUNICATION SKILLS

Communication plays an important role in coping with stress or a mental health problem. There are two types of communication: the internal communication you have with yourself and the external communication you have with others.

Internal communication

Sometimes the way you speak to yourself (self-talk) has as much impact as what others say. The thoughts that go on in your head can be repeated so many times you might start to believe them.

Figure 4.15 Stop the negative chat!

FAST FACT

Athletes are encouraged to use positive self-talk as a strategy to help them achieve their personal best. For example, a netballer saying, 'I'm going to get this goal in' will have more success than one who says, 'I'll never get this goal in'. The aim is to connect the statement with the belief in what is being said.

Figure 4.16 If you believe, you can achieve.

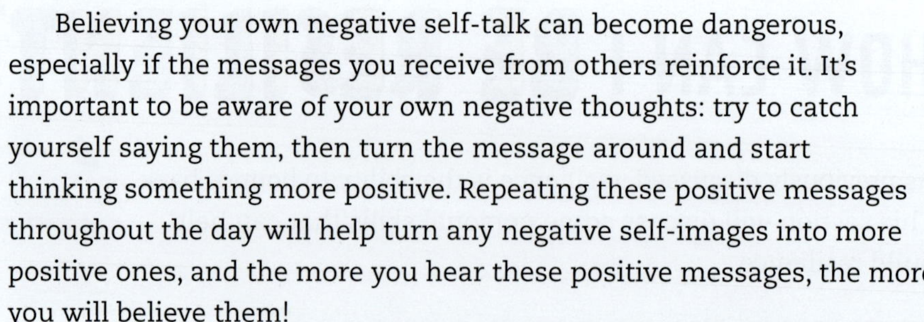

Believing your own negative self-talk can become dangerous, especially if the messages you receive from others reinforce it. It's important to be aware of your own negative thoughts: try to catch yourself saying them, then turn the message around and start thinking something more positive. Repeating these positive messages throughout the day will help turn any negative self-images into more positive ones, and the more you hear these positive messages, the more you will believe them!

We need to be mindful of what we are saying to ourselves. You wouldn't tell a friend that they are stupid, or a loser, so why do we think it is okay to say these things to ourselves? Sometimes we do things we think are stupid, but it is our action, not ourselves, that we are really talking about. It's important to get that clear – yes, we sometimes do stupid things, but that doesn't mean we are stupid. Our self-talk needs to change. Catching yourself saying something negative can be hard; it takes practice to notice when it is happening, stop it and change your talk.

Worksheet 4.5

For example, if you hear yourself think 'Why did I say that in front of everyone? That was stupid!', try to stop that thought and replace it with something like 'I was brave to speak up in front of everyone'. The more we stop the negative self-talk, the more we have a positive focus on our life and ourselves. Like anything, we need to practise this in order to get better at it.

When we are hearing a lot of negative self-talk, it can be hard to remember what we are good at, or where our strengths lie. So it is also important to identify our strengths, so we can remind ourselves of these. Make a list of all the words that describe you in a positive light. These will help you when you need to come up with some positive self-talk. The activity below might open your eyes to some new strengths you didn't realise you had.

FACE TO FACE I'm a star!

Sit in a large circle for this activity. Your teacher will give you each a worksheet and a pen. Write your name at the top of the worksheet, then pass it to the person on your right. They will write down one reason they think you are a star, then they will pass the worksheet on to the person on their right. Eventually each person will get their sheet back and read what their classmates feel makes them a star!

External communication

Worksheet 4.6

External communication is the way you communicate with other people. You do this in a number of different ways, including:

- what you say
- the words you use
- the way you say things, for example, fast and loud, or slow and soft
- your body language.

There are many important skills you need to master for good communication. We are going to look at three of them – listening, negotiation and **empathy**. Effective communication can assist greatly with enhancing mental health.

empathy the ability to understand someone else's situation

Listening

By listening carefully, you are showing the speaker that you think what they are saying is important. Sometimes people just want someone to listen to them, to help them feel better.

There are many ways you can show that you are listening, such as:

- lean in
- repeat what the speaker has said
- nod or smile
- ask relevant questions
- make eye contact.

If you are listening to your own internal conversations, you will not be listening to the speaker effectively. You may be thinking about what you need to do when you get home from school, or how much homework you have. You may want to tell the speaker something and are just waiting for an opportunity for you to talk. Often this begins with 'That happened to me and I ...' or 'You should ...'. This can be really frustrating for the speaker, who is trying to tell their story, not hear about your experiences. If you do this a lot, you might find that people don't talk to you very often.

Listening is a skill that takes patience; not everyone is good at it! Try to focus on what the speaker is saying. If you find your mind beginning to wander, try some mindfulness – refocus by nodding, repeating what they have said or encouraging them to continue their story. Use these tips to help improve your listening skills.

Figure 4.17 Are you really listening?

Tips for being a good, active listener

1 **Let them talk**

If someone's telling you something difficult or important to them, don't interrupt with a story about yourself, even if it's relevant. Let them finish and try to work out what it means to them.

2 **Remain judgement free**

If someone comes to you with a problem, try to be a friend without saying things that might hurt them. Work through whatever they're dealing with and suggest options rather than passing judgement.

3 **Let someone disagree**

If someone comes to you for advice, tell them what you might do in their situation. Remember, though, that your advice might not work for them and they may disagree with you. That's okay; just help them by supporting their ideas, as long as they are safe.

4 **Ask open questions**

Good questions are open questions that let the speaker go anywhere. Instead of asking yes/no-style questions, think about 'how does that work' or 'tell me about …'.

5 **Show them you're listening**

People will trust you more if they can see you're really listening to them. Ask questions about what they're saying, and relay it back to them in different words to see if you've got it right.

6 **Check your body language**

Having open, relaxed body language – facing a person without being too close, and making occasional eye contact – will make whoever's talking to you more comfortable.

Source: '3 steps to better communication', ReachOut, http://au.reachout.com/Listening-skills-worth-having. Adapted with permission.

Negotiation

negotiate
work to find a solution that everyone agrees to

When we **negotiate** with someone, we are both more likely to be happy with the outcome. Use the following steps to guide you through the negotiation process.

1 The first step is to find out what the issue is that you need to negotiate (for example, you might want to go to the movies but your friend wants to go to the swimming pool).
2 Each person then needs to explain their reasons for wanting that option (there is a movie showing that you've wanted to see for ages; your friend wants to check out the new waterslide at the pool).
3 You can both then brainstorm all the options available (you could do something else that you both want to do; you could go to the pool today because it is going to be hot, and go to the movies on Friday because it is supposed to rain; you could flip a coin to decide, etc.).
4 Decide on an outcome that you are both comfortable with.

Sometimes you will not be able to reach a mutually agreeable solution; that is alright, as long as the negotiation has been fair. It is okay to agree to disagree! Next time you might choose something different.

Sometimes there are situations where you disagree with your parents or carers. These negotiations can be difficult for both parties. Your parents want to keep you safe, but you want to become more independent and try new things, which can lead to clashes. Perhaps you have friends who are allowed to do more than you; this can feel unfair, or make you feel as if your parents don't trust you. The following activity is an opportunity to put yourself in your parents' shoes and try to negotiate a solution. You still may not agree with what they say, but you might be able to see why they make the decisions they do.

FACE TO FACE Negotiation

Turn to the person next to you and negotiate whether you will go to the movies or the pool. Then explain to the class how you decided. Think about another scenario where you would need to negotiate an outcome (or your teacher might give you one).

Use the checklist provided to work through that negotiation. How would the negotiation change if someone was really upset, angry, frustrated, happy or excited? How do our emotions affect the way we react? Try the negotiation again with one of you being angry, upset, tired or scared.

1	Issue	✓
2	Explain	✓
3	Options	✓
4	Outcome	✓

Figure 4.18 Checklist for negotiation

FACE TO FACE Keeping everyone happy

Pair up with someone. One of you will take the parent/ caregiver role and the other will take the child role, then you will swap. Choose one of the following scenarios. Your challenge is to negotiate an outcome that both parties are happy with. Use the checklist above if you get stuck.

Figure 4.19 Let's talk this through

Scenario 1 Your best friend has asked you to come to their brother's 18th birthday party. You really want to go because there is a really cool band going, and you don't know when you will you ever get that kind of chance again. However, you know your parents won't let you go because there will be alcohol at the party.

Scenario 2 You want to start catching the bus after school and going to the local youth centre to hang out with your friends. You know your parents won't want you to go because you usually look after your younger sister until one of your parents gets home.

Scenario 3 You've just got a new iPhone and have access to your household plan of 200GB of data a month; you can message your friends using social media or video chat with them at any time of day! Your parent or carer doesn't want you to be on your phone all day, but you don't want to miss out on any of the socialising that is happening between your friends, especially on the weekend.

Scenario 4 Your favourite artist is coming to town and you really want to go to the concert. You've talked about it with your friends and you are all going to ask your parents if you can go. You want to catch the train or bus to the show, and you wouldn't get home until about midnight. Your parents do not think you are ready for this kind of adventure, but they said if you really want to go, they will come with you. You couldn't think of anything worse; they would embarrass you if they came.

1 Was it harder to be the parent or child?
2 Can you see how difficult it might be for a parent to give their child freedom while still keeping them safe?
3 What could you do now to make negotiations with your parents easier?

Empathy

To show empathy, you try to put yourself in someone else's shoes and attempt to understand how they are feeling and why they might be feeling this way. Empathy is an important skill that helps to develop and maintain relationships. It allows you to understand other people without judgement, and it can often provide assistance when difficult times arise.

People express a range of emotions, and some are easier to pick than others. For example, an emotion like embarrassment might be expressed as anger, fear or even uncontrollable laughing! Being able to read emotions is a great skill to have. Sometimes, however, we have to ask what is going on and how our friend is feeling. Some people show a lot of emotions and others show very few. Everybody is different, so reading emotions can be difficult.

FACE TO FACE Emotions

1 Come up with a list of all the emotions you can think of.
2 For each emotion, list three ways people might express that emotion. For example: Sad – cry, withdraw, go quiet.
3 Discuss with a partner how you would express each emotion.

Understanding where an emotion is coming from and having an empathetic response are high-level skills, particularly if we are also filled with emotion. This can make situations like arguments and negotiations difficult – emotions can get in the way. Sometimes when we stop and think about why someone is reacting in a certain way, this can help us understand where the emotion is coming from. How do you feel when someone tries to understand your emotions and shows empathy towards you?

9780170463096

CASE STUDY  READING INTO EMOTIONS

Identify

Being empathetic means understanding why someone is having an emotional response.

Understand

Your teacher caught your friend looking at your test paper during an exam. She brings both of you into her office and says she is going to call both your parents/guardians. You are angry; you have done nothing wrong. Your friend is also angry, and is trying to tell the teacher that you did it first. You are confused. Why is your friend angry? You are not sure why they are lying. Perhaps they are angry at themselves for cheating or for getting caught. What else might be going on that they feel such pressure to do well that they had to cheat?

Discuss

1 List three other emotions might your friend be feeling.
2 Consider why your friend might be feeling angry.
3 Consider two reasons why your friend might not be taking responsibility for their actions.
4 What emotions might they be feeling at the idea of their parents finding out?
5 How would you showing empathy help in this situation?

Sometimes we have things going on that we don't deal with fully. We can push difficult emotions down and try not to think about them. Often when our emotions get too much for us to handle, these ones we've pushed down can also surface, and we can feel overwhelmed or confused about how we are feeling. There are also some emotions that are more 'appropriate' to express in public than others. Where do we get the chance to express those 'inappropriate' ones, so that we don't carry them around with us? Some people use physical exertion to get rid of emotions, e.g. punching a boxing bag or their pillow, or going for a walk/run. Some people listen to sad music and cry. Some people like to talk about it, some don't. Everyone deals with it differently. It is about finding a safe way to express emotions without hurting yourself or someone else. How do you express those emotions?

When someone is experiencing stress they might become angry, sad or frustrated. They may isolate themselves from their friends or just act differently. This can be hard for their friends and family, who may not understand what the problem is and so don't know how to help. Friends and family can also become frustrated by this behaviour. When you notice people you know behaving differently, think about why they might be acting this way and ask them what they need. Sometimes they won't even know, but just showing that you are there to support them might be all they need.

Worksheet 4.7

CASE STUDY  WHO'S EMPATHETIC HERE?

Identify

Working in groups, consider the following five scenarios and who is displaying empathy.

Tyler thinks his parents are about to split up and he is worried that they will make him choose who he wants to live with. He doesn't need this kind of pressure and it is stressing him out. He is overreacting to things his friends are saying and not concentrating in class. His worries seem to be taking over his brain – he can't think about anything else!

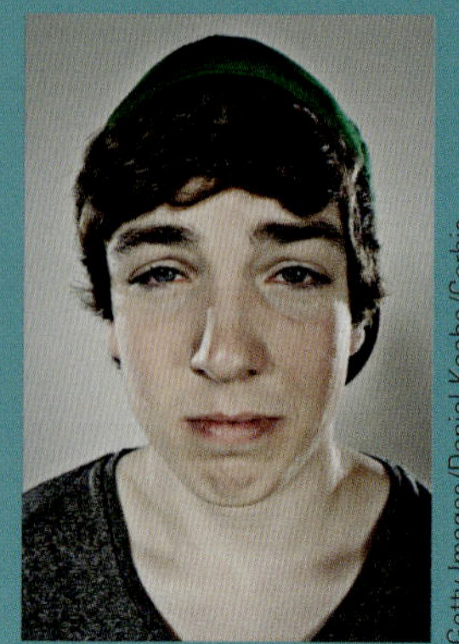

Figure 4.20 Tyler is having a hard time.

Understand

Scenario 1 Luke is in most of Tyler's classes and Tyler is really slang him out at the moment. Tyler won't even come and kick the footy with him after school anymore. He tells Tyler that he'll have to find a new friend if he doesn't stop being so moody of this mood soon.

Scenario 2 Miss Grey is Tyler's Maths teacher. She notices Tyler talking quietly to Sara. Tyler has been waiting for an opportunity to talk to Sara because her parents split up recently and he wants to hear about her experience. Miss Grey asks Tyler to stop talking and get on with his work. He obeys for a few minutes, but then their conversation starts up again. 'Right, Tyler, I'm sick of this. Come down the front here away from Sara, maybe you'll be quiet now. And I want you to come and see me after school to apologise for your behaviour.' Tyler tries to object, as he knows he has to go straight home with his dad. 'No excuses, Tyler. My office after school!'

Scenario 3 Mrs Peace is the Student Wellbeing Coordinator at school and Tyler's parents have told her about their situation. She has been keeping an eye on Tyler and worries that he is withdrawing from school. When she sees Tyler and Luke at lunchtime having what looks like a serious discussion, she tells Tyler that she needs his help with something and asks him to come to her office. She really just wanted to get Tyler away and see if things were okay. She lets Tyler spend the next period in her office, just reading and catching up on some homework.

Scenario 4 Geoff, Tyler's dad, picks him up from school every day. Tyler was late coming out today and Geoff was getting angry. Tyler got in the car and Geoff shouted at him, 'You know things are really stressful at the moment, do you want to make them worse?'

Scenario 5 Kane, Tyler's brother, notices that he is really quiet tonight, and asks Tyler if he's okay. Tyler shakes his head; Kane can see tears in his brother's eyes. 'Don't worry bro, go to your room and pretend to be doing your homework, I'll keep Dad occupied,' says Kane.

Discuss

1 Which characters are being empathetic to Tyler or his situation?
2 List the ways that they are being empathetic.
3 If they are not being empathetic, discuss ways they could treat Tyler differently in order to show more empathy.
4 Suggest what Tyler could do next to help his situation.
5 **Identify** who you believe is responsible for Tyler's mental health. Create a table to include the multiple people this might include and the role that they play in Tyler's mental health. Consider both positive and negative roles.

1 How does communication impact our resilience?
2 Recall why it is important to use positive self-talk on a regular basis.
3 **Explain** what you believe 'empathy' to be and how it can best be included in day-to-day interactions with others.

1 Outline how being a good listener helps develop empathy.
2 State how you express empathy by providing two recent examples.
3 Consider and discuss the way self-talk affects the way we view the world around us and express emotions.

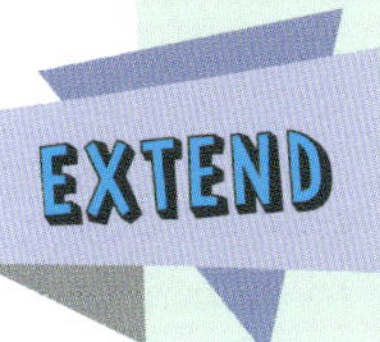

1 In many professional sporting events, we see someone losing their cool and acting out due to being angry, e.g. smashing their tennis racquet on the ground. Is this an appropriate response?
2 What if everyday people did that at their workplace? What would happen to them?
3 Debate as a class whether this is appropriate or not.
4 Have you ever been taught empathy in school? You are the teacher, design an activity you could use to teach your students about empathy.

HOW CAN I ACCESS SUPPORT IN MY COMMUNITY?

Quiz
Anxiety

It is important to know where to access the best type of help in your community. However, there can be many barriers to accessing help. We need to overcome these barriers to ensure that help is available to all.

BARRIERS TO ACCESSING SUPPORT

Getting help for a mental health issue can sometimes be difficult. There are many factors that influence how easy it is to get help. There are several issues that need to be considered when understanding why getting help may be difficult. If people don't seek help when it is needed, their mental health and wellness can suffer, and this can affect all aspects of their lives.

The stigma of mental health issues

stigma a mark of shame or disgrace attached to something

If you have a cold, or even a more serious illness, you would usually feel comfortable telling people, and perhaps even feel better just by talking about it. Mental health can be different because in the past there has been a **stigma** attached to it. In the past, there were strong stereotypes placed on men, telling them they needed to be tough and deal with issues quietly by themselves. This made it difficult for them to ask for support. It's important to remember that everyone gets upset during their lives and has the right to ask for help, without being judged. Often people are uncomfortable discussing mental health issues, or don't understand, so the person with the issue feels alone and unsupported.

When you look at the following barriers to getting help, think about how people might respond emotionally if these barriers are in place. For example, if you didn't speak English very well, and you didn't understand what the doctor was saying but didn't feel like you could ask questions, how might that make you feel? How might your mental health be affected?

Access to help

Access to services can be limited by lack of transport or money, disability or knowledge of available services. Where you live can also affect your ability to access support. For example, in some rural towns everyone knows everyone else, even the health professionals, so you may not feel comfortable discussing mental health issues with them. The distance patients and healthcare providers must travel also contributes to the overall expense of treatment, meaning some people cannot afford the services they need.

Rural and remote access

In some remote places in Australia, the doctor may only be available on a fly-in, fly-out basis. In rural and remote Australia there are many accessibility issues due to distance, cost and availability. Phone and internet services can be limited and unreliable. Natural disasters, such as fire, drought and flood, can also impact the availability of services in times when they are desperately needed.

Language barriers

Most healthcare professionals will speak to their patients in English if they don't have immediate access to an interpreter. This means that a non-English-speaking patient must do their best with the limited English that they have. Sometimes an adult patient might take a child with them to appointments so that the child can interpret for them, but this can be problematic. The child may not understand or have the right words to accurately translate the information. Telephone interpreting services are also available to non-English speakers.

A third of young Australians live in regional and remote areas	Mental health issues are more common outside of cities	Suicide rates are twice as high for young men outside of major cities
We're talking huge numbers: one in three young people, in fact. That's one million young people who are living away from the convenience of city support services, and facing unique issues that call for unique interventions.	Compared to those in major cities, young people living in regional and remote areas show heightened levels of psychological distress. Plus, mood disorders like depression and anxiety are twice as prevalent in remote areas as they are in cities. Young people are doing it really tough in Australian towns.	Devastatingly, the rate of suicide among young men who live outside major cities is almost twice as high as it is for their metro counterparts. Getting young men help sooner is the best chance we have of preventing suicide.
There's less access to specialised mental health care	**People outside of cities are less likely to access mental health services**	**It's harder to get mental health help in regional and remote areas**
Inner regional areas only have 37 per cent of the psychiatrists and 61 per cent of the psychologists that major cities have. That means if a young person decides to get some face-to-face help, they're going to have fewer choices than their peers in the city.	The reduced choice probably has a negative effect on service uptake: for every $1 spent per capita (person) on Medicare mental health services in major cities, just 77 cents is spent in inner regional areas. That reduces to a miniscule 10 cents in very remote areas.	Waiting lists. Long travel times. Stigma. Social isolation. Imagine waiting six months for a first session with a psychologist, then having to drive two hours (each way), all the while not wanting to talk to anyone about it, and also having literally fewer people to talk about it with. It doesn't sound easy, does it?

Figure 4.21 Mental health in rural and remote areas

Time

Often it takes more time to get the help needed for a mental health problem than for a general health issue. There are several reasons for this:

- You need to visit a general practitioner (doctor) first to get a referral for a mental health plan.
- Time is needed to establish trust between the person being treated and the health professional.
- It takes time to get family members involved to help with decision-making and/or treatment.
- It takes time to organise interpreters, ensure understanding and arrange any necessary follow-up care.

Weblink
Access the Rural Adversity Mental Health Program website and click on the Glove Box Guide to Mental Health Volume 8. Choose an article to read and discuss what issues it has brought up for rural and remote mental health. Then discuss what it suggests to help, or to keep mental health strong.

Cultural considerations

People understand, talk about and find help in different ways. When talking about mental health with others, it is important to be respectful of their culture. Aspects of verbal and non-verbal communication such as eye contact, some types of body language and the use of certain words can be seen as offensive or judgemental, causing communication to break down. It is always best to ask the person if they are comfortable talking about these issues.

Worksheet 4.8

relaxation techniques methods used to calm down or relax

Video
Mindful movement activity: Slow motion overarm throw

RELAXATION TECHNIQUES

It is useful to know some **relaxation techniques** to use when you feel yourself becoming stressed. Sometimes, stress can be so overwhelming that you can't think of anything productive to do. If you are able to calm yourself down and clear your head, you will be able to make better decisions about the situation that's creating the stress. With a clear head you can see things from different points of view and come up with solutions.

Some relaxation techniques to try:

Shutterstock.com/Dudarev Mikhail

Figure 4.22 Try one of the relaxation techniques!

- **Breathing** – just concentrate on breathing in and out. Don't listen to anything except the air coming into your lungs and then going out.
- **Meditating** – concentrate on one thing, or perform a repetitive action so your mind is focused on that action alone. Some things that can help with meditation include repetitive exercise, drawing, or playing or listening to music.
- **Practising mindfulness** – focus your mind on one thing at a time and take the time to enjoy what you are seeing. Consider the Dadirri deep learning we looked at earlier. If you find your mind is wandering, bring it back to the object.
- **Listening to music** – choose a slow or relaxing song to listen to. Be careful not to play sad songs if you are stressed and upset or sad.
- **Interacting** – smile at someone or give someone a hug; pat your cat or dog.
- **Visualising** – think about yourself being happy and relaxed, or think of a place you would like to be and all the sights, sounds, smells and colours you would be surrounded by.
- **Laughing** – this can make you feel calmer, and maybe even happier!

iStock.com/DipakShelare

Figure 4.23 Interacting and relaxing

FAST FACT

According to Dr Song from the University of Chicago Medical Centre, smiling uses 12 muscles and frowning uses 11 muscles. Despite this, smiling actually takes less effort.

Even though smiling uses more muscles, it is believed that it takes less effort than frowning. This is because people tend to smile more, which means the relevant muscles are in better shape. When muscles are in better shape, they require less energy (effort) when used ... Humans are born with the ability to smile, it is not something that we learn. For instance, even blind babies are able to smile.

Source: 'Does It Really Take More Muscles To Frown Than To Smile?' ByZidbits, September 20, 2011, https://zidbits.com/2011/09/does-it-really-take-more-muscles-to-frown-than-to-smile/

9780170463096

FACE TO FACE Tongue twister

Can you read this tongue twister to another student without smiling or laughing? This can be a fun way to make yourself or a friend laugh, which will help you relax.

Mr See owned a saw.
And Mr Soar owned a seesaw.
Now, See's saw sawed Soar's seesaw before Soar saw See,
which made Soar sore.
Had Soar seen See's saw before See sawed Soar's seesaw,
See's saw would not have sawed Soar's seesaw.
So See's saw sawed Soar's seesaw.
But it was sad to see Soar so sore,
just because See's saw sawed Soar's seesaw.

Shutterstock.com/Ron Leishman

Figure 4.24 Tongue twisters can make people laugh.

WELLBEING

PROGRESSIVE MUSCLE RELAXATION

This activity has been developed in collaboration with Dr David Bakker of MoodMission

Identify

It's hard to 'just relax', but there are some ways we can make our muscles relax.

Understand

When we're feeling upset or stressed, our muscles tend to tense up and our breathing gets faster and shallower. It's part of the fight or flight response, where our body prepares us to deal with a threat by running away from it or fighting it. However, this response often gets triggered by a false alarm and we actually don't have to run away or fight. In these moments, we can trick our bodies and brains into feeling calmer by relaxing our muscles and breathing more slowly and deeply.

Practise

Find a quiet spot to sit or lie down. Now take a few minutes to slow your breathing and progressively tense and relax different parts of your body. Tense the body part for a count of five and then relax it for five. It can be good to start with your hands and forearm, by making a tight fist and then relaxing it slowly. Then tense other parts of your body one after another, as seen in Figure 4.25. Here's a suggested sequence:

1 Hands and forearms. Make fists with hands.
2 Upper arms/biceps. Bring your forearms up to your shoulder.
3 Neck. (Take care tensing these muscles.) Slowly pull your head back to look at the ceiling and then tense.

4 Shoulders. Bring your shoulders up towards your ears.
5 Shoulder blades/back. Push your shoulder blades back, pushing your chest forward.
6 Stomach. Tense up your abs and then push your belly right out when you relax.
7 Upper legs. While sitting down, go to stand up and tense your thighs and buttocks.
8 Lower legs. Point your toes like a ballerina to tense your calves.
9 Face. Close your jaw tight and screw up your eyes, just like you've eaten a lemon.

Reflect

Do you feel any more relaxed after doing this? When might you be able to do this, or parts of this? For example, some people just tense their fists or their stomachs if they get stressed while they're out and about and can't do the whole sequence.

Worksheet 4.9

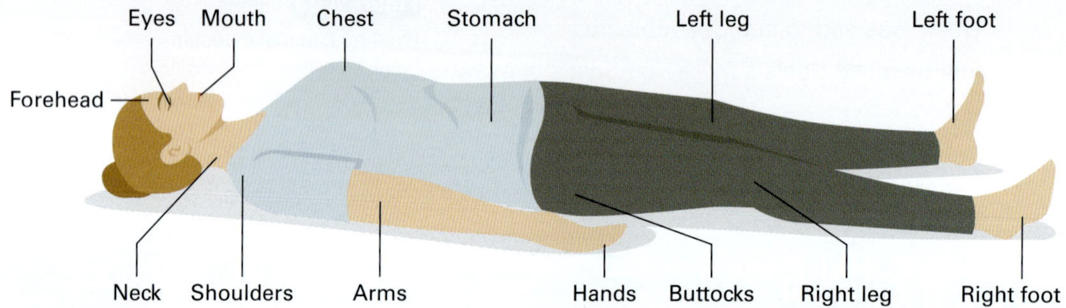

Figure 4.25 Tense, then relax, each of the indicated areas individually from the bottom of your body to the top.

WHERE CAN I GO FOR HELP?

It is important to know how to find people or places (including trustworthy websites) that will provide good, accurate and timely information to help you with the things that are causing you stress. Seeking help is an important life skill. If you can get help quickly, you are less likely to be stressed out!

There are lots of good sources of information available, including:

- family
- friends
- teachers
- student wellbeing coordinator or school counsellor
- nurse
- doctor
- youth worker
- relatives
- internet
- books
- pamphlets
- health centres.

Some of these may not be good sources for all the information you need. For example, your friends may not know any more than you do. You may need to refer to multiple sources.

FACE TO FACE Finding sources of help

1 Do this activity in groups of three. Each person should contribute one example to each of the three questions.

   a What kind of family situation might cause a young person stress? (e.g. an older sibling moving out of home)

   b What kind of decision might a young person have to make that causes them stress? (e.g. choosing between doing the elective that you want to do or the one all your friends are doing)

   c What stressful things might happen to young people during puberty?

2 Using one of your own examples from question 1, name three different options (people, places, websites, etc.) where you could go for help if you were experiencing that particular type of stress.

3 How could you help one of your friends if they were stressed? What advice would you give them?

FACE TO FACE The helping hand

On a blank piece of paper, trace a picture of your hand. You can decorate your hand in any way you like. Think about who you can go to for help when you are stressed. Share your ideas with the others on your table. On each finger and your thumb, write one person, place or source where you could find good information that could help you.

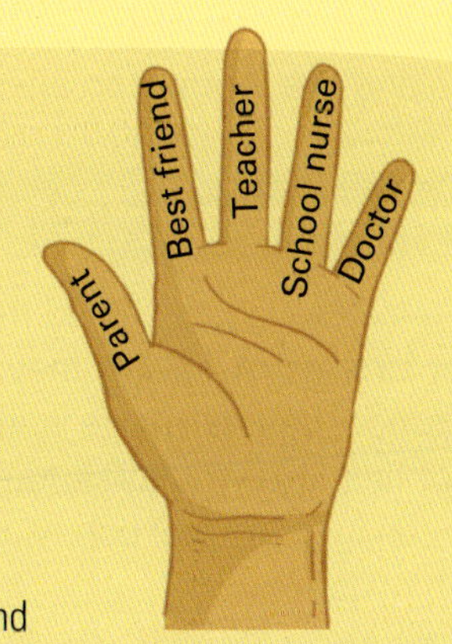

Figure 4.26 The helping hand

Worksheet 4.10

The internet as a source of help

The internet provides a huge amount of information, but not all of it is accurate! How do you know whether the information you have found is correct?

First, look at the web address. If it includes '.gov' or '.org', the site should provide accurate information that has been written by professionals in the field. If the site has '.com', it may have been written by a company to convince you to buy a product, therefore the information may be one-sided. Also check which country the website is written for or targeted to; it is often better to use Australian information. You also need to think about the reason the website has been created.

FACE TO FACE Reliable information?

In pairs, discuss how you can tell if a website is giving good health information. Make a list of things that you could use to work out if a website is reliable.

Local resources

Knowing your local area is an important part of seeking help. Knowing what is available and where makes it easier to get help when you need it. You may never use this information yourself, but it may mean you can help a friend or family member in the future.

INVESTIGATION  PLACES TO GO FOR HELP IN YOUR LOCAL AREA

Purpose

To be able to find reliable sources of information in your local community.

Materials

In pairs or groups, find a map of your school and the surrounding local area and print it out. Referring back to 'Where can I go for help?' on page 176, identify where the health-related resources are located on the map, e.g doctor, nurse, health centre, youth worker. Using your local knowledge and the internet to answer the following questions.

Method

Ameen and his family have just arrived from overseas. They have moved into the house next door to you. Your families met while they were moving in and you found out that Ameen will be starting at your school next term. He is feeling anxious about starting a new school, and worries that he might not fit in. He would like to meet some people his own age before school starts.

1 What resources in your local community could Ameen utilise?
2 What might be Ameen's main worries?
3 What emotions might Ameen be feeling?
4 What might his self-talk be?
5 What mental health supports can be found in the community for Ameen's anxiety?
6 Are there any support services for culturally and linguistically diverse (CALD) people in your community?
7 Where could Ameen go to meet other young people?
8 What could you do to help Ameen feel welcomed into your community?

Getty Images/Jose Luis Pelaez Inc

Figure 4.27 What resources in your community could help Ameen?

Discussion

As a class, discuss your responses to the investigation. Are there similarities and/or differences between the responses? How could your class welcome a new student to the school?

FACE TO FACE Make a mental health first-aid kit

Collaborate with your group while doing this activity. Others may have great ideas that you haven't thought of, and vice versa.

1 Bring in a cardboard box from home and decorate it with pictures, words, colours or objects that you love.
2 Fill the box with things that you can rely on or do to help improve or cope with the daily stressors of life. For example, you could include photos of friends/family you can talk to; a list of phone numbers you could call if you needed help; the names of songs you love or that relax you; ideas of activities that make you feel better (exercise, spending time with friends, relaxation, etc.); or a childhood toy that brings back good memories and a feeling of safety.
3 Keep your kit somewhere nearby, where you can look at the contents when you need a bit of a lift.

Figure 4.28 Some of the things that help us cope

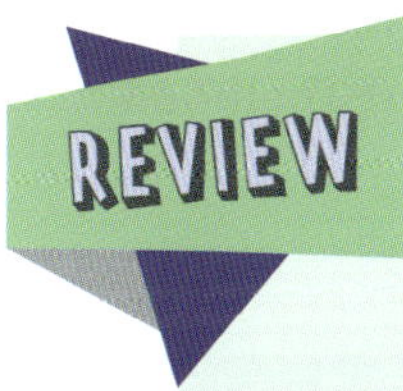

REVIEW

1 Consider why some people find it hard to access support for mental health issues.
2 Determine how you would know if a website is reliable and that you can trust the information.
3 Investigate the availability of mental health support services in your local area. Provide the details including their address, contact phone number(s), website as well as the services they offer.

REFLECT

1 Why might a young person not talk to their friends about their mental health issues?
2 If you were worried that a friend had mental health issues, and they wouldn't talk to you, would you ask someone else for help? If so, who would that be? If not, why not?
3 What services for mental health does your school promote? Are there posters around the school, in the wellbeing office, in the office, etc.?

EXTEND

1 Imagine you are a parent and you are worried about one of your children because you think they might be depressed. Where could you go to get more information about how to help your child?
2 What services does your school have for people with mental health issues?
3 **Evaluate** your school's approach to mental health issues. Consider people, places, services, support, subjects, information to parents, etc.

Worksheet 4.11

HOW CAN I ANALYSE HEALTH INFORMATION?

MENTAL HEALTH PROMOTION

Mental health promotion is any action that is taken by individuals or groups to improve mental health and wellbeing.

The things that can be done to help strengthen your mental health are called protective factors. The things that put you at risk of having mental health issues are called risk factors. When parents, carers and schools work together to help put protective factors in place, young people are more likely to have good mental health and wellbeing. These protective factors help to build resilience by providing opportunities to grow and develop in a safe and supportive environment.

Shutterstock.com/Mega Pixel

Figure 4.29 Promoting mental health: the green ribbon signifies mental health awareness.

Table 4.1 contains part of the Melbourne Charter for promoting mental health and preventing mental and behavioural disorders. These protective factors are included in the planning stages of health promotion strategies.

Video
Wellbeing: What do you think of when you hear the word 'wellbeing'? What is most important to your own personal sense of wellbeing? Watch the video and join the discussion.

Table 4.1 Protective factors and risk factors for mental health and wellbeing

Protective factors	Risk factors
Positive family functioning	Poor family functioning
Social support (including online)	Lack of social support and loneliness
Resilience and self-esteem	High screen time
Community support	Alcohol, cigarette and other substance abuse
Physical activity	Poor diet and sedentary behaviour
Cultural factors (ethnic belonging and identity)	Social media and cyberbullying
	Chronic illness and obesity
	Out of home care
	Stressful life events and factors related to refugee status
	High demand academic environments

Source: 'Evidence Check – Mental Wellbeing Risk & Protective factors', Sax Institute for VicHealth

Worksheet 4.12

FACE TO FACE Protecting students at school

In pairs, have a look at the protective factors outlined in Table 4.1. Can you think of anything your school already does to help build these protective factors into your life? An example might be promoting diversity through a Harmony Day event. Come up with a list for as many as you can think of.

Promoting social participation can help increase mental health and wellbeing. One of the protective factors listed in Table 4.1 is social participation, which helps prevent **social isolation**. Mental health promotion includes helping people to connect and remain connected with others in their local area. Being a part of a friendship, team, workplace, social group, etc. can bring enjoyment and improved mental health.

social isolation when people feel lonely, and not connected to friends or family; this can have an impact on our mental health

Schools are mandated to provide policies that address mental health and wellbeing in their schools. These include the Student Wellbeing and Learning Policy, the Student Engagement Policy and the Student Attendance Policy.

Worksheet 4.13

FACE TO FACE Wellbeing policy

Analyse your school's student wellbeing policy for protective factors. Do you think your school's policy does enough to build these into a student's day-to-day life?

World Mental Health Day and Bright Futures Challenge are examples of health promotion strategies that are designed to improve the awareness of mental health, provide information and/or build skills in dealing with mental health problems. These types of strategies all try to include protective factors to help build mental health and wellness in their communities.

World Mental Health Day

World Mental Health Day is an example of international health promotion. Held on 10 October each year, the aim is to educate and raise awareness about mental health issues. The organising body is the World Federation for Mental Health (WFMH). Participating countries organise various events and activities to promote mental health and wellness and to try to reduce the stigma around mental health. Each year is based on a special theme related to mental health; previous themes have included mental health and older adults; depression; diabetes and depression; and caring for the caregiver.

Weblink Check out the Australian website for World Mental Health Day.

Weblink
Bright Futures Challenge

Worksheet 4.14

Local health promotion

Organise a stress-awareness week at school to help students, teachers and parents understand and cope with stress. Decide how your class will go about organising this week. Here's a list to help get you started.

- Research and **identify** ways of dealing with stress, e.g. exercise, good sleep habits, talking about worries, time management skills, having fun with friends, etc.
- Research and **develop** a display that can be left out for the week, highlighting your ideas for dealing with stress, with pamphlets and handouts available for students to take home. This could include you making a version of the '10 Tips to Stress Less' poster on page 161.
- Make use of school newsletters (write an article for students and/or parents), daily bulletins and announcements.
- Organise a lunchtime event based on exercise, food, study skills, relaxation, etc.
- Make up a skit, rap or poem that you could perform or read at assembly.
- Organise a speaker for assembly and invite parents to come.
- See if the local newspaper could write an article about your health promotion activities.

If this event falls around the time of Mental Health Week in October, don't forget to register your activity with the Mental Health Foundation of Australia. Make sure you evaluate the week to see what went well and what could be improved next time!

Quiz
Chapter Review

1. Contrast the differences between protective factors and risk factors.
2. Discuss how your school addresses some of the protective factors.
3. Summarise at least three benefits associated with health promotion strategies.

1. Looking back on your primary school experience, can you think of any things that the school did to promote/strengthen your mental health? Refer to the list of protective factors to help.
2. Reflect on your transition to secondary school – what did your primary and secondary schools do to help promote/strengthen your mental health during this big change?

1. Try to recall any mental health promotion campaigns you have seen in the media.
   a. Who are they targeting?
   b. **Evaluate** how effective you think they are, consider their target group.
   c. Could you improve the campaign based on the protective factors?
2. Create a mental health promotion clip aimed at young people. Begin by writing down the messages that you want to come across and develop a creative way to do that.

CHAPTER 4 REVIEW

1. What is the definition of mental health?
2. Why should we strive for good mental health?
3. What is the difference between a mental health issue and a mental illness?
4. Name two effects good physical health will have on your mental health.
5. Name two effects good social health will have on your mental health.
6. As people grow up, who or what are the influences on body image?
7. What role does the media play in defining the ideal body?
8. What causes stress?
9. How does stress impact our day-to-day life? Consider the physical, social and emotional effects.
10. Does the way that you communicate with others have an impact on your mental health? Why or why not?
11. How does diversity impact how we access mental health care?
12. Why is it important to have resilience?
13. List four barriers to someone finding help for a mental health issue or illness.
14. **Identify** three relaxation techniques.
15. How could you incorporate these relaxation techniques into your everyday life?
16. Name three people you know who you could go to for advice on a mental health issue.
17. Name three organisations you could seek help from about a mental health issue.
18. List an example of an international, national, state and local health promotion. Can you think of others that are not mentioned in this chapter? How do these help people with mental health issues or illnesses?

PUBERTY AND RESPECTFUL RELATIONSHIPS

5

IN THIS CHAPTER

You will learn about puberty and the changes that occur during this time. These include physical, emotional and social changes. These changes often have an impact on your identity as you develop into an independent young adult. You will have an opportunity to explore how your values, beliefs and relationships with others also influence your identity, actions and decisions.

You will learn about respectful relationships and the people and places you can access if you, or someone you know, needs help or has further questions.

Shutterstock.com/Nicetoseeya

By the end of this chapter, you should be able to:

- discuss the changes that occur during puberty and how they impact on your identity
- evaluate strategies that will help manage changes during puberty
- discuss the factors that can influence your emotions
- identify how empathy can help in strong relationships
- identify all the types of relationships you have in your life
- recognise qualities and understand the importance of positive relationships in your life and how relationships impact health and wellbeing
- discuss why young people may not seek support, and what can be done to improve this
- examine the consequences of an imbalance of power in a relationship
- identify good communication strategies
- identify support services in your own environment
- evaluate strategies for managing conflict
- evaluate health information and support services
- recognise the role of the media in promoting respectful relationships.

PUBERTY

Quiz
Pre-chapter

Before you start, take the pre-chapter quiz to find out how much you already know.

puberty a short time of rapid physical growth when the body begins its transition from childhood to adulthood

hormones substances produced by the body that have an effect on growth or development

Puberty is the stage in life where a child's body changes as it develops the ability to reproduce (create a baby). It is not known what causes puberty to begin, but when it does, the brain will signal the body to produce the required **hormones**. Puberty is a time of significant growth for a child and a time when many changes take place. These include physical changes, emotional changes and social changes, among others. Some young people seem to fly through puberty with no worries at all, while others have a more difficult time. Most experience something in between. Everyone has to go through puberty; it is a normal part of growing up.

Puberty usually occurs between the ages of 9–16 for those with ovaries and 12–18 for those with testicles. It takes several years for the changes associated with puberty to be complete. Most people have a lot of questions about puberty, so hopefully this chapter will help answer them! In this section we will talk about the physical and emotional changes that occur during puberty. The next section will address the social changes that occur.

peer group a group of people that you associate with, e.g. class members at school or people who work with you. They may or may not be classified as friends, but they do similar things to you.

hereditary passing of genetic characteristics from parents to children

PHYSICAL CHANGES

The factors that have an impact on how people grow and develop over time include the environmental aspects that you don't have much control over, such as where you live, the climate, your community, your **peer group** and the media. Other factors that can't be controlled are **hereditary** features. These are the characteristics that are passed on from parents to their children, such as hair colour, eye colour and height.

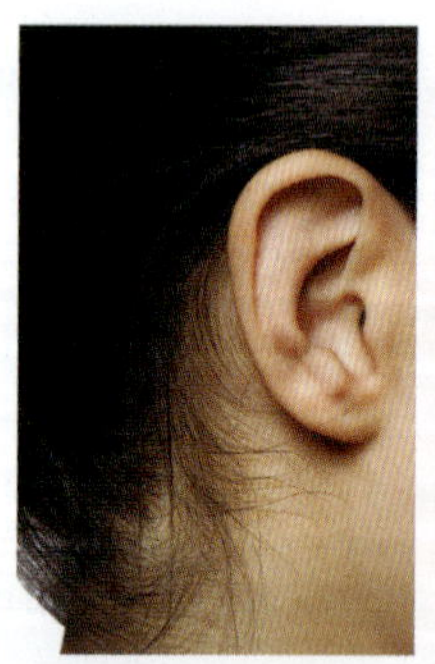

Attached earlobes

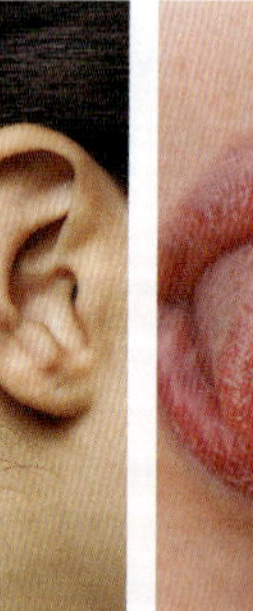

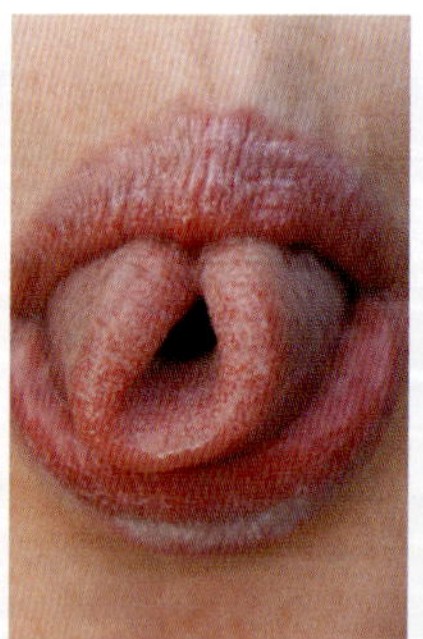

Can roll tongue

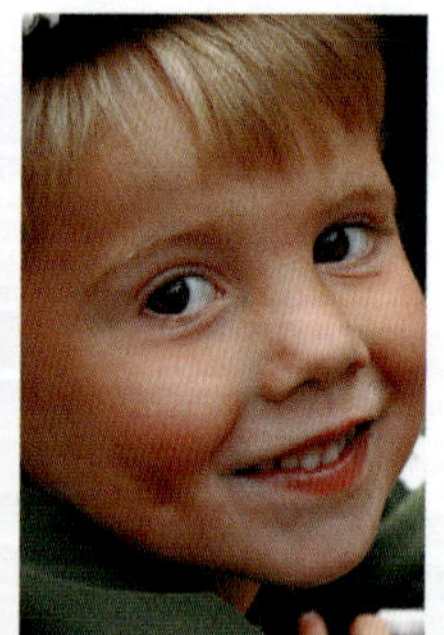

Dimples

Right-handed

Freckles

Naturally curly hair

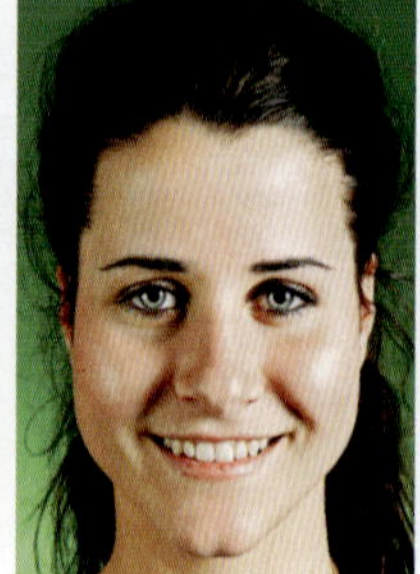

Cleft chin

Allergies

Cross left thumb over right

Can see red and green

Clockwise from top left: Shutterstock.com/Seanika; Shutterstock.com/Mateusz Kopyt; Shutterstock.com/Tracey Helmboldt; Shutterstock.com/Novelo; Shutterstock.com/Savanevich Viktar; Shutterstock.com/Djomas; iStock.com/RapidEye; Shutterstock.com/Eric Isselee; Shutterstock.com/Vankad; Shutterstock.com/FeelIFree

Figure 5.1 Which inherited traits do you have?

9780170463096

Once they are through the childhood stage, after walking, talking and social skills have been learnt and developed, young people experience puberty.

While the **primary sex characteristics**, which are the parts of the body responsible for reproduction (including the ovaries and testes), have been present since birth, during puberty the **secondary sex characteristics** begin to appear. This occurs because the **pituitary gland** at the base of the brain starts to produce hormones, which cause physical and emotional changes in the body. These hormones send a message to the ovaries and the testes, promoting the development of secondary sex characteristics.

primary sex characteristics
sexual reproductive organs such as the penis, testes, ovaries and vulva

secondary sex characteristics
physical features, other than the reproductive organs, that appear during puberty and distinguish bodies with testes and bodies with ovaries

pituitary gland
a small gland at the base of the brain that produces several hormones

Functions of the reproductive system in bodies with testes

Structure	Function
Testes	Responsible for the production of sperm and testosterone. Located in the scrotum sac.
Sperm	The reproductive cell that fertilises an egg.
Scrotum	Loose pouch of skin that hangs below the penis and contains the testicles. The scrotum keeps the testes at an optimum temperature for sperm development.
Vas deferens	Tube used to transport sperm and semen to the urethra and penis.
Urethra	Carries urine and sperm to the outside of the body.
Seminal vesicles	Small glands that add fluid to the sperm to produce semen.
Epididymis	A sperm storage area that links the testes with the vas deferens.
Prostate gland	Surrounds the urethra under the bladder and secretes a fluid that assists with movement of the sperm.
Penis	The organ used in sexual intercourse to release sperm.

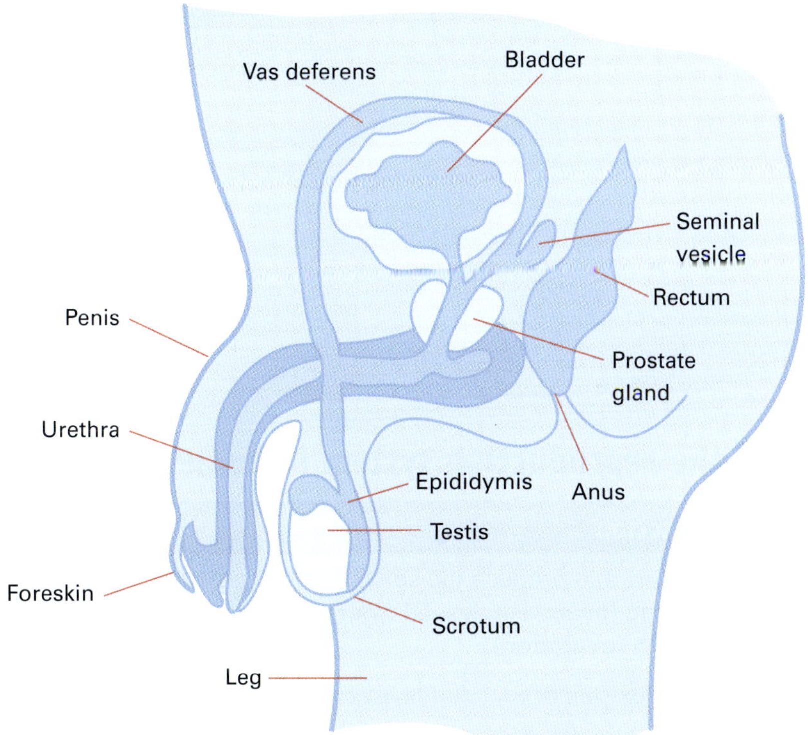

Figure 5.2 The reproductive system of a body with testes

The secondary sex characteristics that may develop in a body with testes during puberty include:

- bigger muscles
- voice 'breaks'
- shoulders widen
- hair growth on the body, around the pubic area, under the arms and on the face
- the penis, scrotum and prostate enlarge
- sperm production will begin.

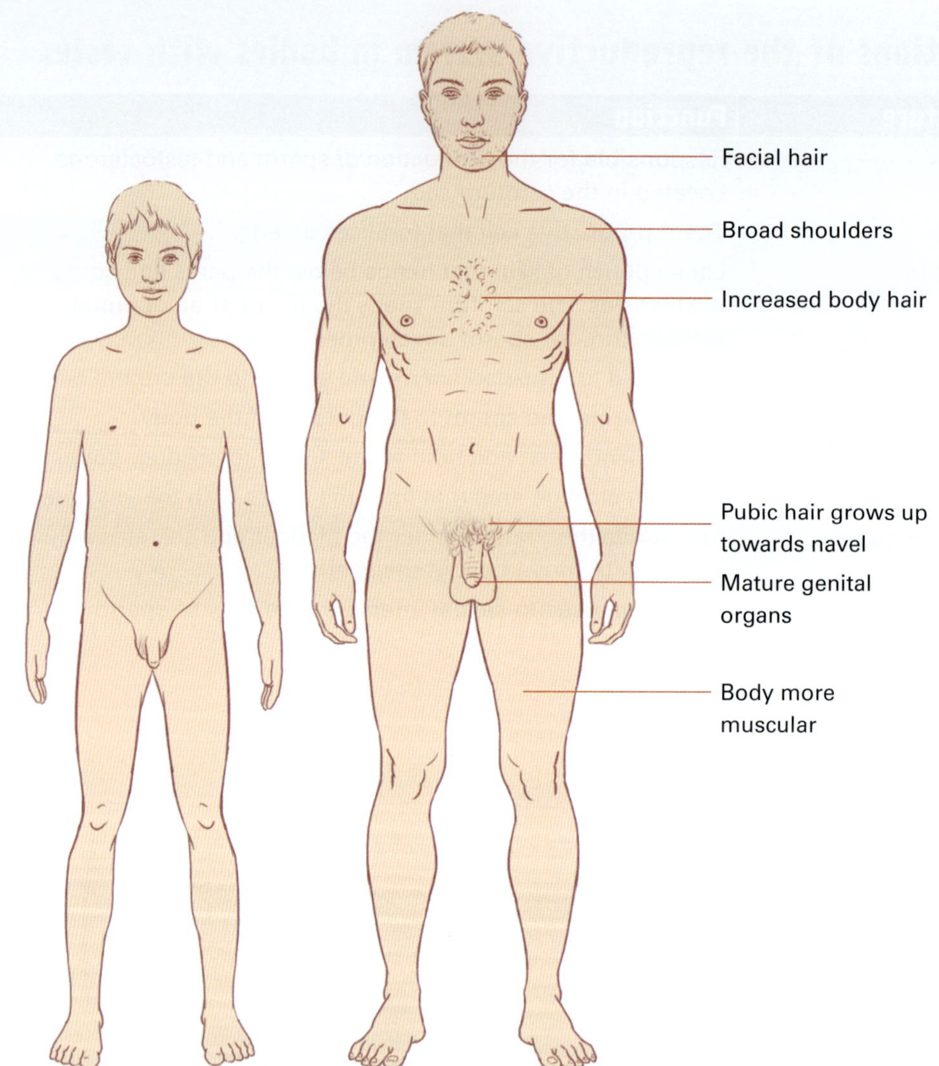

Figure 5.3 Secondary sex characteristics of a body with testes

Functions of the reproductive system in bodies with ovaries

Structure	Function
Ovaries	Glands that produce the sex hormones oestrogen and progesterone. An egg (ovum) is released from the ovaries approximately once a month (**ovulation**).
Ovum	The sex cell, or egg, fertilised by a sperm at **conception**.
Fallopian tube	Extends from the ovary to the uterus to carry the egg after ovulation. This is usually where fertilisation takes place.

ovulation when a mature egg is released from the ovary

conception when the sperm fertilises the egg and the embryo is created

menstruation the process through which the lining of the uterus is expelled as blood from the body

Structure	Function
Uterus	A muscular pear-shaped organ where a developing baby will grow for 40 weeks. The lining of the uterus is shed during **menstruation** when an egg is not fertilised.
Endometrium	This is the lining of the uterus; it builds up every month in preparation for conception. If no conception occurs, the body releases the lining in the form of a period or menstruation.
Cervix	Joins the uterus and the vagina. This is the part that opens when a baby is ready to be born.
Vagina	Joins the uterus to the outside of the body. This is where menstrual blood travels out of the body. This is where the sperm can be released during sexual intercourse.
Urethra	A tube that runs from the bladder to the outside of the body to release urine.
Clitoris	A sensitive organ located just above the vaginal opening that extends under the skin.
Labia	Folds of skin that protect the opening to the vagina.
Vulva	The whole outer area, including the labia.

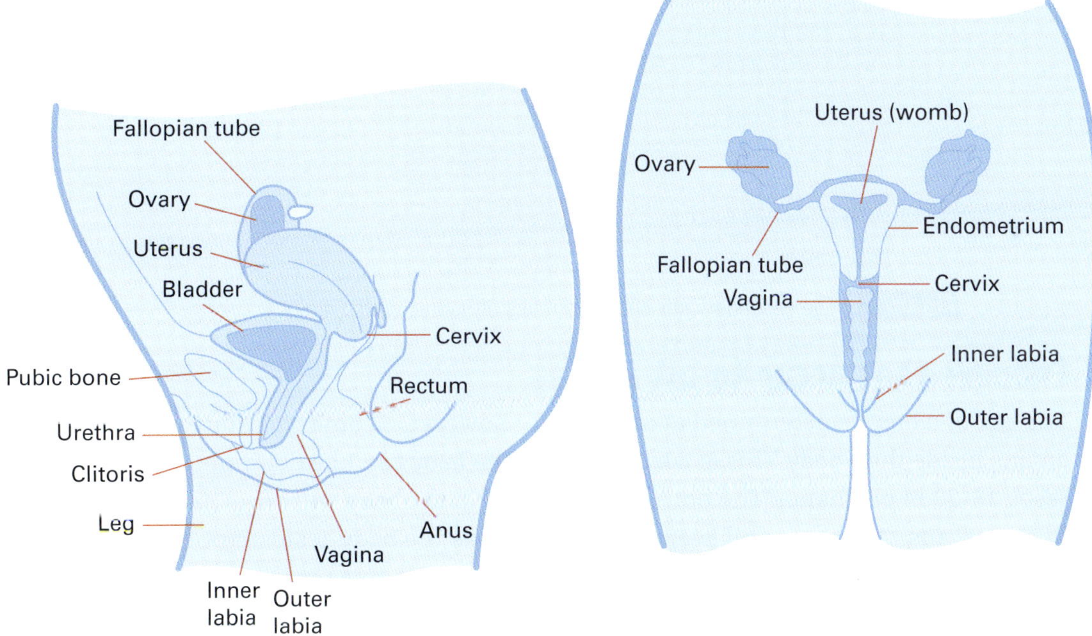

Figure 5.4 The reproductive system of bodies with ovaries

The secondary sex characteristics that may develop during puberty in a body with ovaries include:

- breast development
- hips become bigger
- body hair may appear around the pubic area and under the arms
- reproductive organs such as the vagina, fallopian tubes and the ovaries enlarge
- ovulation and menstruation will begin.

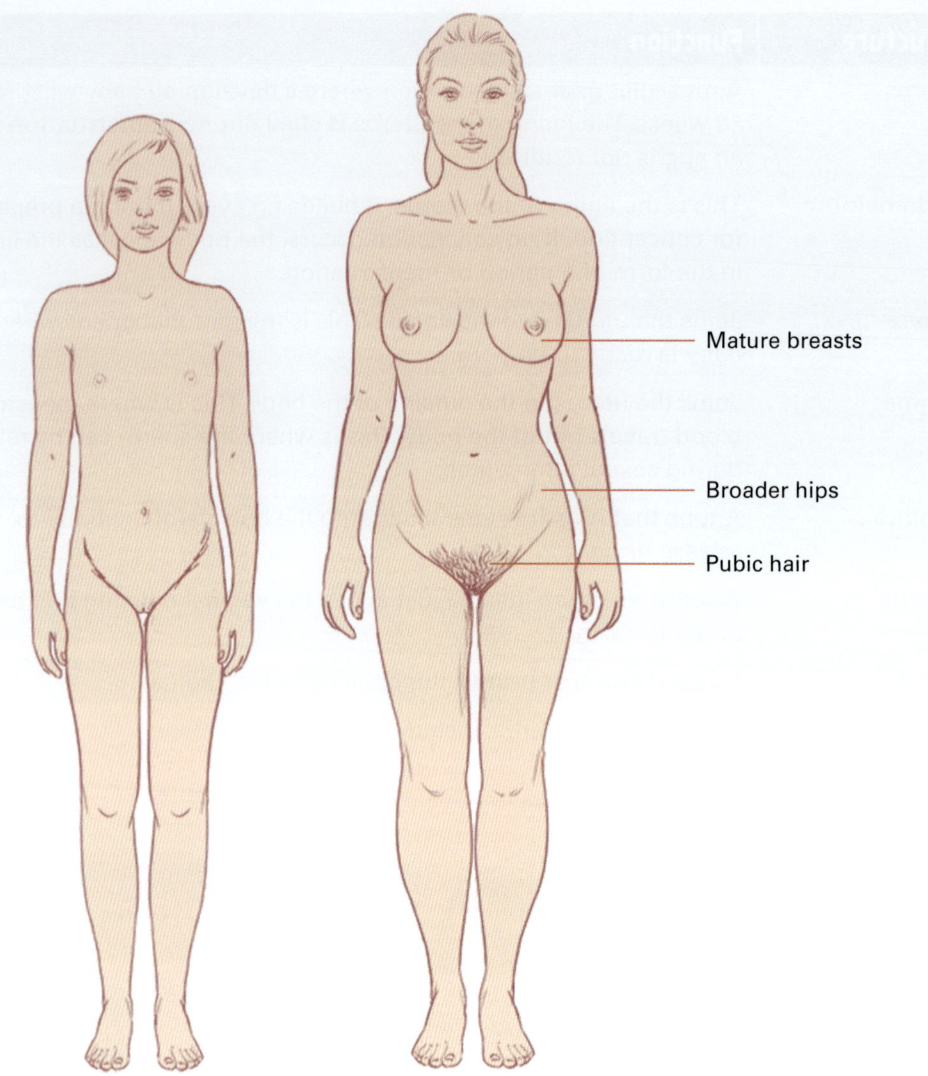

Figure 5.5 The secondary sex characteristics of bodies with ovaries

Changes that occur to all bodies

sex whether a person is biologically male or female

embryo the fertilised egg for the first eight weeks of development

intersex people born with variations of sexual characteristics, e.g. physical sexual characteristics, hormones, chromosomes, etc.

transgender when the physical sexual characteristics of a person don't match the gender identity of that person

stereotype a widely held view or perception of a person or thing based on appearance, gender, race, culture or religion

The **sex** of a baby is determined at conception; however, all **embryos** start with their sex organs inside their body. This is why the sex of the baby can't be determined until around Week 16–20 in pregnancy. The hormones released in the baby will develop the required sexual organs. Oestrogen will keep the organs inside the body, testosterone will allow the penis and the testes to drop outside the body. Sometimes this process doesn't happen, or isn't complete. When this happens, the baby might be born with variations of hormones, genitals or sex characteristics; this is called **intersex**. About 1.7 per cent of the population is born intersex; this is nothing new, and will continue to happen.

Puberty can be confusing for children who are born intersex, as their body parts may not develop as expected for their sex. For example, they may develop breasts when they have a penis. There is nothing physically wrong with these children.

People who are **transgender** don't identify with the body they have been born into. As you can imagine, this can cause a lot of stress for them and their loved ones. Imagine looking at your genitals and feeling like you have been given the wrong ones. People who are transgender may choose to express themselves as a different sex or gender. This can include changing their name and the way they dress. Can you think about how harmful gender **stereotypes** might be to people who identify as transgender?

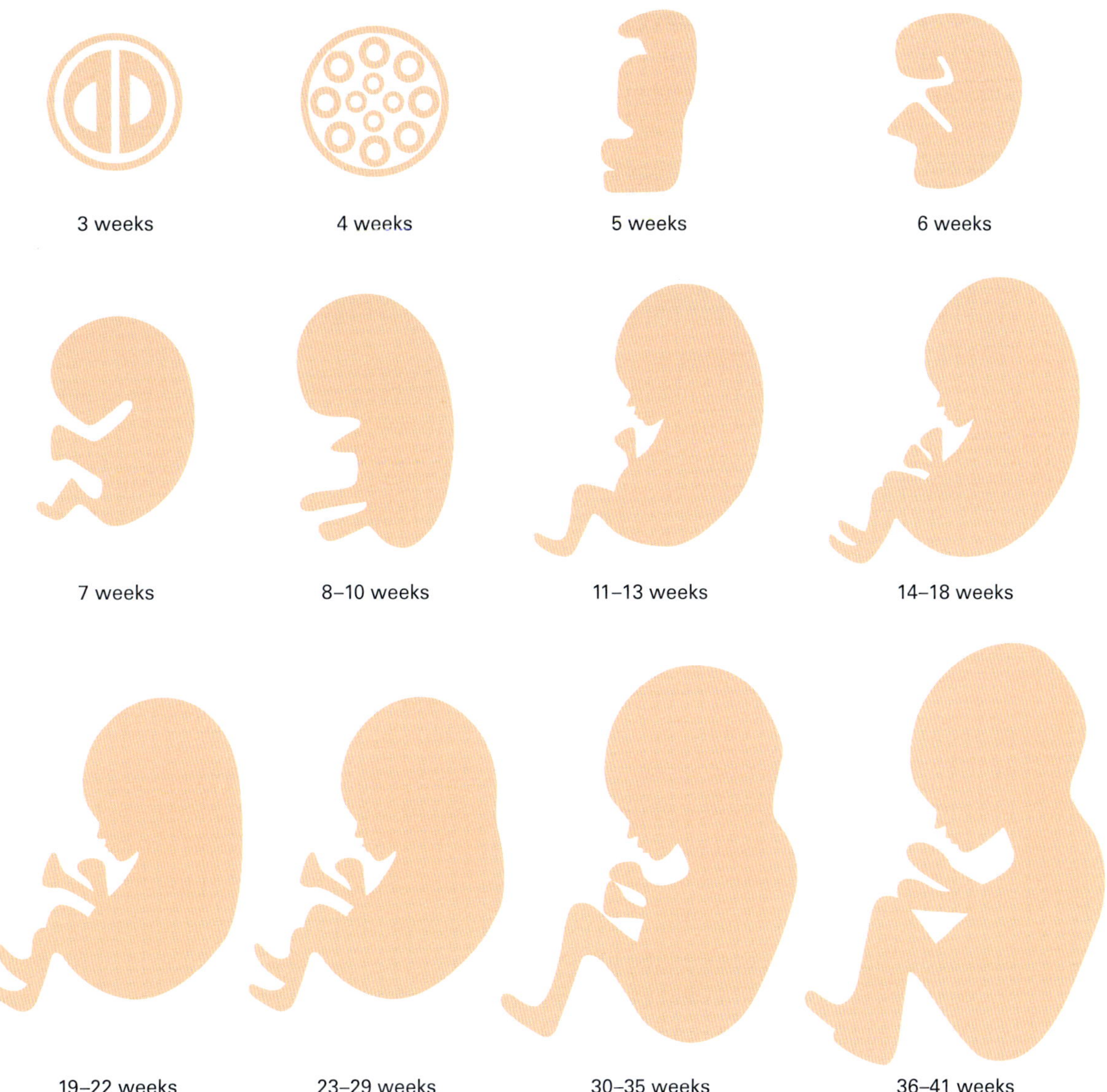

Figure 5.6 Growth from conception to birth

> **FAST FACT**
> All people have nipples, even though not everyone can use them. This is because all babies start off the same way!

For the most part, most physical changes occur both to bodies with ovaries and bodies with testes. These include getting taller and therefore putting on weight, voices deepening, developing body hair, body odour, acne and an interest in **intimate relationships**, etc. There are also other changes that are less well understood, such as **erections** and **nocturnal orgasm**, commonly known as 'wet dreams'.

People often assume that only bodies with testes experience these, but as all sexual organs start out the same, the clitoris is also very sensitive, just like the penis. Both can harden in an erection as they are both made out of the same erectile tissue. All bodies can also have wet dreams. For bodies with testes this occurs when semen, the fluid containing sperm, is discharged from the penis during an ejaculation while sleeping. The person may find they have more erections during this time.

intimate relationship a deep emotional connection with another person

erection when the penis becomes enlarged and fills with blood, causing it to stand upright; when the clitoris becomes enlarged and fills with blood

nocturnal orgasm a night-time discharge of semen, containing sperm; or vaginal lubrication and discharge

Bodies with ovaries can have vaginal lubrication and orgasms during the night. Erections are thought to happen only when someone is sexually excited, but they may also occur for no particular reason. A person may feel embarrassed or upset about this, but it is perfectly normal and part of growing up. As the production and release of hormones becomes more consistent, the frequency of wet dreams and unexpected erections will decrease.

Choose the changes

Worksheet 5.1

Stand up and move to the centre of the room. Your teacher will identify a change that occurs at puberty. If you think it is a change that happens only to bodies with ovaries, move to the left of the room. If you think it is a change that happens only to bodies with testes, move to the right of the room. If you think it is a change that happens to both, bob down.

Menstruation

The one thing that only bodies with ovaries experience is menstruation, or a period. Menstruation occurs every month once the person reaches puberty. It can take up to a year for menstruation to become regular – this is totally normal. The onset of menstruation is triggered by the production of oestrogen and other hormones in the ovaries. When periods start to occur regularly, this is called the menstrual cycle, and is the monthly preparation of a body for pregnancy. The average cycle is 28 days, but it can range from 21 to 35 days. Everyone is different!

The start of the cycle is confirmed with the release of the period blood. This is a substance rich in nutrients that was built up to house the baby. There is nothing dirty about this substance. A period usually lasts from 3–5 days; again, everyone is different, so whatever is 'normal' for you is normal! However, if you do find your period lasting for more than seven days, it would be a good idea to visit a doctor, just to check everything is working okay and that you are retaining enough iron in your body. Ovulation, when the ovary (usually one each month) releases an egg, occurs in the middle of the cycle. If you have a regular cycle, subtract 14 days from the length of your cycle and that will *usually* be your ovulation day. So, a person who has a 30-day cycle will ovulate on *approximately* day 16 (30 days – 14 days). During this time the lining of the uterus has been growing and thickening with blood and tissue in anticipation of nurturing a developing baby, or embryo.

Weblink
Days for Girls

FAST FACT

Days for Girls is a charity organisation aiming to help all people access sanitary facilities. There are schools in Australia who are helping disadvantaged communities around the world by making sanitary products for them. The packs have been distributed in more than 125 countries since 2008, including to remote Australian communities. Check them out at the link on Nelson MindTap!

The egg travels down the fallopian tube, which joins the ovary to the uterus. If a sperm cell fertilises the egg during this journey, it will attach into the thickened lining of the uterus. If the egg is not fertilised, the lining of the uterus is not needed and breaks down, passing out of the body through the vagina.

Sanitary pads, tampons, menstrual cups and various types of period underwear can be used to absorb the menstrual flow. Choosing which is right for you is a personal choice, and your parents, the school nurse or your doctor can help you decide. You should follow instructions on the packaging and change pads and tampons regularly for hygiene purposes.

Signs that the menstrual cycle is about to begin can be both physical and emotional. You may experience cramps, feel tired, have tender or sore breasts, or be more sensitive emotionally. Not everyone will feel the same during menstruation, and not all menstrual cycles will be the same. Your periods may vary in the amount of blood lost and the length of time they occur. It is important to maintain a nutritious diet and get plenty of exercise and sleep during your period. Exercise and a heat pack can often decrease the menstrual symptoms and make them more manageable. If you are concerned that your periods have stopped or are not as regular as they should be, you should seek advice from a trusted person or your family doctor.

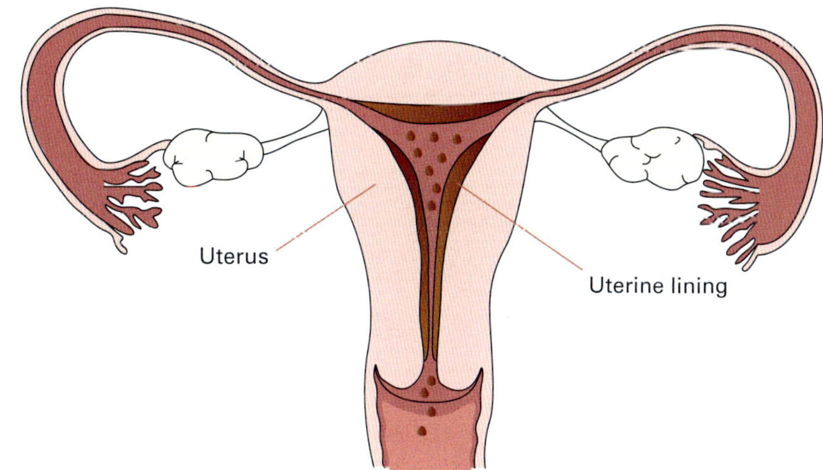

Figure 5.7 The uterus lining breaks down to form part of the menstrual flow.

Menstruation continues until you are around 45 to 55 years old. The end of menstruation is called **menopause**.

menopause the end of ovulation and menstruation

Conception, pregnancy and birth

Human **development** is a continual process that begins at conception when, after sexual intercourse or **invitro fertilisation**, an egg unites with a sperm. This is known as conception, and the single cell that results is called an embryo. The embryo grows and develops into a baby in the mother's uterus until it is ready to be born after approximately 40 weeks.

Usually after about 40 weeks, a baby is born. The baby is delivered either by the mother pushing it out of the uterus through the vagina, or via a caesarean (commonly called a c-section) where the lower abdomen is cut open to the uterus and the baby is lifted out. There are specific reasons why both delivery methods might be recommended; the most important thing is that both parent and child are healthy.

Sometimes the baby might come early, or a little later. If the baby is born too early they may need some help to finish maturing their organs. Sometimes a miscarriage can occur; this is a natural death of the embryo or **foetus**. A miscarriage can happen up to 20 weeks **gestation**. After 20 weeks, the death of the foetus is called a still birth, as the mother will need to deliver the baby.

development how you learn to handle the physical, social, emotional and intellectual changes through life

invitro fertilisation also referred to as IVF; the process where a sperm and an egg are joined together outside the human body. If the process results in fertilisation, the embryo will be manually planted in the uterus to grow

foetus follows the embryonic stage, from eight weeks post conception

gestation the period in which the baby is developing in a mother's uterus, usually approximately 40 weeks

What can you do now?

To simulate the feeling of being pregnant, get your school bag and put it on your front, instead of your back! Try tying your shoelaces/buckles, walking around the room/outside without bumping into anyone, jogging on the spot, stretching or dancing. Does the bag get in your way at all? How might it affect your day-to-day activities?

Table 5.1 The stages of human life

Conception to birth	(40 weeks)
Infancy	0–2 years
Childhood	3–12 years
Puberty/adolescence	13–18 years
Adulthood	19–39 years
Middle age	40–65 years
Old age	65 years+

Birth, growth and development continue through numerous stages over the course of your life. These stages include infancy, childhood, puberty, adolescence, adulthood, middle age, old age and finish at death.

Puberty and hygiene

Good hygiene is particularly important during puberty. Your skin and hair can become oilier, and not washing regularly increases the chance of acne or pimples appearing on your face, neck, shoulders and back.

Worksheet 5.2

Acne is bumps on the skin in the form of whiteheads, blackheads and red pimples. Your skin is covered in pores that contain sebaceous glands, which produce sebum, an oily substance that moisturises your skin and hair. Sebum is produced in increasing amounts during puberty. Without regular washing and cleansing, acne and pimples are more likely to occur as the pores become blocked. Eating a good diet and using face wash when showering will reduce the likelihood of severe acne.

The sex hormones also trigger an increase in body odour. This can be managed by washing and change your clothes regularly, particularly after physical activity, and using a deodorant or anti-perspirant.

Worksheet 5.3

During menstruation it is important to change pads and tampons as recommended. It is necessary to dispose of these sanitary products in an environmentally friendly way. There are special bins for these in most public toilets. Never flush them down the toilet, as they can cause blockages.

EMOTIONAL CHANGES

responsibilities obligations that ensure rights are maintained

Coping with these physical changes can be difficult for everyone. You may feel self-conscious about your changing body, or embarrassed if you look different from your friends. People may start treating you more like an adult because you look like one, but you may not feel comfortable with these additional **responsibilities**. Communicating and associating with people of the same or opposite sex may become more of a priority, and you may take more of an interest in having a partner. The release of hormones during puberty may also cause mood swings and increased tension between you and your family. Not all parents feel comfortable discussing puberty and the physical, emotional and social changes that accompany it. It's important to remember that you are not alone – everyone goes through puberty at some stage. Talk to a parent or caregiver, an older sibling, a friend or your family doctor to help you understand and deal with how you are feeling.

During puberty, the brain is developing, particularly the parts that feel emotions – sometimes quite intensely. Unfortunately, the part of the brain that helps with regulation of emotions and making decisions is one of the last to develop – sometimes not until around 25 years old. This means it can be very difficult for young people to deal with their emotions, and they can often experience mood swings.

Mood swings are often caused by increased hormone levels during puberty. Your mood can change in a second, which can be very confusing. Mood swings can be frustrating for both the young person and their family, teachers and friends. Sometimes there are other reasons for mood swings that are not hormone related. These could include lack of sleep, anxiety, demands from school/home, physical developments and stress, to name a few. As you grow, you get used to your mood swings, and they become easier to manage. If a feeling of sadness or feeling down continues for a couple of weeks or more, it is important to get checked out by a doctor, as this may be a symptom of a mental health issue rather than a mood swing.

Figure 5.8 How do I really feel?

WHAT DRAINS MY BATTERY?

Identify

Controlling our emotions can be really tricky, especially when we're experiencing hormonal mood swings. It's good to show some emotions, but others are more useful to try and calm before we do or say something we'll regret.

This activity has been developed in collaboration with Dr David Bakker of MoodMission

Understand

The brain uses a lot of energy when controlling emotional responses. That's why we sometimes lash out when we're hungry, tired, or emotionally exhausted. At these times, you could say that your battery is drained. When your battery is drained, you need to do things to recharge it. But first you need to realise the signs of it being drained so you can proactively take steps to prevent big, uncontrolled emotions.

Practise

Note down some **physical signs** you might notice as warnings that your battery is being drained. Signs might include:

- Feeling tired or hungry
- Feeling tense, hot, shaky, or highly alert.

Note down any events that might drain your battery. These might include:

- Long days at school
- Stressful tests or assignments
- Being around lots of people
- Having to control your feelings.

Note down some **reactions** you might have that show your battery is drained. These might include:

- Snapping at someone
- Wanting to hit/throw something or slamming doors
- Not wanting to do what other people are asking you to do.

Reflection

Now that you know what might drain your battery, what things could you do to recharge it before your brain goes into low-power mode? Recharge strategies might include having a snack, doing something you enjoy, having some alone time, or just sitting down and breathing for a minute.

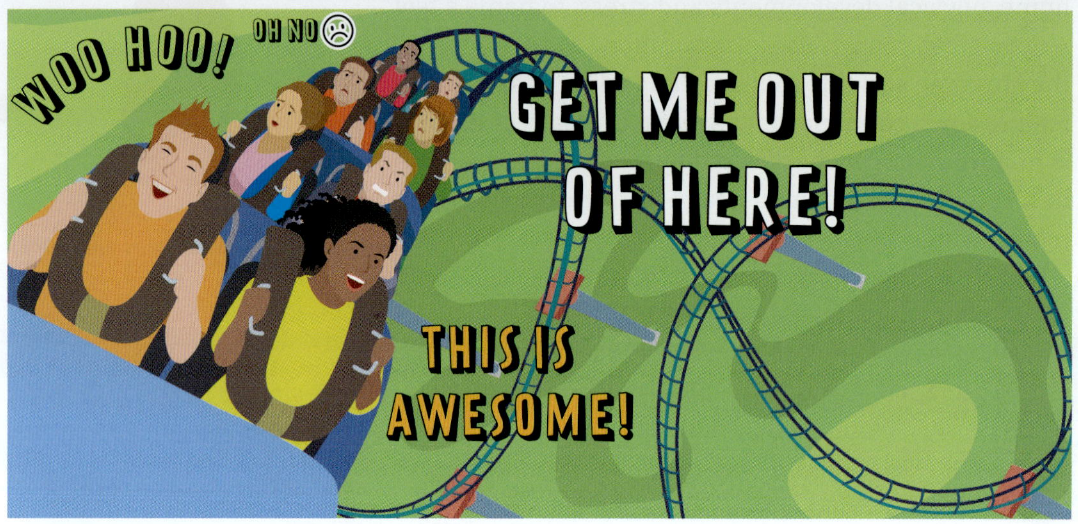

Figure 5.9 Roller coaster of emotions

Along with the above emotional changes, we also have to consider the impact on our identity as we move from child to young adult. Our values and beliefs have a significant influence over our thoughts and actions and therefore how we feel about ourselves. These will usually drive our behaviour; however when placed in a difficult position, such as peer pressure, it is up to us whether we can hold true to these. For example, if you have a strong belief that smoking is bad for you, and a 'friend' attempts to pressure you into smoking with them, you have to think about whether you stick with your beliefs or do what your 'friend' is asking. Sometimes it depends on how strong your value or belief is as to whether you challenge it. Acting in accordance with your values and beliefs usually makes you feel good inside, however you still have the social world to consider.

FACE TO FACE My values and beliefs

1 In a small group make a list of the values and a list of the beliefs you each hold about growing up. These could include ideas you hold about family, independence, rights and responsibilities and relationships.
2 Identify where these values and beliefs come from, e.g. did you learn them from your family, friends, school, etc.?
3 Choose one belief and propose three behaviours that would support that belief and three behaviours that you would not do because of that belief.
   a For example: Young people have to prove they are trustworthy.
      i Supportive behaviours include: being where they say they will be, coming home at the time they are asked to and calling a parent if something happens that makes their plans change.

ii Behaviours you would not participate in: sneaking out of the house, lying about where they are and blaming others for their mistakes.

4 Reflect on how strong your values are and what would it take to go against them.

Our values are generally learned over time and often we don't question them. Values are like a code that we live by. It is important to understand why we think the way we do so that we can always act in an ethical manner. These will change and develop over time as we are exposed to more challenging experiences. Usually the truer you stay to these values, the easier it is to live with yourself!

FACE TO FACE How can you support a friend?

Pair up with another student and think about how you might be able to help a friend who is confused or worried about the emotional changes they are experiencing.

1 How will you recognise a friend who needs help?
2 What could you say to help them feel better right now?
3 What could you do or say to help them with a long-term solution?
4 Is there anyone you might need help from?

WELLBEING CHECK IN: TALKING ABOUT FEELINGS

Identify

Sometimes it's hard to talk about how we feel, but it can be one of the best things we can do to deal with our emotions.

This activity has been developed in collaboration with Dr David Bakker of MoodMission

Understand

Talking about our feelings helps with several things. Firstly, it gives us a chance to figure out how we're feeling. Sometimes it's hard to know how we feel until we open our mouth and start talking. Secondly, it helps other people empathise with us so they can help us feel better. Empathy is the process of understanding how other people feel, which is a bit different to sympathy. Sympathy is when you actually feel the same feelings and emotions as someone else, e.g. you see someone who is sad and you start feeling sad too. Empathising and understanding someone else's feelings is super helpful for close relationships.

Practise

1 Take a look at the emotions on the wheel. Are there any you don't understand? Look up definitions for three words you don't know.

2 Think of a name for how you feel. Now use the wheel to pick how you feel right now.

3 Think of the reason you might be feeling that way. For example, you might be feeling nervous because you have a test tomorrow.

4 Put it into words, e.g. 'I feel nervous because I have a test tomorrow'. You could also add the thoughts going through your head, e.g. 'I keep thinking, what if I forget the answers?'

5 Share this with someone close to you. It might be a parent, teacher, family member or friend.

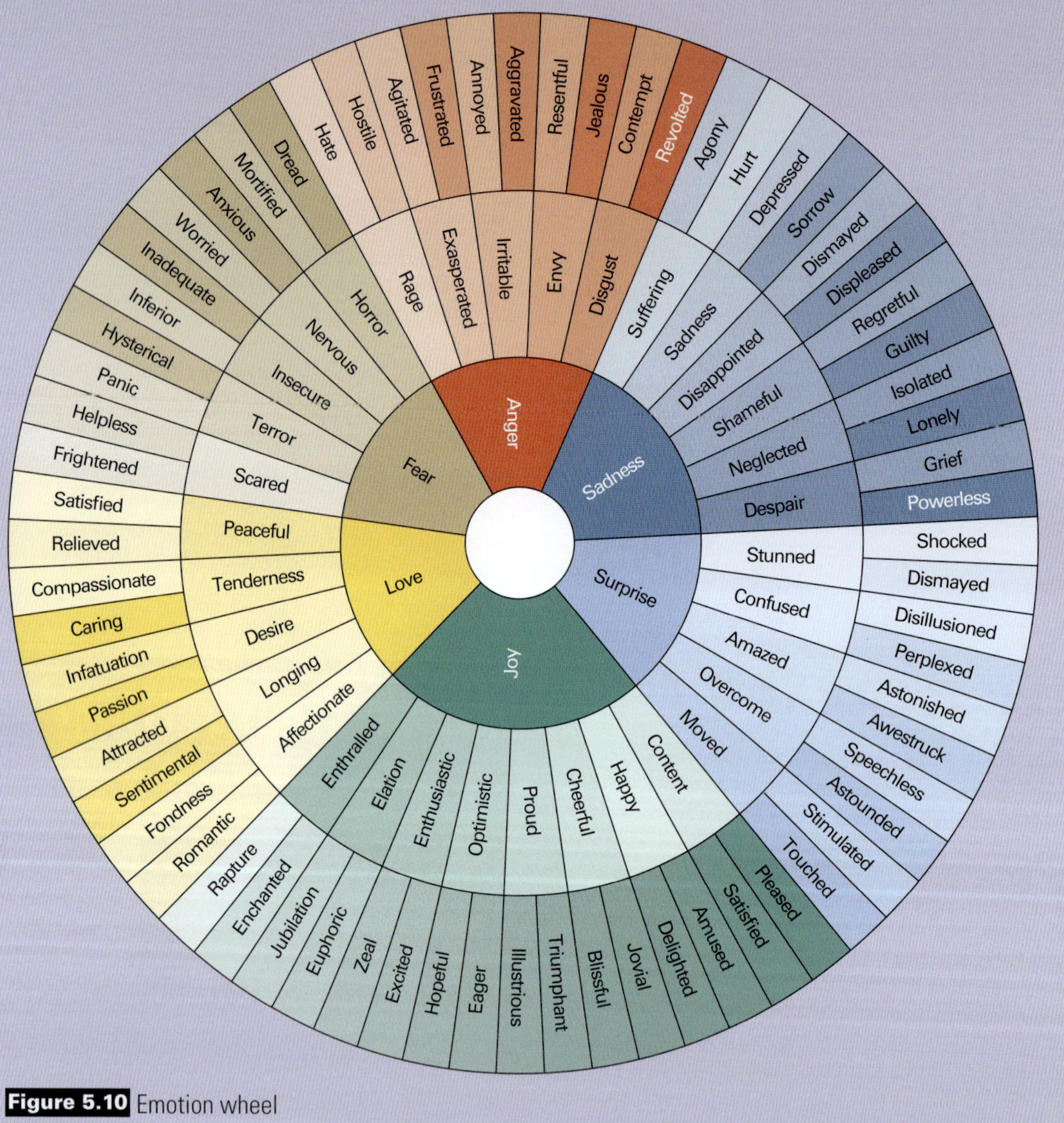

Figure 5.10 Emotion wheel

Reflect

If a friend came to you and told you the same thing, how would you react? And did the person who you actually shared this with react in a similar way? Did you feel like they understood? They might have shown they understood by giving you advice, telling you about a time they felt a similar way, or reassuring you.

Parents/carers are usually good people to talk to about puberty because they've been through it as well. Genetics might have an impact on how your puberty plays out, and your parents might be able to give you some insight into this. Alternatively, teachers, the school nurse, older siblings and friends who have started puberty can all be of help. There are also many books and reliable online sites where you can gather information. Sexuality education at school is set up to answer your questions and get discussions going.

1. Identify five physical changes that can happen during puberty.
2. Identifty five emotional changes that can happen during puberty.
3. What is the difference between someone who is born intersex and someone who is born transgender?

1. We all go through puberty, so why is it so hard for some people to talk about it?
2. Describe three ways that puberty has an impact on your identity.
3. Reflect on your values and beliefs and write a few lines explaining why you think you may or may not be doing things that are beneficial (healthy) for your growing process and development.

1. Puberty is a long process of growing, learning and changing. What do you think you will learn about yourself that will help you as an adult?
2. If you have a friend who is not coping with the changes of puberty, how could you help them initially?
3. How could you use the many resources available for puberty (e.g. books, websites, phone lines, etc.) to support yourself or others through all the changes?

Quiz
Puberty

RELATIONSHIPS

Relationships are the social connections we have with others. These can change during puberty as you start to look at the world differently. There is usually a need for greater independence, which means negotiating with parents, friends, team members, etc. (see Chapter 4). Romantic relationships might also be of interest, as might be work and the relationships you develop with colleagues.

Shutterstock.com/Rawpixel.com

Figure 5.11 Social health is just one part of your wellbeing.

From the time you were born, you began to connect and interact with the people around you. There are so many benefits of having relationships in your life. Good relationships provide us with support, care, guidance, understanding, love, honesty. Good relationships should make us feel good, and want to put in effort to maintain and grow them. As we have discussed in Chapter 4, the aspects of health compliment each other. If your social health is good, i.e. you have good strong relationships, this can positively impact your mental health by making you feel good about yourself. Therefore the ability to maintain effective relationships will lead to an improved sense of wellbeing and assist your overall **holistic health**. During COVID-19, relationships have been even more important as many of us have been in lockdown, or had periods of isolation. Being isolated at home and not being able to see friends and family face to face can put a strain on these family relationships. Online communication with friends and extended family members have become critical for social interaction. It is important to understand the significance of positive relationships, what makes them work and how to cope when they change throughout your life. Forming, maintaining and developing relationships helps to satisfy the human needs of feeling loved, accepted and a sense of belonging.

holistic health the physical, social, emotional, cognitive and spiritual wellbeing of a person

Every relationship formed is different. Some will be very close, such as those with your parents and friends, and some will be intimate relationships, such as a sexual partner. Others will be more distant and casual, such as those with your next-door neighbour or family friends. The type of relationship you have with someone may depend on factors such as how long you have known each other, your interests and your age. We are going to look at some of the relationships you might have.

RELATIONSHIPS WITH FAMILIES

Relationships with family members are very important and help to satisfy many physical, social, emotional and cultural needs. When people are younger, they are dependent on family for food, clothing and shelter. Families make important decisions and shape **values**, **morals** and social behaviours that are accepted in the community. Families help to develop communication skills, **self-esteem** and a sense of safety and security in the members of their household. They establish the roles, **rights** and responsibilities that help children learn how to relate to people in different ways. They also pass on cultural values and traditions to their children, such as cultural celebrations/holidays, food, dress, rituals, etc.

values things that are considered to be good, appropriate and important in people's lives

morals behaviour and attitudes towards what is right or wrong

self-esteem how people feel about themselves; their sense of worth

rights entitlements that everyone should have

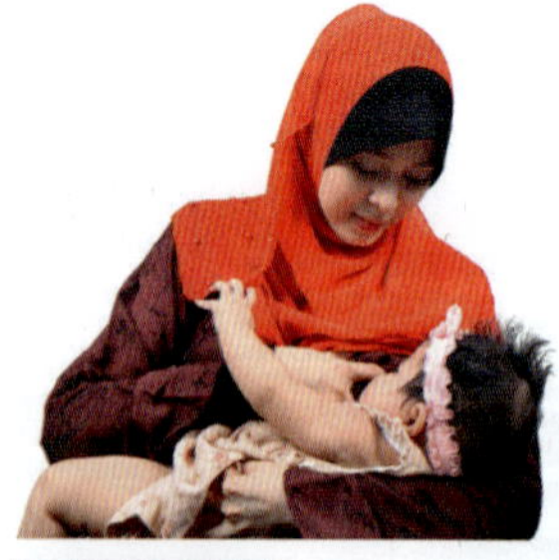

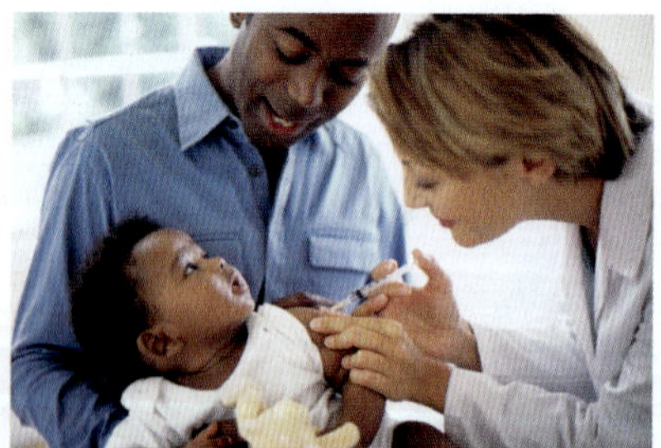

Shutterstock.com/Ery Azmeer; Getty Images/Science Photo Library – IAN HOOTON; iStock.com/CasarsaGuru; iStock.com/VM Jones

Figure 5.12 We start off being dependent on our families for many things.

Worksheet 5.4

UP AND MOVING — What makes a family?

On the board, brainstorm a list of things that make up a family. Don't forget all the positives and negatives about being part of a family!

As a class, determine whether each of the following is a family or not, based on your list on the board:

- a single dad with three children
- a couple with six children, and grandparents living with them
- a couple with no children
- a woman and her dog.

Usually things like love, care and shelter are what determine a family, rather than things like being related by blood.

9780170463096

Types of families

A family is usually a group of people who live together and love, care for and support one another, keeping each other safe. They may or may not be **biologically related**.

biologically related genetically related to the parents

The Australian Bureau of Statistics defines a family as people who are related by blood, marriage or through adoption and who usually live in the same household together. In today's society there is no longer a 'typical' type of family. There are many different family types, including:

- **nuclear** families, which consist of a couple and their children. These children can be biological or adopted. The couple may or may not be married
- **sole-parent** families, where only one parent is raising the children. This may be due to divorce, death or adults choosing not to marry
- **extended** families, where adults from different generations of a family live together. These may include parents, children, cousins, grandparents, aunts and uncles
- **same-sex** families, which consist of two adults of the same sex in a relationship; they may or may not have children, and may or may not be married
- **blended** families occur when two people who already have biological or adopted children from a previous relationship form a family. This family type may also be called a step-family
- **couple-only families**, when a couple live together and choose not to have children
- **communes**, which consist of a number of adults and children who are not all related but choose to live together. The group share resources and tasks and may live together because of a common purpose, such as their religion.

Worksheet 5.5

Shutterstock.com/Romrodphoto

Shutterstock.com/ARENA Creative

iStockphoto/Juanmonino

Getty Images/Shaw Photography Co.

Figure 5.13 There is so much diversity in what can make a family!

Functions of families

No matter what type of family you come from, all families have a purpose. There are many functions a family provides, including:

- physical: care and safety, financial, food and water, hygiene needs, shelter
- social: experience of co-existing, teaching of relationship skills
- affective: love and care, emotions and control of
- cultural, religious and spiritual: handing down of family beliefs, morals and values
- authority: rules and regulations, both in society and specific to the family
- day to day: how the family functions each day, expectations of roles and responsibilities.

FACE TO FACE Families with children

Today in Australia, the number of couples with dependent children is declining, while couples with no children are on the rise. Why do you think this is?

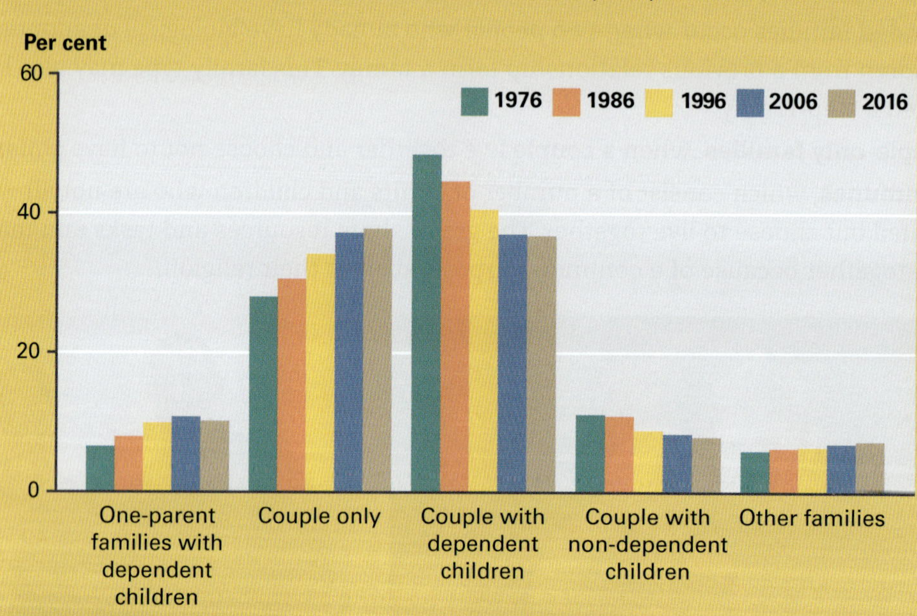

Figure 5.14 Types of family in Australia, 1976–2016

The Australian Bureau of Statistics has predicted family types through to 2041. At this point in time, mothers make up the majority of parents in sole-parent families. Why do you think that sole-father families are predicted to increase by 2041?

In 2041, there are projected to be between 9.2 million and 9.4 million families in Australia (up from 6.7 million in 2016).

Couples with children are projected to make up 43% of all families in 2041, slightly down from 44% in 2016.

Couples without children are projected to be the second most common family type, up from 38% in 2016 to 39% in 2041.

Single female parent families are projected to make up 13% to 14% in 2041.

Single male parent families are projected to increase the fastest of any family type, increasing by between 44% and 65% by 2041.

Figure 5.15 Projected family types in Australia in 2041

INVESTIGATION

FAMILY FUNCTIONS – ARE THEY THE SAME?

Purpose

To find out whether other families have the same functions as yours.

Materials

Come up with a questionnaire, either on paper or on a tablet/computer. You will need access to others to discuss the functions that their families perform.

Method

1 Use the functions of families list to write a questionnaire that will help you find out what functions each family performs.
2 Make a list of your family's functions.
3 Interview at least three of your peers and make a list of their family's functions.

Discussion

1 Compare the functions each family has.
2 Do any of the following have an impact on the types of functions the families perform? Consider: family size, age of children/parents, whether parents are working or not, type of family, any extended family living with them, volunteering, extra-curricular events, etc.

Influence of families

Individual characteristics and features make people unique, or different from each other. These qualities help to establish each person's personal identity, which is made up of their physical features, individual qualities, skills, values and beliefs. As you grow, you will develop your own feelings, ideas, values and beliefs about different aspects of life, based on your upbringing, friends, religion, culture and the environment around you. These influences then help you establish what is important to you and who you feel you are.

Values are established by your family relationships from a very early age. They are ideals that provide guidance on what to do and how to behave in certain situations. Children often take on their family's views on particular issues from a young age. As they grow and mature, some of those views will change, according to their life experiences. Values will vary from person to person because of their individual backgrounds.

UP AND MOVING Family values

This activity will start to look at some of the values that your family sees as important. Can you think of any off the top of your head?

The class will form two circles, one inside the other, with students facing each other in pairs (an odd number can be a group of three). Your teacher will ask a question and you have one minute each to discuss this question in your pair. Your teacher will then ask one circle to move to the left or right a certain number of spaces. You will then have a new partner and a new topic. This can get pretty noisy!

FACE TO FACE Your family

Discuss the following points with the person next to you:

1 Describe why families are important.
2 List three values your family taught you.
3 Are there any values in your family you don't agree with?
4 Explain how your relationship with your family changes as you develop.

Families also have an influence on perceptions of gender roles. Gender roles or stereotyping refers to views on how people should behave, act or dress, based on whether they are assigned male or female at birth. Children learn very quickly what it means to be a male or female through activities they are involved in, opportunities they are given, responses from those around them and parental guidance on appropriate behaviours. As children grow and develop into adolescents, gender stereotypes are often reinforced by other factors in the environment, such as culture, friends and peers, and the way that gender issues are presented in the media.

UNSUPPORTIVE FAMILIES

Sometimes babies are born into families who do not provide enough support for them to grow and develop to their fullest. This could be for a variety of reasons: they do not know how to, they can't afford to (including money, time, etc.), they may have other children/a partner who demands more of them, they might have a mental illness, they may be drug dependent or they simply may not have formed a bond with the child.

We encourage most young people to seek out their families for support, but this might not be the best source of help for some. When seeking help, it is important to turn to someone who you trust and who you can talk to. Just because they are older doesn't mean they have your best interests at heart. If there are members of your family who are not supportive, perhaps a teacher or a friend's parent might be a helpful alternative.

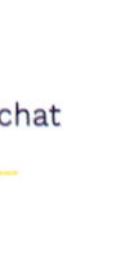

Weblink
Kids Helpline webchat

Sometimes it feels as if there is no one to help. Families are usually private and don't want their issues broadcast to the public so it can be hard to take the first step to tell someone about what might be going on. If there isn't anyone to talk to that you know, there are organisations that can help. Please know that if someone you tell believes that you are in danger of being hurt, they might have to tell an authority.

This is called mandatory reporting and is there to keep children and young people safe. This person can be the voice of the child or young person, as often they don't feel like they have a voice. There are help lines that you can ask for advice, Kids Helpline is specifically for children and young people. It is a 24/7 national service and their number is 1800 55 1800. If you don't want to talk in person you can also have a webchat with a counsellor on the Kids Helpline website.

INVESTIGATION  HOW CAN WE HELP OUR PARENTS OR CAREGIVERS HELP US?

Purpose

Often parents or caregivers want to be able to give you all the information you need about growing up and relationships, but feel too nervous, or think you already know. Sometimes they will wait until you approach them, when you are waiting for them to come to you!

The purpose of this investigation is to ask both friends and family about how parents or caregivers can do a good job of supporting their teenagers. This research will inform a clip you will make giving parents or caregivers advice on how to have a good relationship with their children. This will help those parents or caregivers who find it difficult to know how to help their children.

Materials

In groups, watch the video about teens and their relationships with their parents or caregivers.

- Paper and pens or computer/tablet to record your questions on

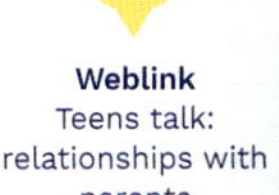

Weblink
Teens talk: relationships with parents

Method

1. Write a set of questions that will help you present a clip to help parents or caregivers. Consider the following examples from the 'Teens talk' clip:
   - How do I build trust with my teen?
   - How do I show my teen I am interested in them and want to help?
   - How do I start a conversation with my teen?
   - How to know if I need help?
2. Interview your friends and family, including parents, caregivers, teachers, etc., using the questions you have come up with. Plan a clip that will help parents talk to their teens about growing up and relationships. Be as creative as possible.
3. Record your clip.
4. Present your clip to the class so you can all evaluate each other's.
5. If you feel comfortable doing so, take your clip home and show it to your parents/carers. Ask them how effective they think it is and why.

Discussion

1. Was it easy to get advice about how to help parents or caregivers?
2. Why do you think this is the case?
3. How effective do you think your clip would be in helping your own parents/carers? Why do you say this?
4. What did you think about the other clips that were presented in your class?
   a. Did you see anything other groups did that might have improved your clip?
   b. How did other groups creatively approach parents or caregivers in their clips?
   c. What did other groups do well in their clips? Why do you say this?
   d. What could the other groups improve in their clips? Give clear examples.
   e. If you felt comfortable showing your clip to your parents/carers, what did they think of it? How effective did they think it would be and why?
5. How can your class use these clips to help parents talk to their teens about growing up and relationships?

Intercultural understanding

ethnic people of the same race or nationality who share the same culture

census official count of the population of a country

Australia's population is made up of people from many different cultures and **ethnic** backgrounds. According to data collected by the Australian Bureau of Statistics during the 2016 **census**, about 67 per cent of people living in Australia were born in Australia. Approximately 34 per cent had both parents born overseas.

Getty Images/Jasmin Merdan

Fairfax Syndication/Jeffrey Chan

Figure 5.16 Australia's population is extremely multicultural, the population having grown due to immigration after the British colonisation and genocide from the 1700s to the 1970s.

First Nations People make up 3.3 per cent of the population of Australia, but tend to have a larger number of children than other Australians. They are more likely to live in multifamily and multigenerational households, and are also more likely to live in regional and remote areas.

iStock.com/Thurtell

Figure 5.17 Many First Nations Australian families live in rural and remote communities.

Diversity of families make for rich communities. Each family has their own cultural or religious beliefs and activities they participate in. Discussing these differences in class, gives you a wide understanding of why families behave the way they do. These personal stories help create harmony rather than relying on media stereotypes of certain cultures and/or religions. You see each family as a group of people doing the best they can with what they have just like every other family!

FACE TO FACE Some benefits of families

1. What qualities help make strong, happy families?
2. Identify the qualities you have gained from your family.
3. Describe some specific examples of how your family has passed these qualities on to you.
4. How do these qualities help to build positive relationships?

RELATIONSHIPS WITH FRIENDS

As you move through adolescence, your friends and peers play a greater role in your life. They can have an influence on what you wear, what music you listen to, what activities you like to participate in and how you feel about yourself. Peer relationships develop through common interests such as school, sport or other leisure activities, and are usually dynamic, meaning that they continually change as you mature, have new experiences and move through life. Friends usually come from your peer group and are the people you feel the most comfortable sharing your thoughts and feelings with.

Worksheet 5.6

Belonging to a peer or friendship group can allow you to feel safe and secure, supported and respected. Having friends encourages the development of qualities such as respect, trust, **tolerance**, good communication skills and honesty. The sense of belonging and being connected with friends or peers plays a part in developing self-esteem, **self-confidence** and self-identity, and further develops social and emotional health.

tolerance willingness to accept attitudes different from your own

self-confidence belief in one's own abilities

FACE TO FACE Friends

1. Complete the sentences:
   - **a** A friend is …
   - **b** Three things I like about my friends are …
   - **c** It is good to have friends because …
2. Some people find it difficult to make friends. Why? How could this be overcome?
3. Discuss some problems that can arise from belonging to a group.
4. Describe some rules that exist in your group of friends.
5. Write an acrostic poem using the word FRIEND, with each letter of FRIEND as the starting letter of a new line of the poem.

Shutterstock.com/Denis Kuvaev

Figure 5.18 What do you like about your friends?

Peer group

During adolescence, your peer group plays a larger role in establishing what you feel is important. The influence of your peer group during this time can be either positive or negative. A positive influence might be when your friends encourage you to try something different, such as a new sport, or they may help you to make decisions that benefit you in some way, such as setting yourself a challenge. Here are a few more examples of positive peer group influence:

- studying for a test together
- volunteering at a local community centre to build your skills and confidence
- talking to someone you wouldn't normally have felt comfortable approaching.

Worksheet 5.7

An example of a negative influence might be if your friends persuade you or push you into making a decision that is harmful to yourself or others, such as trying a cigarette. There are other examples of negative peer group influence:

- skipping school
- vandalising something
- letting someone copy your homework
- bullying.

peer pressure the strong influence of a group to act in a particular way

This is called **peer pressure**. Common peer pressure techniques include:

- dares or threats: 'if you don't do this, you won't be a part of our group anymore'
- guilt: 'if you really wanted to be my friend, you would go along with it'
- generalisations: 'everyone else is doing it'
- poor logic: 'no one's going to find out'.

It can often be difficult to stay true to the values and standards you have learnt from your family when your friends feel differently and you want to fit in and be part of a group. Feeling like part of a group satisfies a need to belong, so it is understandable that you may feel pressured to go along with what your peers are doing in order to be accepted. If you are being encouraged to try things that don't feel right or that make you feel uncomfortable, listen to your body. If you feel sick, sweaty or shaky, for example, your body is probably trying to tell you that these actions are not right for you, or that you are not ready for them yet.

iStock.com/sturti

Figure 5.19 Peer pressure can happen in many different environments.

It is important to do what you feel is right and commit to actions that represent what you believe in. If your friends are constantly pressuring you, they may not be the right group of friends for you. It can be difficult to stand up to them, particularly if you are the only one doing so. Here are some simple strategies that could be used to avoid negative peer pressure:

- hang out with people who have the same interests as you, as they will probably have similar values and goals

- use humour – a funny one-liner may help to take the pressure off
- avoid risky situations where peer pressure is more likely to occur
- be assertive – this means being able to say no in the nicest possible way!
- seek help from your support networks – this might be your parents, your siblings, a teacher, the school counsellor, or more formal organisations such as Kids Helpline.

UP AND MOVING **Friendship cake recipe**

If you had to make a recipe for a friendship, what ingredients would you include and in what amounts?

1 Work in a group of three or four to design your own recipe.
2 Compare your friendship cake with another group.
3 Are both recipes the same? Why or why not?

Friends on the web

Having lots of friends online can make you feel special, particularly if they're a group of people you have a special connection with or who have similar interests to you. However, there are also risks in forming relationships in this way. It is very easy to remain anonymous online. It's important not to give away your personal details. Never arrange to meet people you do not know. Ask yourself the following questions about your online friends:

Shutterstock.com/Sarawut Aiemsinsuk

Figure 5.20 Social media allows online contact to be made 24/7.

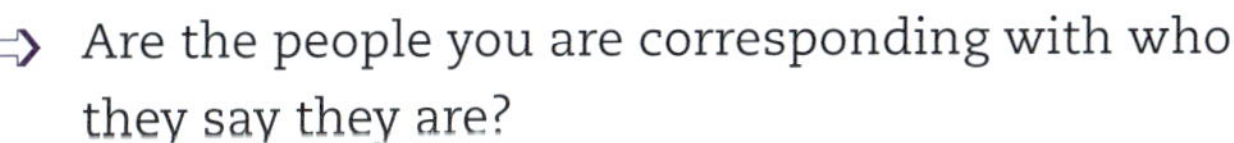

- Are the people you are corresponding with who they say they are?
- Are they really your friend?
- Do they value the special qualities that make you unique?

There are nasty people who spend all day trying to lure young people out of the safety of their homes into the community where they aren't as comfortable. There are also scammers who try to lure people into relationships online, never intending to fall in love, so they can ask that person for money once trust has been established. It feels good when someone pays attention to you, especially if you don't have many really good connections or people who you feel understand you. These people tell you what you want to hear – they are very good at it. Each person they do it to teaches them new things about what people like and don't like.

These people usually ask for your phone number or email address very quickly, or try to get you to sign up for a communication app like WhatsApp or Kik or one of the hundreds of others around. Make sure you stay on the original site – you don't want to provide them with any more information about you than they already have. Sending pictures of yourself in a school uniform tells them which school you go to; photos of yourself at a shopping centre tells them the area you might live in. Ensure they don't have access to your location on the apps. In particular make sure your photos don't reveal your address, many selfies are taken at home. It is very hard to trust people online; you need to make sure you keep yourself safe and don't provide strangers with any of your personal information.

Worksheet 5.8

It's important to tell someone if things are getting too stressful or if they are causing you problems like anxiety, difficulty sleeping or losing touch with your friends. There are people who can help – you do not have to go through anything alone. Even if you think you might have instigated something, it's important to seek help.

1 What have you learnt about relationships?
2 How do families impact your relationships with others?
3 Why do people sometimes pressure their friends, and keep doing so even though they have said no?

1 Reflect on your own friendships. Would you say they are strong? Why or why not?
2 Identify the five most important things you need in a friendship (this might come from your friendship cake), and ask an adult in your life what the five most important qualities were that they wanted in a friendship when they were your age. Then ask them if those things have changed over their life and why.
3 Consider if there is anything else you would like to know about building and maintaining strong relationships.

1 You have just been given the job of being a peer mentor to next year's new year seven students. You have to come up with a resource, an activity, a story, a play or a movie that you can do or show them during their transition to high school. If it is an activity, you will lead the group in doing it. If it is something that you tell, sing, perform or show, you will need to develop some questions to get the new year sevens involved and thinking.
   a Reflect back to when you were starting high school. Was there anything that worried you about making new friends? What sorts of things helped you? What would have helped you?
   b Create a draft of your resource, and questions to ask the students if you are not doing an activity.

RESPECTFUL RELATIONSHIPS

Worksheet 5.9
Think Pair Share

The types of relationships that you have now may change several times as you progress through secondary school. Relationships that are unhealthy can hurt emotionally, socially and sometimes physically. Respectful relationships, on the other hand, promote a sense of safety and caring. People in respectful relationships generally:

- feel good about themselves
- freely practise their own cultural and religious beliefs
- feel safe and secure
- feel free to express their opinions and thoughts
- feel accepted for who they are
- feel they have a right to be heard
- feel trusted.

iStock.com/101dalmatians; iStock.com/kali9

Figure 5.21 Respectful relationships are needed to feel safe.

Safe relationships exist when young people not only expect to be treated in certain ways, but also carry out their responsibilities regarding the needs of others. They show respect for other people's needs by:

- being respectful
- listening to what people are saying, not just hearing what they are saying
- providing support
- accepting different points of view
- exercising empathy
- protecting others from harm, or alerting them to danger
- being considerate of other people's feelings.

Shutterstock.com/Solovyova Lyudmyla

Figure 5.22 Listen to what people are saying by paying attention not only to their words, but also to their body language and other cues.

Young people have the right to express themselves and their opinions, but they should also be prepared to put themselves in others' shoes when listening to what they have to say. In positive relationships, both people have an equal say in decision-making and should feel free to be who they are and say what they think without fear, anxiety or risk of punishment. In this type of relationship there is a balance of power, with both people equally respecting each other's feelings, their right to be heard and their right to feel safe.

RIGHTS AND RESPONSIBILITIES

Every positive relationship involves being treated with understanding and respect. In order for this to happen, you need to be aware that you have rights that need to be satisfied and responsibilities to fulfil.

Worksheet 5.10

Rights are the things every human being deserves to have, such as the right to clean drinking water. Responsibilities are the things that you are required to do in order to maintain other people's rights and entitlements, such as the responsibility to cross at a pedestrian crossing in order to keep ourselves, other pedestrians and car drivers safe from accidents.

FACE TO FACE Rights and responsibilities

1 Which of the following are rights and which are responsibilities?
   a To learn without being disrupted by other students.
   b To listen to others when they are voicing their opinion.
   c To clean up after yourself.
   d To work in a safe environment.
2 As a class, develop a Class Charter with a list of 10 rights and responsibilities that you could maintain in your classroom to ensure the best possible learning environment.

CONSENT

Consent is an agreement to do something. Some people call it a free agreement or giving an enthusiastic 'yes'.

Consent can be difficult to give, it is not as easy as just saying 'no'. This is an oversimplification of a very complicated action. The decision is based on a lot of things, these include, but not limited to, values, beliefs, self-esteem, communication skills, relationship status, expectations and social pressures. Exploring different situations and experiences and practising negotiating consent is a good way to improve your skills.

FACE TO FACE Consent in the media

In small groups, make a list of the television shows, series or movies you watch. Whenever someone is kissing or participating in other sexual activities there is not usually any consent involved, so we do not see examples very often. Choose one off your list and discuss the following:

1 What messages is the media telling us about giving and getting consent. How do they do this?
2 Design a different way the media could portray the realities of giving and gaining consent that young people could relate to.
3 If you could influence the writers of a movie, what would you tell them about including examples of consent in their script? Would it be different for a show or series?

Think about all the times you need to consent each day, how many of them do you actually say 'yes' or 'no' to? How many of them involve implied consent? Some examples of this might be eating whatever dinner is on the table, or joining a group in class to complete an activity. When it comes to relationships and sexual activity, there must be a 'YES' to be able to legally participate with someone. Everyone involved must consent. A nod, a shrug, a wink, a shy glance is not considered consent. Neither is what the person is wearing nor what they are doing. If you are the one attempting to gain consent, sometimes it is hard to wait for the other person to say 'yes', even if they want to. There are usually a lot of decisions going on in their mind that might make them unsure as to whether to say 'yes'. If this ever happens, stop and give the person enough time so that they can make the right decision for

5

them. They will appreciate your patience and that you didn't rush them. If you are the one giving or not giving consent, you have every right to say exactly what you want. It is against the law to coerce someone into participating in sexual activity; this includes threatening, tricking and pressuring someone to participate. There is never a time when you should consent if you do not want to.

WB Worksheet 5.11

Consent requires good communication skills, which we will talk about later in this chapter. It also requires you to be level-headed and able to make a decision. Being tired and/or emotional can interfere with being able to give consent. If you are angry, sad, lonely, grieving, excited, etc. you might make a different decision than if you weren't any of these. It is impossible not to be emotional at all, so ensuring you can make decisions when you are put in that situation is key. Alcohol and other drugs can get in the way of making a good decision. Age is another important factor when you are contemplating sexual activity with someone else. In Australia the law requires you to be at least 16 or 17 to be able to consent. States and Territories in Australia all have different laws that you need to abide by.

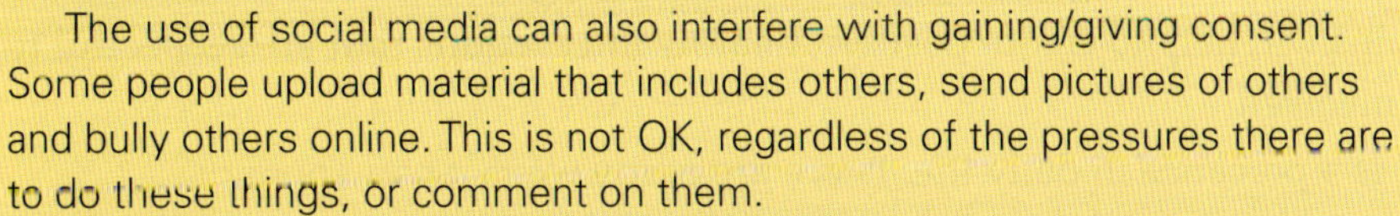

The use of social media can also interfere with gaining/giving consent. Some people upload material that includes others, send pictures of others and bully others online. This is not OK, regardless of the pressures there are to do these things, or comment on them.

Weblink
Sex, dating and the law

Consent is a component of a respectful relationship, when friends/partners want to know their friend/partner is participating in behaviours they want to. In a respectful relationship friends/partners want each other to tell the truth and be honest about how they feel. Showing respect involves being open, honest and trustworthy; it also includes being empathetic and understanding the thoughts and feelings of the other person.

EMPATHY

Empathy is a wonderful skill to have when dealing with people. Empathy helps you think about what someone else might be going through, even if you aren't experiencing it yourself. Being empathetic helps you take a step back, so instead of getting frustrated by someone's behaviour, you can try to understand how they are feeling and help them find a solution. Being empathetic is a skill, and like all skills, it takes practice to develop. When someone comes to you with an issue, the best things to do are to listen without interrupting, avoid making any judgements, tell them you can understand that it must be frustrating/worrying/scary (use the words they use), and ask if there is anything you can do to help.

empathy being able to put yourself in someone else's shoes and think about how they might be feeling

Empathy toss

Stand in a circle. Your teacher will give one student a ball (or soft object). Everyone must think of a time when they have been empathetic, or seen someone else be empathetic. When you have the ball in your hands, discuss the example you have thought of and how it made the other person feel. Then throw the ball to someone who hasn't had a turn yet.

POWER

Figure 5.23 Achieving a balance of power

During puberty and adolescence, teenagers often feel vulnerable. This is because puberty is a time of exploration and uncertainty about identity and the future. Teenagers are afforded more independence than they had previously, and sometimes have to make choices on their own. It is also a time when teenagers are most influenced by people around them, often because others seem to have more power.

Power has many meanings. In terms of relationships, power can mean the ability or capacity to do something or act in a particular way, or to direct or influence the behaviour of others or the course of events. Power can also play a significant role in the gaining and/or giving of consent. It is one of the reasons consent can be so complicated.

Everyone has some degree of power. Finding the right balance of power within relationships will help to create positive, healthy interactions that are respectful to everyone involved.

Power is a presence in daily life. There are many ways it can appear:

- the power of knowledge, such as a parent teaching a child right from wrong
- the power of authority or position, such as a school principal
- the power of expertise in a particular area, such as a doctor
- the power of an institution, such as the law
- the power of being physically or emotionally strong, such as an older brother or sister
- the power given by customs and traditions, such as the power held by a First Nations elder.

People are at risk of being emotionally and physically hurt when they have relatively little power in a situation. It is important to maintain a balance of power so that positive, fair relationships can be formed.

Positive use of power involves acting or influencing others in ways that show respect for their rights. Everyone has rights, no matter who they are, what culture they come from, where they live or what they believe in.

prejudice a judgement or opinion formed without knowledge of facts

discrimination treating people differently based on their personal characteristics, race, religion or beliefs

Knowing yourself – your attitudes, needs, values, beliefs and interests – will contribute to your sense of identity. This allows you to have power and feel confident in who you are and what you believe in. It is also important to be empathetic towards others who hold different values and beliefs. This develops tolerance and reduces the possibility of **prejudice** and **discrimination** taking place.

9780170463096

FACE TO FACE Power

Discuss the following with a partner:

1 How is power used positively in our legal system? Give some examples.
2 How could power be used negatively in the way brothers and sisters treat each other?
3 Define the following uses of power as positive or negative. Explain your choice.
   a Your sports coach asks you to play in a position you don't normally play.
   b A friend at school asks you to lie to a teacher for them.
   c The Student Representative Council enforces rules about wearing the school uniform correctly.
   d A teacher sends a student to the principal for being disrespectful in the classroom.
   e A group of senior students make jokes about a younger student's hairstyle.
4 Explain how the misuse of power can be destructive to relationships.

Privilege, or assumed privilege, can lead to someone feeling powerful or having power over someone. There are groups in society who hold privilege over others. These people are looked at as being the best choice, the right fit, the smartest, etc. just because of the group they belong to, not because of the person they might be. Can you think of any groups that might hold privilege in our society? Think of some examples of the privileges that they receive.

privilege an advantage given to someone because of a particular group they belong to

Worksheet 5.12

Power in relationships

Power exists in relationships when someone tries to control a situation in order to bring about change. Power can be used positively when the outcome is also positive. Examples include helping a friend complete a task in which you are more skilled; helping grandparents complete tasks around their home that you find easy but they tend to struggle with; speaking up when someone is being bullied because you have the confidence to do so; and helping a mate when they have received bad news about a team selection. In all of these examples, the outcome is positive in terms of physical, mental and social health.

In some relationships the balance of power is skewed in favour of one person. This can be harmful because it can lead to an increased possibility of abuse. The types of abuse that can result from abuse of power in relationships are emotional, physical, social, sexual, psychological, financial, spiritual and cultural abuse. Let's talk about the first four, remembering that abuse can sometimes be difficult to talk about, especially if you are experiencing or have experienced it. If you need a break, please ensure you let your teacher know and go somewhere safe.

123RF/Graham Oliver

Figure 5.24 A balance of power allows all involved to benefit and feel positive.

Emotional abuse

vilification criticism or abuse directed towards someone or something

Emotional abuse includes put-downs, non-inclusion, racial or religious **vilification**. It happens when a person attempts to exert control over another person. If your friend or partner displays any of the following behaviours, then your relationship would be considered to be emotionally abusive:

- being possessive
- being jealous
- telling you who you can and can't see
- telling you what you can and can't wear.

FAST FACT

Neglect is the failure (usually by a parent or caregiver) to provide for a child's basic needs, including failure to provide adequate food, shelter, clothing, supervision, hygiene or medical attention. Neglectful behaviours could be physical, emotional, educational or environmental. This is also considered a type of abuse.

Physical abuse

Physical abuse includes behaviours such as punching, kicking and hitting, and occurs when someone uses their power to be physically violent towards another, less powerful person. It is also important to remember that it is not always the bigger person (physically) who is the attacker. Typical examples of physical abuse include:

- kicking, hitting, striking, etc.
- smashing personal belongings
- threatening to physically hurt someone.

In Australia, physical abuse is a criminal offence and carries serious consequences and penalties.

Social abuse

Social abuse includes someone telling you who you can and cannot see or talk to. The abuser usually uses their power to ensure that you become reliant on them alone. This usually leads to the victim being isolated from their usual support systems. Common signs of social abuse include:

- needing to know where you are and who you are with all the time
- telling you they want to spend all their time with you
- telling you negative things about your friends/family
- changing your plans so you can spend more time with them.

Sexual abuse

Sexual abuse includes unwanted sexual activity, unwanted touching, suggestive behaviours or comments. It occurs when a person uses their power over another person for sexual reasons. Common sexual abuse involves the following actions:

exposure showing your sexual organs to other people in public; also called indecent exposure or flashing

- forcing others to have sex
- **exposure**
- sending unsolicited pictures to someone, or sending naked pictures without permission
- suggestive behaviours or comments.

Sexual abuse involving young people always includes elements of emotional abuse, and can have damaging effects that last well into the future. It is important to remember that young people who have been sexually abused are not responsible for the abuse, and should not feel guilty or at fault. Sexual abuse is against the law.

Other types of abuse in relationships

Along with the other types of abuse listed there are other ways that abuse can be seen in relationships. These include:

- psychological abuse – deliberate attempt to hurt, frighten, confuse or manipulate someone
- financial abuse – using money or resources to control someone
- spiritual abuse – controlling someone's ability to practice their spiritual rituals or beliefs
- cultural abuse – controlling someone's ability to practice cultural rituals or beliefs.

Abusive relationships are not safe relationships because of the negative effects they can cause, which include:

Worksheet 5.13

- distress
- anxiety
- lack of trust in others
- fear
- withdrawal
- stress
- lack of confidence
- bullying and harassment.

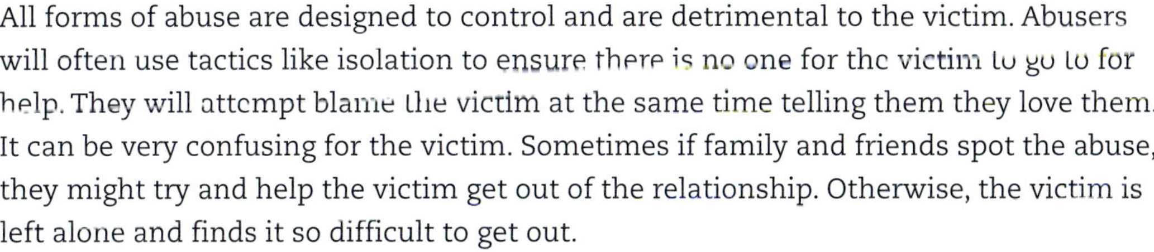

All forms of abuse are designed to control and are detrimental to the victim. Abusers will often use tactics like isolation to ensure there is no one for the victim to go to for help. They will attempt blame the victim at the same time telling them they love them. It can be very confusing for the victim. Sometimes if family and friends spot the abuse, they might try and help the victim get out of the relationship. Otherwise, the victim is left alone and finds it so difficult to get out.

All types of abuse need to be stopped; this often requires the intervention of a trusted person. Supporting family and friends in abusive relationships sometimes means sticking with them, even if they push you away. Letting them know that you are there, regardless of whether you are in contact often, might be the only reassurance/support they have. Just being there for them when they are ready is very important. If you think a family member or friend is in danger, it is important to get some help. School counsellors are often really good resources if you are not sure what to do as along with advice and information, they can refer you to specialist services.

National helplines

1800 RESPECT (1800 737 732) National Sexual Assault and Domestic Violence hotline

1800 176 453 Australian Childhood Foundation

1800 55 1800 Kids Helpline

13 11 14 Lifeline

COMMUNICATION IN RELATIONSHIPS

Interpersonal communication is sending and receiving messages between two or more people. Effective interpersonal communication is essential for forming positive relationships with family or friends because it allows you to express how you are feeling, your needs and your emotions. We spoke a bit about communication in Chapter 4 and looked specifically at internal and external communication skills, listening and negotiation (see pages 164–67).

To communicate effectively, messages about thoughts, feelings and emotions need to be sent in a clear, easy-to-understand way. For the communication cycle to be complete, you also need to listen to find out how your messages are being interpreted by the person you are communicating with.

verbal the use of spoken language, including words, letters or numbers

non-verbal non-spoken, including facial expressions, gestures, posture and emotions

Communication can be through words, text, facial expressions or even posture. These are known as **verbal** and **non-verbal** forms of communication.

Verbal communication can be face-to-face, over the phone or via video chat, through the radio or TV, or through social media, websites and blogs.

Non-verbal communication includes eye movements, tone of voice, speed and volume of talking and body position. Be aware of your body language when you are speaking to someone; even your hand gestures and the way you are speaking will affect whether your message is received in the way you intended. We often rely on our facial expressions for non verbal communication. With COVID-19 restrictions and face masks, people can only see our eyes. This makes this type of communication difficult. Even seeing a smile can be reassuring and start a conversation off on a positive note.

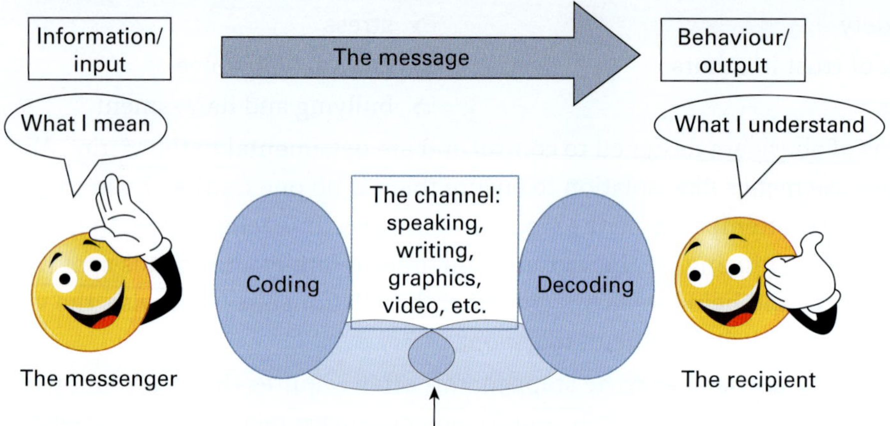

Figure 5.25 The communication cycle

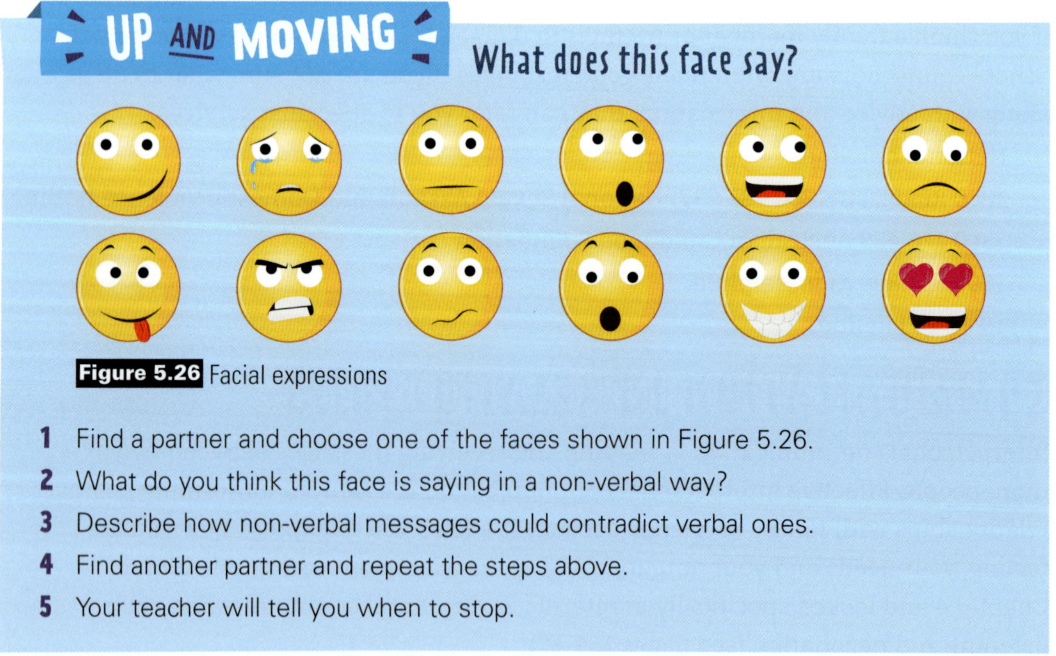

What does this face say?

Figure 5.26 Facial expressions

1. Find a partner and choose one of the faces shown in Figure 5.26.
2. What do you think this face is saying in a non-verbal way?
3. Describe how non-verbal messages could contradict verbal ones.
4. Find another partner and repeat the steps above.
5. Your teacher will tell you when to stop.

Effective messages

Remember that being a good listener is also a vital part of communication. Sometimes there can be barriers to communication. Potential barriers include:

- poor listening and speaking skills
- background noise
- one or both people being upset or angry
- confusing messages
- not having enough time to explain
- individual differences.

To reduce these barriers, you may need to try some of the following strategies:

- Reduce or eliminate background noise.
- Repeat or rephrase the message you received if you are not sure you heard or understood it correctly.
- Select the best method for communication.
- Select the words you use carefully.
- Avoid difficult communication when you are in a hurry, emotional or tired.
- Try to reduce the physical distance between yourself and the other person.
- Try to see things through the eyes of the other person – this gives you a different perspective.

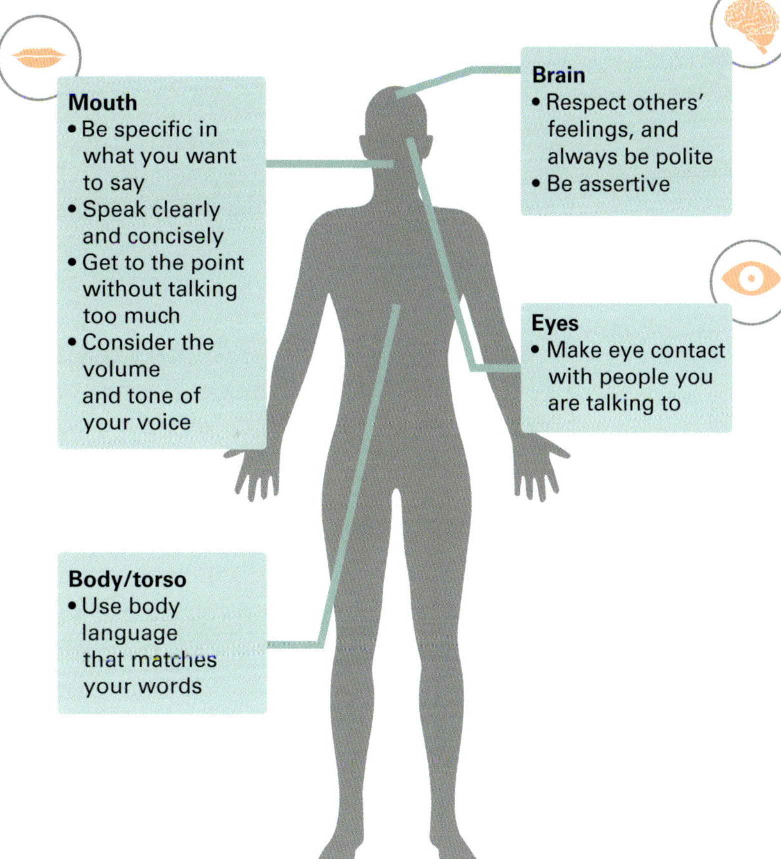

Figure 5.27 Here are some tips to help make sure your messages are sent and received in the way you intended.

UP AND MOVING — Communication games

1. Play a game of Grapevine. Stand in two lines. The first person in each line thinks of a message to whisper to the person next to them. Pass the whispered message down the line until it reaches the end. Was the message at the end the same as it was at the beginning? Identify some barriers to communication that may have led to the message being changed.
   a. Start the game again from the other end of each line and make it a competition. Who can get the message down the line the fastest? Did the competition make it easier or harder? Why? Were there any additional barriers to communication?
   b. Try doing the grapevine game in the language that you learn at school. What barriers now exist?
2. Work with a partner to do the following activity.
   a. Stand as close to each other as possible and talk about your favourite music.
   b. One person sits while the other stands. Talk about what you are doing during the next school holidays.

c Touch each other's shoes and describe your family.
d Stand on opposite sides of the room and talk about what you want to do when you leave school.

3 How did each situation in question 2 affect the ability to communicate? From these activities, what can you determine to be the most effective way to communicate?
4 Research a different form of communication, such as Morse code, sign language, semaphores or the signals used by a referee to control a particular sport. Describe the method of transmission and the skills required to effectively send and receive messages in this form.
5 Play charades as a class. Charades is a word-guessing game where one person mimes the title of a movie, TV show or book without speaking. How does communication change without the use of verbal cues?

Communication styles

The way you communicate will have an impact on the way you can relate to people. The style of communication you choose will affect whether the message is sent or received successfully. There are three main styles of communication: passive, aggressive and assertive.

Assertive communication

This is the most effective style of communication. It allows all ideas and feelings to be expressed while respecting other people's rights. Others are treated with dignity, and the communicator believes that each person has something worthwhile to say. Being assertive allows you to say 'no' and feel in control of your choices. Assertive communicators state their needs, wants and feelings clearly, appropriately and respectfully. They listen well without interrupting, and use 'I' statements such as 'I feel hurt when you speak to me in that way' or 'I would like you to ask for my opinion before making a decision'.

Passive communication

Passive communicators often don't stand up for what they believe in. They ignore their own rights and can allow other people's ideas and actions to take over conversations and decisions. They may do this in order to avoid confrontation, but may then become frustrated at not being able to get their message across. They may often feel anxious, resentful and confused, leading to a lack of self-esteem and self-confidence, saying things like, 'People never consider my feelings' or 'My needs don't matter'.

Getty Images/Fuse

Figure 5.28 Aggressive communication is a barrier to getting your message across.

Aggressive communication

Aggressive communicators express their feelings and opinions and satisfy their needs in a forceful way that doesn't acknowledge the rights of others. They may be verbally or physically abusive. Aggressive communicators may try to blame, humiliate or criticise others. They may interrupt frequently and not listen well to other people's opinions.

Conflict in relationships

When communication messages become jumbled, misunderstandings can occur and conflict may arise. Conflict is when individuals or groups disagree about a topic or decision that needs to be made. It is a natural result of human interaction and occurs because of differences in values, opinions, ideas, goals or beliefs. Conflict can be positive, by improving people's understanding and perspectives around the issue, or negative, if there is no agreement or negotiation and a resolution cannot be reached.

iStock.com/skynesher

Figure 5.29 Conflict can occur when people have different perspectives.

Conflict can often occur in families as children grow up and want to become more independent, particularly if the parents hold a different point of view. Conflict can also occur in friendship groups where there may be a variety of personalities and beliefs.

Sometimes during conflict, you say things you don't really mean. To avoid this, there are some conflict resolution strategies that you can use in these situations.

Conflict resolution strategies

- Try to remain calm. Walk away for a few moments if you have to. Often counting to 10 (in your head) before responding helps.
- Think of a compromise. This is when both sides decide to give up some demands and meet in the middle to come to an agreement. Use the negotiation skills you learnt in Chapter 4.
- Get a different perspective on the conflict. A conflict map might help you understand how the other person sees the same situation. Figure 5.30 shows an example of a conflict map.
- Write down the issue, then think about the people involved in the conflict and what their concerns and needs might be. This helps you to gain some insight into how others might be feeling and thinking.
- Communicate honestly.
- Be respectful in how you speak to the other person and in the type of language you use.
- If no solution can be found, you may have to agree to disagree.

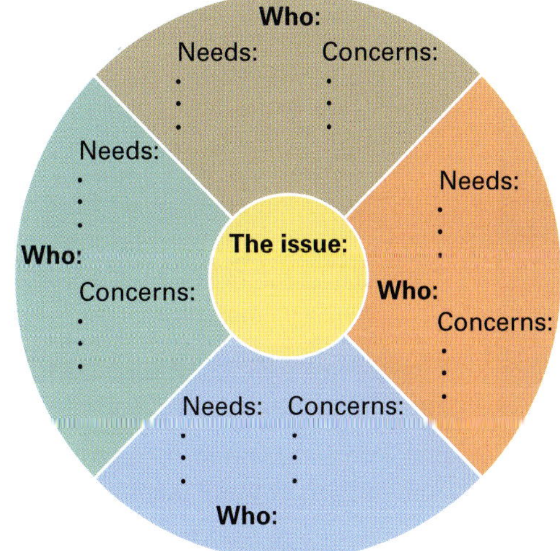

Figure 5.30 Conflict map

Identify

ReachOut.com is an Australian non-profit organisation with a mission to help young people lead happier lives. It provides fact sheets, stories, forums and videos on mental health issues.

REACHOUT

Figure 5.31 ReachOut.com by Inspire Foundation

Understand

ReachOut.com is an online youth mental health service. It provides information, support and stories on almost any topic for people aged 15 to 24.

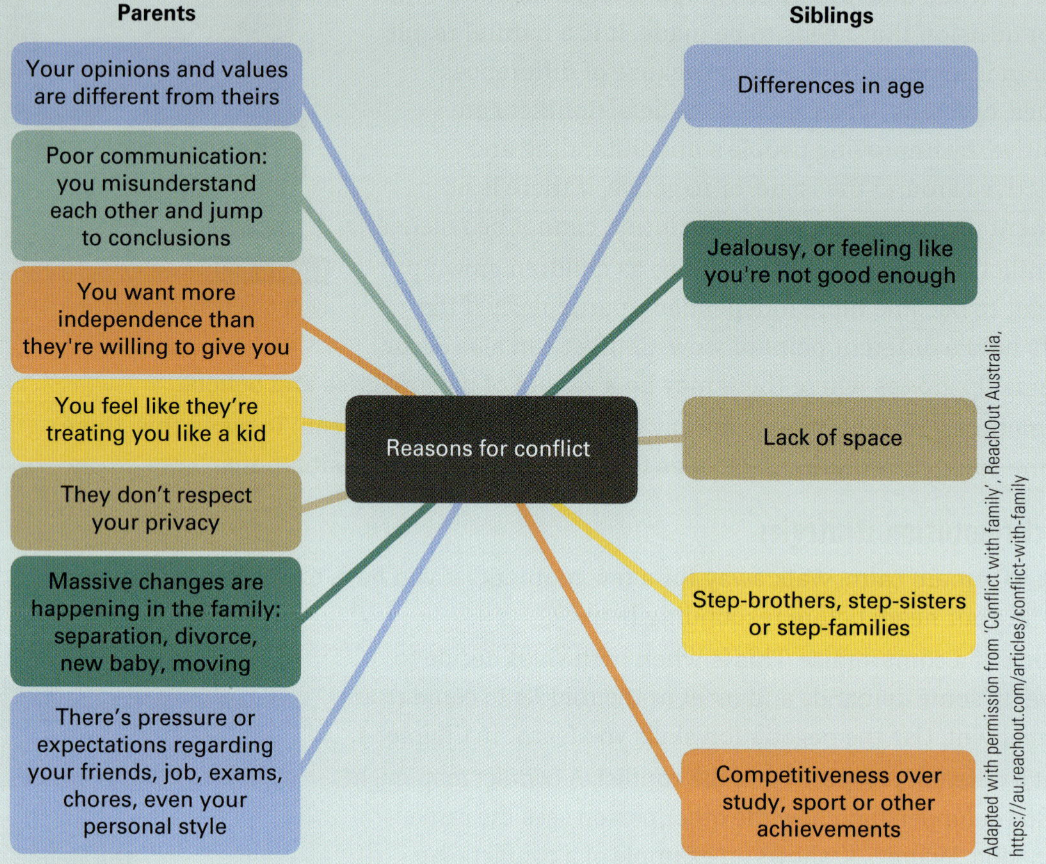

Figure 5.32 ReachOut discusses several common reasons that conflict can occur among family members.

Reachout.com suggests some ways to talk out conflict with those around you. If you're fighting with your parents, you might try having a calm conversation with them about what's going on. They'll probably be impressed to see you take such a mature approach to the problem, especially if you initiate it. Even with annoying siblings, clear and calm communication will almost always be the best way to sort things out and come to an arrangement that works for all of you.

- Pick a time when no one is angry, upset, stressed or tired.
- Choose a place where you can sit and talk without being interrupted.
- Be willing to compromise, and come up with options you're willing to accept.
- Avoid being sarcastic or verbally attacking the other person.
- Be honest. If something really upsets you, let the other person know.
- Listen to what the other person has to say, and accept that their point of view might be just as valid as yours. (This is easier said than done, but it's well worth it!)
- Once you've settled on something you can agree to, stick to it – maybe for a set period of time.
- If talking feels impossible, try writing an email or a letter, explaining how you feel.
- If you can't reach a compromise, you might have to 'agree to disagree'. Remember that you can have your own opinions, based on your personal experience, beliefs and values, and you don't always have to agree with your family.

Weblink Check out the rest of the article on the ReachOut website

Adapted with permission from 'Conflict with family', ReachOut Australia, https://au.reachout.com/articles/conflict-with-family

Discuss

1 What do ReachOut say are common reasons for conflict with parents/guardians?
2 What do they say are common reasons for conflict with siblings?
3 What are their tips for talking it out? For each one, discuss whether it is reasonable.
4 What do they suggest if things aren't getting better?

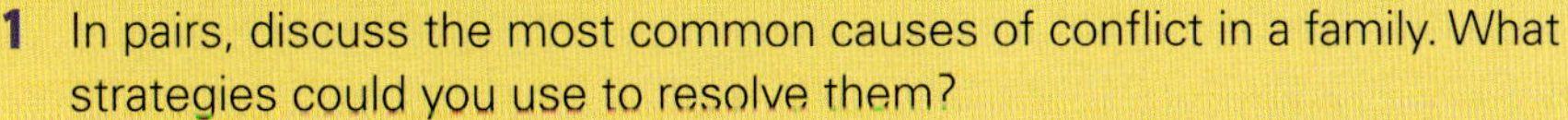
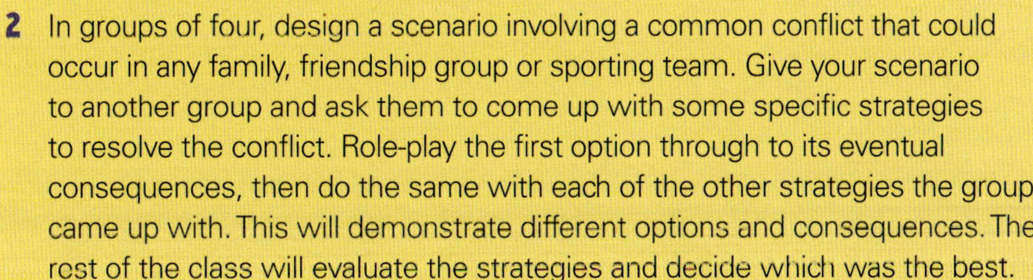

FACE TO FACE Conflict

Worksheet 5.14

1 In pairs, discuss the most common causes of conflict in a family. What strategies could you use to resolve them?
2 In groups of four, design a scenario involving a common conflict that could occur in any family, friendship group or sporting team. Give your scenario to another group and ask them to come up with some specific strategies to resolve the conflict. Role-play the first option through to its eventual consequences, then do the same with each of the other strategies the group came up with. This will demonstrate different options and consequences. The rest of the class will evaluate the strategies and decide which was the best.

Knowing where you can find help when you need it and having this information on hand allows you to have a strong support network. Being **proactive** in getting some help is an important step to resolving any issue that you may be experiencing.

proactive to act, rather than react to events

There are many sources you can turn to depending on the type of help you need. Adults should be included in your support network, as they will often know the best way to deal with the issues you may be experiencing. School counsellors or wellbeing teachers may be able to advise you further on other support options. Year level coordinators, teachers you can trust or school nurses will also listen to your needs and assist you where possible. You may feel most comfortable talking to your friends or your family, as they might know you best and understand your perspective. These are known as informal support networks. There are also formal support networks available online or by phone, which specialise in dealing with young people and their health.

WELLBEING CHECK IN

SAYING THANK YOU

Identify

We all have people in our life who we are grateful for in some way. But we don't always thank them …

This activity has been developed in collaboration with Dr David Bakker of MoodMission

Understand

Saying thank you has lots of good effects. It makes both us and the people we say it to feel good, so it's really a win-win situation. It's especially powerful when the person we're thanking isn't expecting it. Maybe they don't get thanked much for what they do, or maybe there isn't an easy opportunity to say thank you. Sometimes we might also think it would be awkward to say thanks. But as soon as you do so, you'll realise it isn't.

Practise

1 Think of someone you are grateful for but haven't necessarily shown a lot of thanks to. They might be a family member, friend, teacher, sports coach, bus driver, etc.
2 Think of what you'd like to thank them for. It can be just a little thing, like picking you up from school, or a big thing, like always being available to talk to if you have a problem. Note down what you might say.
3 Once you've come up with someone and have thought of what you want to say, think of a time you could thank them. It might be a brief moment when it's just you and them. Try to make it as specific as possible, e.g. at 5 p.m. on Thursday at the end of soccer training.

Reflect

How did they respond? What do you think they felt? And how did you feel when they responded? You can refer back to the emotion chart in 'Talking about feelings' on page 198 if you like.

After this experience, do you think you'll try to say thank you more often? Or maybe you'll say thank you in a different, more deliberate way?

1 Define what is meant by a respectful relationship.
2 Identify three ways someone can exert their power in a relationship.
3 Where can you go for help if you feel like you or a friend are in a non-respectful relationship?

1 Why do you think people feel the need to exert power over others in relationships?
2 Propose strategies to make sure you are being assertive when your emotions are strong, e.g. if you are feeling angry, sad, frightened.
3 Reflect on your experiences with conflict. If you are having a disagreement with someone, is it better to ignore it and hope it goes away? What are the pros and cons of this approach?

1 Identify and list any examples of respectful relationships in the English novels you read for school. Do they have a role in teaching us how to have respectful relationships?
2 Identify any examples of non-respectful relationships in the English novels you read for school. Is anything done to address this behaviour?
3 Debate: Power is necessary in a relationship.

Quiz
Respectful Relationships

CHAPTER 5 REVIEW

1 Describe the physical, emotional and social changes that occur during puberty.

2 Design a flow chart that describes the journey of the ovum after ovulation.

3 Describe three different types of relationships you have and analyse the functions they perform.

4 Consider how the values of your family may influence the values that you have.

5 Why might it be hard to stand up to peer pressure?

6 Reflect on how power can be used both positively and negatively. Give specific examples of each.

7 Explain how communication helps to maintain effective relationships.

8 Draw a diagram to represent the communication cycle.

9 Summarise your top three tips for conflict resolution, and explain why.

10 Describe the formal and informal support networks available to young people.

11 What qualities do positive relationships have? Why are they important to your holistic health?

THINK SAFE, ACT SAFE, BE SAFE

IN THIS CHAPTER

You will learn about factors that contribute to your safety and that of others in various settings, including your home, school, social situations and natural environments.

Shutterstock.com/Nicetoseeya

By the end of this chapter, you should be able to:

- identify safe and unsafe situations and environments
- recognise and recall strategies to seek help for yourself and others
- understand and apply strategies for celebrating safely
- plan for safe participation in physical activity in constructed and natural environments
- propose strategies you could use in emergencies
- use basic first-aid principles and techniques
- identify strategies enabling safe use of ICT, digital technologies and associated cyber safety
- think, act and be safe.

HOW DOES BULLYING IMPACT ON SAFETY?

Quiz
Pre-chapter

Before you start, take the pre-chapter quiz to find out how much you already know.

BULLYING

Video
Bullying: What strategies can you employ safely to stop bullying? How could you safely support a friend who is the victim of bullying? Watch the video and join the discussion.

According to the Australian Human Rights Commission, bullying involves a person or a group of people with more power than you, repeatedly and intentionally using negative words and/or actions against you, which causes you distress and has an impact on your wellbeing. It includes behaviours that are repeated and that may cause you harm and contribute to negative social relationships and environments. It is important to remember that bullying is not a one-off incident or fight. Harassment is a type of bullying that is nasty, degrading, frightening and harmful to the victim. It can affect people emotionally, socially and physically. Sexual harassment involves any sexual behaviour that is not invited and that is threatening, embarrassing or offensive.

FACE TO FACE What's going on?

In pairs, discuss what you believe might be going on in the scene shown in Figure 6.1. Does the image portray any unsafe behaviours? Discuss.

Shutterstock.com/Rawpixel.com

Figure 6.1 Bullying can be physical or verbal.

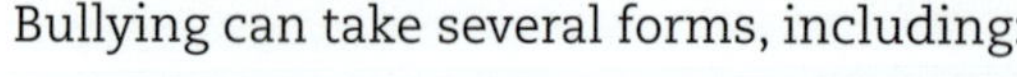

Bullying can take several forms, including:

- verbal behaviour, such as name calling, teasing or threats
- physical behaviour, such as hitting, punching or kicking
- social behaviour, such as keeping someone out of a group
- psychological behaviour, such as spreading stories, hiding or damaging possessions, or sending nasty text messages or social media posts.

Worksheet 6.1

Other forms of bullying include harassment or discrimination based on someone's race, religion, gender, sexual orientation or disability. Remember, bullying is not acceptable in any circumstance, either as a bully or an onlooker. **Bystanders** need to take responsibility and let others know that bullying is not okay.

bystander a witness who sees or knows about bullying happening to someone else

9780170463096

Bullying can happen anywhere – at school, at work, at home, in sporting teams or at the skate park. People often bully others because they feel it may help them fit in or become more popular, will make them seem more powerful, because of prejudices or because they themselves are unhappy.

Figure 6.2 Technology such as mobile phones and the internet allows bullying to occur from a distance.

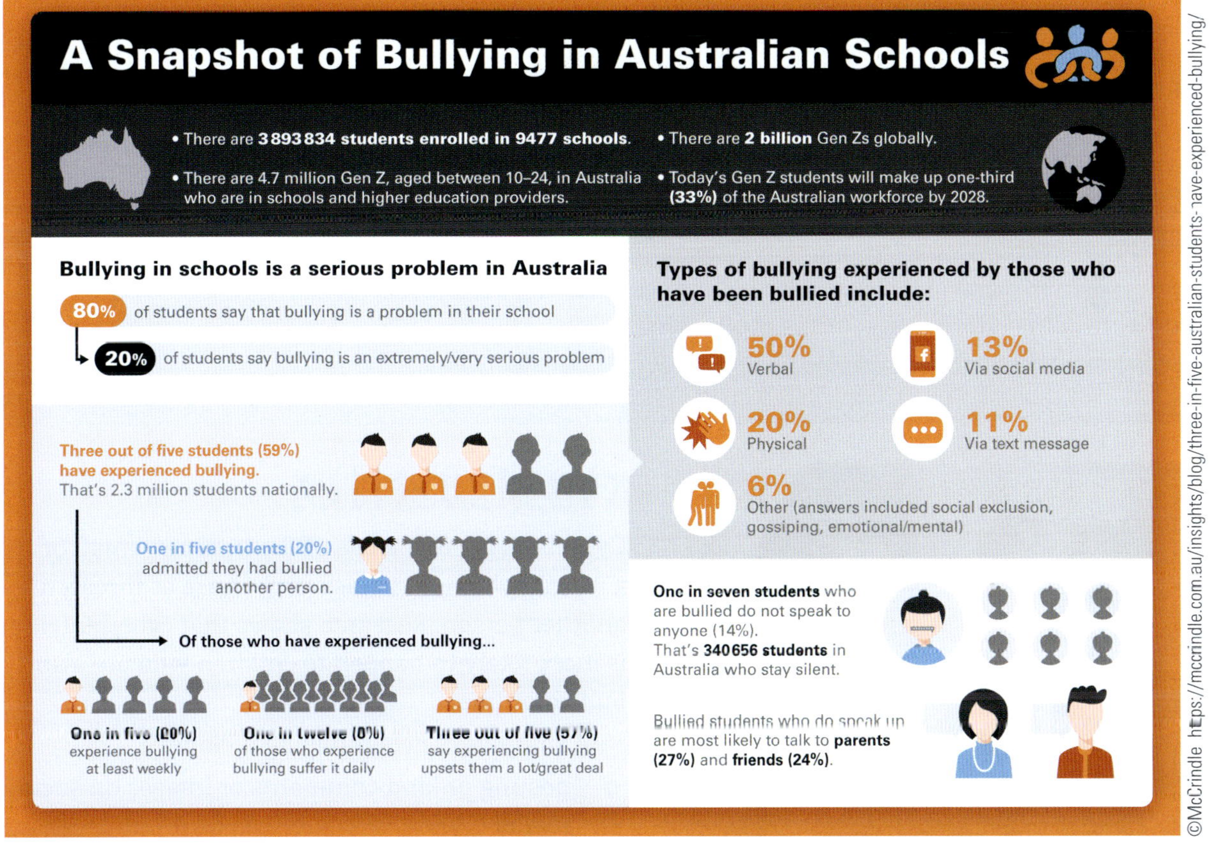

Figure 6.3 A snapshot of bullying in Australian schools

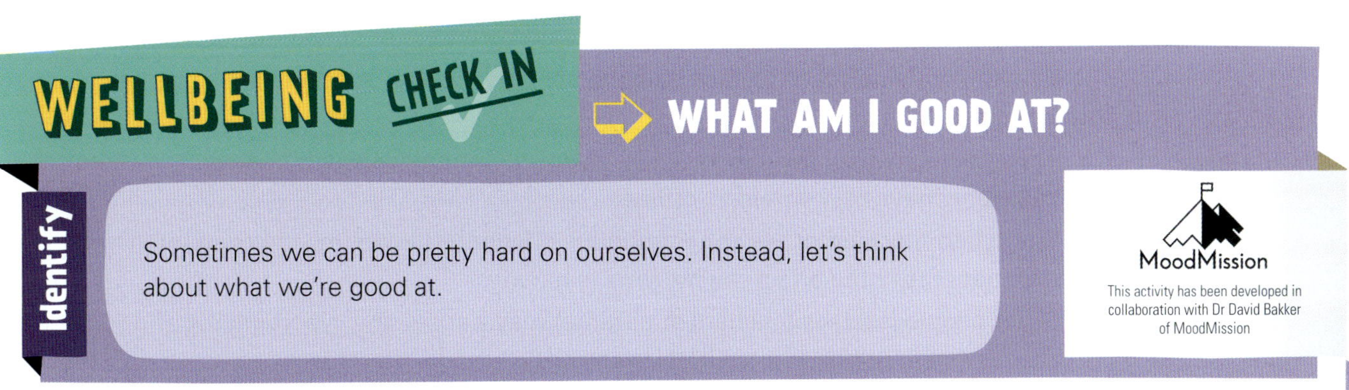

Understand

Research has shown that the way we think about ourselves has a big impact on how we feel. If we think we are confident and capable, we feel pretty good most of the time, but if we think we are powerless and doubtful about our own abilities, we can end up feeling bad. To avoid this, it can be worth reminding ourselves of our strengths. Some people might feel like this is being arrogant, but it's not. We're not saying we're better than anyone else. We're just saying we're proud of who we are.

Practise

1 Take a moment to think back over the past 24 hours. Try to recall what you have achieved or accomplished.

Recall something that you did well during that time. It might not feel like a huge, amazing achievement, and could even just be part of your day-to-day routine. For example, maybe you made yourself or someone else breakfast. Or maybe you got to school on time.

2 Propose why you were able to do that. What are you good at that allowed you to do it? For example, you could say that you are helpful if you helped someone, or that you are careful if you made sure that something bad didn't happen.

Reflect

Was it hard to come up with something? If so, why do you think it was hard? Would it be easier to come up with good things about someone else? If it was tricky, it might be a sign to keep practising, to get better at positive thoughts about yourself.

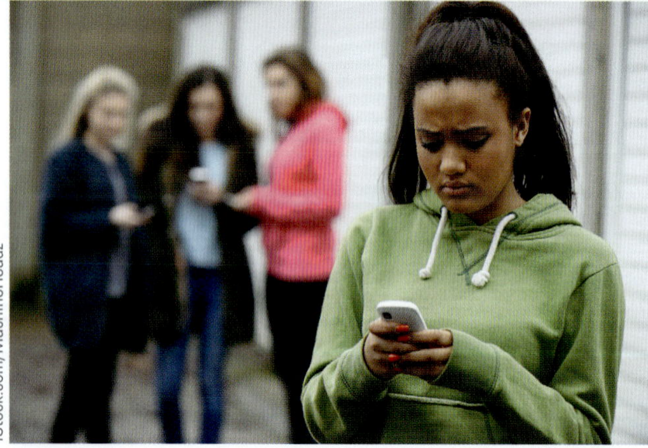
iStock.com/MachineHeadz

Figure 6.4 Cyberbullying occurs online.

CYBERBULLYING

Cyberbullying is the use of the internet, email or a mobile phone to harass, embarrass or threaten another person. Research shows that cyberbullying is more likely to happen to children who are also bullied offline.

Examples of cyberbullying include:

- posting unkind comments or images on social media
- sending abusive texts and emails
- imitating and teasing others online
- excluding others or spreading rumours online.

Cyberbullying and face-to-face bullying are very similar, but there are some key differences:

anonymity not revealing a person's true identity or real name

- The cyberbully may feel protected by a sense of **anonymity**, which may lead them to behave in ways they wouldn't offline.
- The cyberbully cannot see the immediate effect on the victim, and therefore they may not realise the effect the bullying is having.
- Other people (such as teachers or parents) may not see the cyberbullying and therefore will not be in a position to stop it.

- Cyberbullying can take place 24/7, making it hard to escape.
- Cyberbullying can have large audiences when readily shared with groups or posted on a public forum.
- Bullying comments and images posted on the internet are in most cases permanent because they are so difficult to erase.

Cyberbullying safety tips

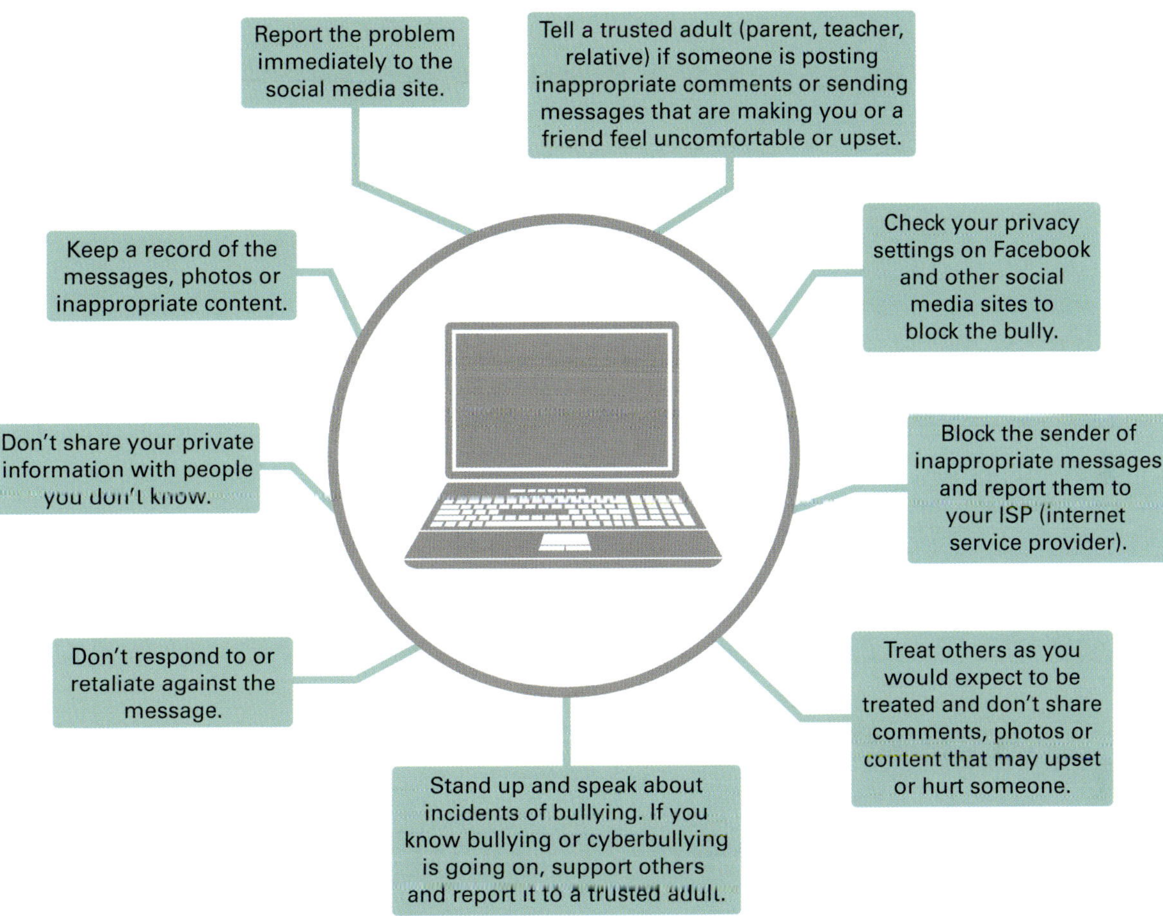

Figure 6.5 Some safety tips to consider when dealing with cyberbullying

FAST FACT

Help is available at Kids Helpline: 1800 55 1800, or online at www.kidshelpline.com.au. In life threatening or critical situations, call 000 or 112.

INSTAGRAM AND SNAPCHAT – FRIENDS OR FOES?

Identify

A Canberra high school teacher says many staff members feel 'powerless' to stop a '24/7 cycle' of cyberbullying and physical violence among their students.

Understand

Last month the ABC revealed private groups on Instagram and Snapchat hosted disturbing videos of Canberra students as young as 12 involved in fights and brutal bashings.

High school teacher Eliot*, who wished to remain anonymous so as not to identify his school or students, said he had witnessed such incidents at his school.

He said it was also common practice for physical violence in the playground to develop into ongoing online abuse, often with devastating impacts. Eliot said breaking up playground fights was possible, but, when students had camera phones, controlling the fallout from the incident proved much more difficult … He said the fighting and filming he had witnessed was often cheered on by other students on the sidelines, and for some, the ability to document and share the violence acted as a kind of social currency …

Nationally, the office of the eSafety Commissioner operates a complaints scheme for Australian children under the age of 18 who are targets of serious cyberbullying. It also has legislated powers to issue civil penalties to social media providers.

Discuss

1 Summarise the forms of bullying are discussed in the article.
2 Discuss the role bystanders (witnesses) play in the spread of bullying. How can they break the cycle of bullying?
3 The article notes that students can complain to the eSafety Commissioner. Discuss what other support options the person being bullied could seek.
4 Does your school have support services or processes for students affected by cyberbullying? Discuss two improvements that could be made to these services or processes.

CASE STUDY  DOLLY EVERETT CHARITY RELEASES ANTI-BULLYING VIDEO

Please note, the following case study discusses suicide. If you do not feel comfortable reading this, please move on to the next page.

Identify

A striking new ad about cyberbullying, directed by a 15-year-old girl, has been lauded as 'brave', 'disturbing' and 'relevant' after the suicide of Amy 'Dolly' Everett shocked the nation. Amy was was attending Scots PGC College in the Queensland town of Warwick.

Weblink
Watch the cyberbullying advert

Understand

Dolly was just 14 when she died early [2018] after being tormented by cyberbullies. Now, stirred by her death, teenager Charlotte McLaverty has created a powerful short film that depicts how modern bullying is more than just sticks and stones in the playground. It follows a teen girl in school uniform holding her phone, which pings with the notification sound every time another girl throws a rock at her. The settings are all at home, including in the bedroom, bathroom and around the kitchen table as oblivious parents sit by …

'I was hoping to relate to how it feels to be cyberbullied for most teens,' Charlotte told the ABC. 'The rocks being thrown was like the visual representation of what it feels like.'

One in five young people report being cyberbullied in any one year, according to research by the advocacy organisation Dolly's Dream, which was established by Dolly's parents last year. The organisation has this week launched a new internet hub to help parents better understand and deal with online safety, including bullying … 'Dolly left us with a message that was "speak, even if your voice shakes",' she [Dolly's mum] said …

Charlotte's ad is being billed as a project that is 'by teens, for teens' and she said part of the problem was cyberbullying was sometimes missed, even by the victims. 'It's not like a punch in the face. It's constant, relentless comments,' she said …

The film ends with the victim catching one of the thrown rocks and staring down her bully. 'Just to give a little bit of hope at the end of the film,' Charlotte said.

Discussion

1 State how many teens report experiencing some form of cyberbullying.
2 Explain the impact that cyberbullying can have on victims.
3 Identify why it is important that this project is 'by teens, for teens'.
4 Propose how the community as a whole can combat cyberbullying.
5 In small groups, create your own short video to combat cyberbullying.

ONLINE RELATIONSHIPS

WB Worksheet 6.2

The Australian Bureau of Statistics states that 96 per cent of 12- to 14-year-olds are spending up to two hours a day online. This kind of accessibility means people have the ability to form new online relationships every day, particularly through chat rooms or social media sites. It is important to remember that all you know about a new person is what you see on the screen – this may be in the form of a photo, an avatar, a username or the characteristics they use to describe themselves, and may not represent who they really are. It is very easy to 'friend' someone and never meet them face to face.

Video
Cyber safety: What digital technologies do you use the most? How can you protect yourself from the risks associated with these digital technologies? Watch the video and join the discussion.

ONLINE AND CYBER SAFETY

Think about how many times you have used digital technologies today – not just at school, but from the time you woke up this morning. You may even be reading this chapter on a computer or tablet, by projection on a smartboard or through the school's own internal learning management system. Can you imagine living without a mobile phone, computer or access to the internet? When was the last time you sent a text, downloaded music, chatted online with a friend or simply used a computer to create and share your ideas?

Shutterstock.com/BigTunaOnline

Figure 6.6 Social media apps are easily accessible on mobile phones.

SOCIAL MEDIA

Social media sites such as TikTok, Snapchat and Instagram, where users can post information, share photos, tell everyone what's happening in their lives and even share videos and/or play games, are very popular with Year 7 and 8 students. You probably have classmates who seem keen to obtain large numbers of 'friends', both people they

WB Worksheet 6.3

know in real life and people they have never met, because they think this will make them seem popular. Online multiplayer games like Fortnite and Minecraft are another way that you can directly interact with many people, including people you don't know and who you are unlikely to ever meet.

'Social Media Statistics for Australia (Updated April 2022)' by Adam Ramshaw, Genroe (Australia) Pty Ltd, https://www.genroe.com/blog/social-media-statistics-australia/13492#source2

FAST FACTS

1. In January 2021, 20.5 million Australians (79.9% of the Australian population), were active users of social media or have social media accounts.
2. Australians spend an average of 1 hour 48 minutes per day on social media.
3. 18.2% of Australian Internet users (16–64) say they follow influencers or other experts online.
4. YouTube (78.2%) and Facebook (77.7%) are tied for the most popular social media platform in Australia.

CASE STUDY  TEENS' SOCIAL MEDIA HABITS AND EXPERIENCES

Identify

81% feel more connected to their friends	**69%** think it helps teens interact with a more diverse group of people	**69%** feel more in touch with their friends' feelings	**68%** feel as if they have people who will support them through tough times
45% feel overwhelmed by all of the drama there	**43%** feel pressure to only post content that makes them look good to others	**37%** feel pressure to post content that will get a lot of likes and comments	**26%** feel worse about their own life

Source: PEW Research Centre

Figure 6.7 Positive and negative thoughts on social media

Amid growing concern over social media's impact and influence on today's youth, a new Pew Research Center survey of US teens finds that many young people acknowledge the unique challenges – and benefits – of growing up in the digital age …

Understand

For some teens, sharing their life online can come with added social burdens: Around four-in-ten say they feel pressure to only post content on social media that makes them look good to others (43%) or share things that will get a lot of likes or comments (37%).

At the same time, the online environment for today's teens can be hostile and drama-filled – even if these incidents may fall short of more severe forms of cyberbullying. Some 45% of teens say they feel overwhelmed by all the drama on social media, with 13% saying they feel this way 'a lot'. And a similar share of teens (44%) say they often or sometimes unfriend or unfollow others on social media. When asked why they've digitally disconnected from others, 78% of this group report doing so because people created too much drama, while 52% cite the bullying of them or others …

Teens generally believe social media helps deepen friendships and are more likely to equate their social media use with positive emotions – but this positivity is far from unanimous.

A central conversation surrounding social media and young people is the impact these platforms may be having on the emotional well-being of teens. A majority of teens believe social media has had a positive impact on various aspects of their lives, the survey finds. Fully 81% of teens say social media makes them feel more connected to what's going on in their friends' lives, with 37% saying it makes them feel 'a lot' more connected. Similarly, about seven-in-ten teens say these sites make them feel more in touch with their friends' feelings (69%), that they have people who will support them through tough times (68%), or that they have a place to show their creative side (71%) …

Others believe social media has had a negative impact on their self-esteem: 26% of teens say these sites make them feel worse about their own life. Still, just 4% of teens indicate these platforms make them feel 'a lot' worse about their life

The survey also presented teens with four pairs of words and asked them to choose the sentiment that most closely matches how they feel when using social media. In each instance, teens are more likely to associate their social media use with generally positive rather than negative feelings.

Majorities of teens believe social media helps people their age diversify their networks, broaden their viewpoints and get involved with issues they care about. Roughly two-thirds of teens say social networking sites help teens at least some to interact with people from different backgrounds

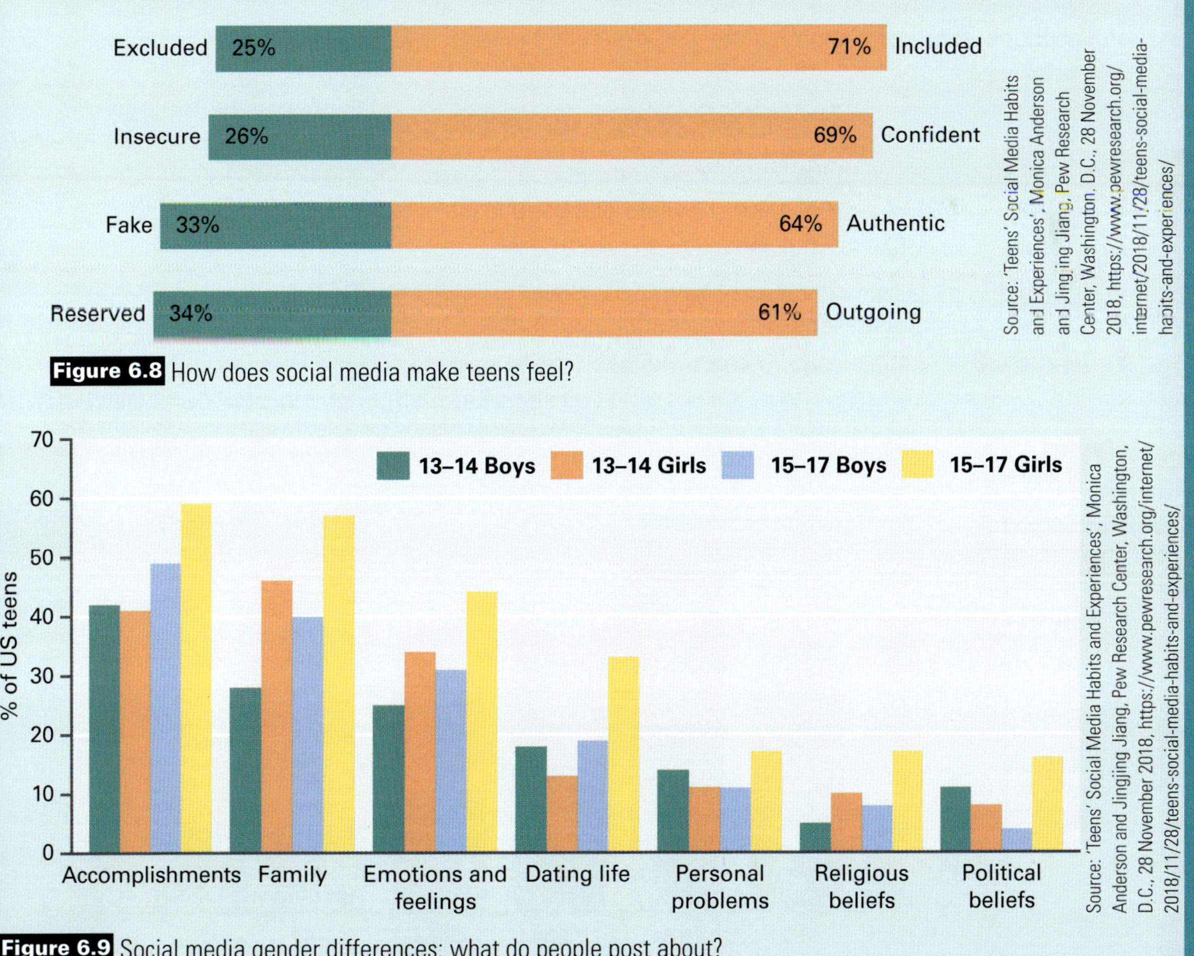

Figure 6.8 How does social media make teens feel?

Figure 6.9 Social media gender differences: what do people post about?

(69%), while a similar share credits social media with helping teens find different points of view (67%) or helping teens show their support for causes or issues (66%).

Source: 'Teens' Social Media Habits and Experiences', Monica Anderson and Jingjing Jiang, Pew Research Center, Washington, D.C., 28 November 2018, https://www.pewresearch.org/internet/2018/11/28/teens-social-media-habits-and-experiences/

Discuss

1 Consider Figure 6.8.

a Do a quick class survey using these same words to find out how your class feels about social media.

b Present the class data in graph form like Figure 6.8.

c Write a short summary paragraph comparing the feelings of the teens in your class to those in the US.

2 Consider the data from the case study, alongside your data from question 1. Do you think social media causes more harm than good? Discuss.

CASE STUDY  AUSTRALIA'S SOCIAL MEDIA USE

Identify

According to the Australian Bureau of Statistics Population Clock there are now approximately 25 million Australians. Facebook's recent data shows there are now 15 million active Australians on Facebook. Therefore, approximately 60 per cent of the total Australian population are active Facebook users.

Understand

The 5 most used social media sites in Australia are:

1. Facebook (82% of Australians (>13 years old) used in December 2021)
2. YouTube (64% of all Australians in December 2021)
3. Instagram (48% Australians (>13 years old) used in December 2021)
4. LinkedIn (31% Australians (>13 years old) used in December 2021)
5. Twitter (28% Australians (>13 years old) used in December 2021)

'Social Media Statistics for Australia (Updated April 2022)' by Adam Ramshaw, Genroe (Australia) Pty Ltd, https://www.genroe.com/blog/social-media-statistics-australia/13492#source2

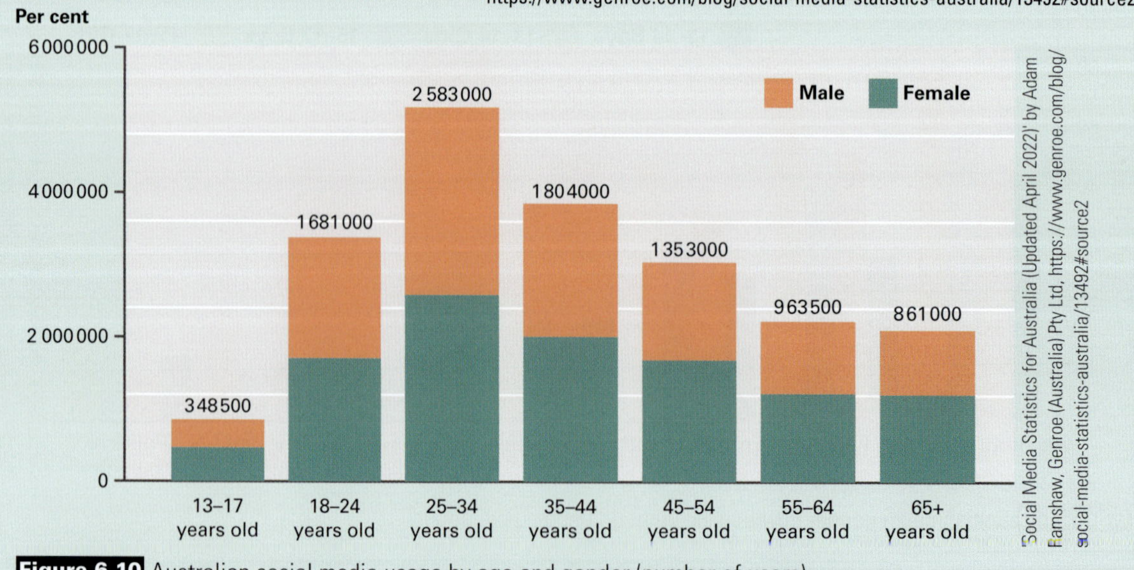

Figure 6.10 Australian social media usage by age and gender (number of users)

Discuss

1. Conduct a quick investigation in your own class of any, or all, of these social media platforms. Present your data as a graph. What was the most popular social media platform in your class?
2. Now investigate other online practices in your classroom.
   a. Use the following table as a starting point to conduct a survey analysing how your class uses the internet. Can you think of any other uses to add to the survey?
   b. Convert your class results into percentages and present them in a graph.

Internet for homework	Internet for socialising/ chatting	Mobile for text messages	Social networking: Facebook, etc.	Internet for gaming/ games	Mobile for talking	Internet for media such as sports, movies, etc.

FAST FACTS

1. Worldwide, there are 1.908 billion daily Facebook users, compared to 1.29 billion daily TikTok users and 265 million daily Snapchat users (Dec 2021).
2. 75% of consumers rely on social media to influence their purchases.

INFLUENCERS

In 2017, YoungMinds and the Royal Society for Public Health (UK) published research singling out Instagram and Snapchat as having the most negative impacts on young people's mental wellbeing of all social networks. Emma Thomas, the charity's chief executive, said that while social media could be beneficial, it also came with increased pressures.

"Being surrounded by constant images of the 'perfect' life and seemingly perfect bodies can also have a big impact on how you feel about your own life and appearance, and it can be really hard not to compare yourself to others", Thomas said.

With the rise in social media use, the number of 'influencers' has also exploded. An 'influencer' is an individual who utilises a variety of social media platforms to express their opinions on products, services or topics consequently influencing their audience/ followers. The larger an influencer's following, the more likely it is that they will be receiving payments for their recommendations.

Do you really know the credentials of the people you are taking health advice from? Glasgow University found that 8 times out of 9, social media influencers give bad diet and fitness advice. Some of these self-appointed wellness 'experts' have more than 80 000 followers!

There are some fantastic health professionals online who have studied, trained and worked hard in their fields to become experts with advice that is sound, scientific and subjective. Remember to check credentials before considering taking advice from an online expert.

FACE TO FACE Who am I?

Some young people belong to several websites on which they post their profiles, mostly real, but sometimes as another person. Discuss the following questions in pairs:

- Propose why you think some young teens would want to pose as someone else.
- List some of the potential dangers of this practice.

Social networking is a great way to stay in contact with friends and family. This can be very important for people who are socially or physically isolated, such as those with a physical disability or students living in rural or remote areas. But social networking needs to be used appropriately because it can be easy to forget who you are communicating with online, as well as who might be able to access or see the information you post. Sometimes, you might do and say things online that you would never consider doing in real life. If this difference between 'real' and 'virtual' worlds becomes blurred, the following problems may occur:

- cyberbullying
- identity theft
- unwanted contact
- exposure to offensive or illegal content
- excessive or compulsive behaviour.

WB Worksheet 6.4

Think about how your online behaviour will affect not only you, but also others. How can your actions ensure that you are cyber safe today, tomorrow and well into the future?

sexting sharing provocative or sexual photos, messages or videos, either via a mobile phone or by posting online

cyber stalking use of technologies such as mobile phones or the internet to intimidate, control, manipulate or humiliate the recipient

Eight cyber smart ways to protect your privacy

Here are some ways you can protect your personal information online.

- Set strong passwords, and use different passwords for each online account.
- Delete cookies.
- Log out of social media sites and email while you browse the web.
- Use private browsing or 'incognito' mode.
- Avoid using social media accounts to sign in.
- Change your smartphone settings to limit advertising tracking.
- Give the least amount of personal information as possible.
- Do a digital check-up of how social media sites are using your data.

Source: eSafetyCommissioner, https://www.esafety.gov.au

Alamy Stock Photo/Cultura Creative (RF)

Figure 6.11 A Year 8 student is upset when a 'selfie' she sent to her boyfriend is forwarded to hundreds of others in her school.

SEXTING AND 'SELFIES'

Taking 'selfies' and sharing them with your friends is something that you probably do without thinking much about it. But sharing selfies of a sexual nature, either of yourself or of others, is known as **sexting**, which has serious legal and social consequences. When an image has been sent or shared, you lose control over who sees it and shares it. Images and messages can spread rapidly. This can have a serious impact on your reputation, both now and well into the future. Images can also potentially be used for cyberbullying, **cyber stalking**, harassment and, in extreme cases, assault.

FACE TO FACE What if ...?

With a partner, discuss the following scenario: A Year 8 boy asks his girlfriend to take a photo of herself topless. Although the boy swears he won't show anyone, the girl says no. He pesters her, saying that if she really loved him, she would do it. After a few days of worrying and being pressured, she takes the photo and sends it to her boyfriend. After a couple of months, they split up and he posts the photo on Facebook with some negative comments about her. The girl tells her mother what has happened.

- Is it a good or bad thing that the girl told her mother about what had happened?
- Should the family inform the school, police or other authorities?
- What long-term impacts could this type of behaviour have?
- What does it mean when you read or hear that 'images last forever and can't be erased'?
- You are friends with the girl. She tells you what is happening and asks for help. What advice can you give her?
- You are friends with the boy, and he tells you about the situation. What can you say to him?

Weblink
Bullying. No Way! Adam Goodes on bullying

minor a girl or boy younger than the legal age for an adult (18)

incidence the occurrence, rate or frequency of an event

FAST FACT

Students may be committing a criminal offence when taking and/or sharing sexual images of themselves or others who are under 18. Creating and/or distributing sexual images featuring minors may be regarded as production and/or distribution of child pornography. This is true even if the people in the image agree to it being taken. Punishments vary by state and territory, and on a case-by-case basis.

Quiz
How does bullying impact on safety?

REVIEW

1 Provide a definition of bullying that would make it easy for someone in Year 6 to know if they are being bullied or not.
2 Discuss why the **incidence** of cyberbulling is increasing much more rapidly than face-to-face bullying.
3 Summarise the benefits social media can bring teenagers, based on the US data provided.

REFLECT

1 When considering whether it is appropriate to post something online, what is a good rule of thumb or self-set guideline to ensure everyone is safe?
2 Discuss the important role bystanders play in breaking the bullying cycle.
3 Briefly discuss how you believe social media can make people feel excluded and/or insecure.
4 Go to the Bullying. No Way! website and research other forms of bullying, such as harassment or discrimination. Create a fact sheet to present to the class on one form of bullying, its effects and some strategies to help deal with this type of bullying.
5 Watch the Adam Goodes clip on bullying at the link provided online in Nelson MindTap and answer the following questions.
   a Describe the forms of bullying that Adam Goodes describes in the video.
   b Research the type of bullying Adam Goodes was subjected to as an AFL footballer.

c Explain how you think Adam Goodes' stand against bullying and films that featured his situation, such as *The Final Quarter* and *The Australian Dream*, might influence others to tackle bullying in a more immediate way.

1 Imagine that you have posted a photo on Facebook or Instagram that shows you at a party behaving inappropriately. How might this have negative consequences for you in the following situations:
   - Back at school
   - Applying for a part-time job
2 Investigate how state and national security officers (police, protective service officers, etc.) go online and assume the identity of someone else in an effort to catch out people who are trying to harm others.
   a Summarise any articles or information you find that outline the identity they took on, who they were trying to 'catch out' and what the outcomes were.
   b Discuss other examples of how people pretend to be other people online and how they try to trick people and compromise their safety.
3 Divide the class into two groups, and then break these into groups of three. One half of the class (in groups of three) will collect information on how social media brings about positive outcomes. The other half of the class will collect information on how social media can be harmful and damaging to our health and safety. Bring the groups of three together to summarise the strongest points and select representatives to conduct a class debate using all of the information collected and discussed within the groups.

HOW CAN I IMPROVE MY SAFETY?

While it's important to be safe online, many other situations may arise in your daily life, whether you are on your own or out with friends, where some., practical skills and strategies for looking after yourself will be important to know.

Identify

When you feel distressed, overwhelmed, or unsafe, it can be useful to take a moment to calm yourself and become aware of your surroundings. This can help us take action to ensure the safety of ourselves and those around us.

This activity has been developed in collaboration with Dr David Bakker of MoodMission

Understand

The psychological technique of grounding involves stopping and paying close attention to our senses. This might be taking note of what you can see, feel, hear, smell, and taste. When we pay attention to these sensations, we can become more aware of our surroundings and less distracted by all the negative or overwhelming thoughts rushing around in our head.

9780170463096

Practise

1 Stop what you're doing right now and look around you.
2 Name 5 things you can see.
3 Name 4 things you can feel, e.g. on your skin or in your body.
4 Name 3 things you can hear.
5 Name 2 things you can smell.
6 Name 1 thing you can taste.

Tip

This is a really helpful strategy for finding calm when you're feeling stressed, anxious, or overwhelmed.

Reflect

1 What did you notice after you had listed your sensations? Did you feel a bit more aware? More calm?
2 When do you think you would be able to use this exercise?

Strategies for personal safety

These strategies, known as your personal safety plan, include developing your confidence so that you are less likely to become a victim and knowing what to do if you find yourself in an unsafe situation. The following section presents some tips for keeping yourself safe in various situations.

Worksheet 6.5

Travel with friends

Where possible and practical, travel with at least two or three friends. The saying 'there's safety in numbers' applies here. If you or your group are threatened, there are more opportunities for seeking assistance.

Shutterstock.com/Artens

Figure 6.12 Safety in numbers

FACE TO FACE Safety in numbers

Discuss situations when walking together might actually decrease safety.

When might a group of teenagers attract negative attention for their behaviour? How might walking through a shopping centre or down the street be considered dangerous?

Stay alert

- If you appear distracted, you are more likely to be targeted. Attackers will target people who look less able to defend themselves.

- Stay awake on public transport – sometimes it is tempting to close your eyes or have a quick sleep, but this will leave you very vulnerable. If you feel yourself dozing off, open a window to get some fresh air.
- Don't listen to music or talk on your mobile as you walk home, as this will reduce your awareness of what is going on around you. Being aware of your surroundings will also help you identify places or people who might be able to help you if necessary.

Look confident

- The more confident you look, the less likely you are to be attacked.
- Walk with your hands by your sides and stride out confidently and with purpose. Look ahead rather than down at the ground so that you are aware of what is going on around you.
- Find comfortable and secure ways to carry your bags, such as wearing the strap over your shoulder and across your body.

Act on your feelings

- If someone is making you feel uncomfortable, or you suspect that you are being followed, don't ignore it.
- Cross the road if you think you are being followed. If the person continues to follow you, go into a shop, petrol station or somewhere where there are lots of people and tell someone what is happening. You should then either phone home and ask to be picked up or call the police.

Stay in touch

- Make sure that people know where you are, where you are going and when you are due home so that they can raise the alarm if you are not home by the time you specified.
- Ensure that you carry your mobile phone, if you have one, at all times, but keep it hidden from sight in a pocket or bag.
- If you and your friends are travelling home late at night, text each other when you get home. If someone doesn't check in when expected, tell an adult.

Stay in the light

- Choose well-lit areas with other people around wherever possible.
- When walking home after dark, use main roads with good street lighting as much as possible. You are a less obvious target if other people can see you.
- Meet friends in a familiar and safe environment that you all know. If one of you is waiting alone, it is much better to do so in a busy, well-lit place than in a dark park or on a street corner.
- Stick to major routes and roads and don't take short cuts.
- Walk facing oncoming traffic and do not walk near the kerb.

Stay close

- When using public transport, stay close to people who can help you. Travel in carriages or sit where there are groups of people who could help you if needed.
- Arrive at the station, depot or bus stop as close as possible to the departure time of the train or bus, and stand in well-lit areas close to other people.
- When using a taxi, give the driver clear directions explaining how to reach your destination. If you are not on the agreed route, stop the taxi. If feeling unsure, insist on being taken to a safe place and end the trip.

Newspix/Craig Greenhill

Figure 6.13 Friends 'sticking together' and looking out for each other

FACE TO FACE Public transport

1 Go to your local public transport website and create a list of the top 10 strategies suggested to increase safety while using public transport.
2 Of the top 10 strategies, choose the five that you think are the most important. Discuss with your classmates why you have chosen these five.

Carry some essentials

There are a few things you should carry with you at all times, just in case things go wrong. These include:

- at least a couple of $1 coins and a $10 note so that you can always make a phone call or catch public transport or a taxi if you are feeling anxious
- emergency contact numbers – these should be stored in your mobile phone, but make sure you also have them written down.

Look out for your friends

At major events such as concerts, parties and outings, look out for your friends by following these guidelines:

- Never leave valuables unattended at the beach, in parks or at major events. Offenders watch the movements of people in public spaces to pick their targets. Make it obvious that you are alert and not alone.
- Stay with your group, especially in large crowds during parties, concerts, festivals, etc. When you decide to leave, do so in groups of two to three.
- Don't take too many belongings and bags with you, as they can be difficult to care for. Some venues may not allow you in with bags. Keep belongings in the middle of your group and always have one person stay with them.
- Avoid getting into arguments, especially if you or your friends are drinking or doing drugs – logic and common sense are affected by alcohol or illicit drugs.
- Call an ambulance immediately if a friend is seriously affected by alcohol or drugs. Police are not necessarily called when an ambulance is requested, so don't be scared about this happening.
- Give yourself enough time to get to and from where you want to be, especially when relying on public transport.

Three steps to keep you safe

The following three steps can minimise your chance of harm in risky situations.

1 **Think** about any potential for a situation to become unsafe or risky. Listen to what your body is telling you: are you experiencing a faster heart rate, increased sweating or uneasy feelings? In most cases, if a situation doesn't feel right, it probably isn't.

2 **Evaluate** the level of risk associated with a situation. Try to think ahead about what might happen. Quickly decide whether the situation is risky but under control, or if it is too risky and potentially harmful for you and your friends.

3 **Act** quickly to remove yourself from situations where you think your safety could be at risk and things could go wrong for you or your friends. Planning ahead will help you to respond in a safe and positive way.

CATCH A THOUGHT

This activity has been developed in collaboration with Dr David Bakker of MoodMission

Identify

Sometimes it feels like we have emotions in response to events. Something happens and then we feel good or bad about it. Sometimes this emotion is so strong that we might not be able to think of anything else. But research shows it's actually the thoughts we think that give rise to our feelings. If we catch our thoughts, we can keep ourselves from being overwhelmed with emotion.

Understand

Cognitive behavioural therapy (CBT) is a type of psychological therapy that has been shown to help people feel better. CBT shows us that thoughts come before feelings, and if we change our thoughts, we can change our feelings for the better. However, first we've got to catch our thoughts. It can sometimes be hard to know exactly what we're thinking when we're feeling bad. This takes practice!

Practise

1 Think back to the last time you were feeling sad, anxious or another strong negative emotion. Note the situation briefly, e.g. what you were doing, where you were, who was there.
2 Now write down the emotions you were experiencing, e.g. sadness, anger, etc.
3 Now write down the thought that was probably going through your head, e.g. 'They're all going to laugh at me', 'I can't do anything right'.
4 The next time a similar situation happens, take a second to note what's going through your head. You might even like to write the thought down.

Reflect

Now that you have the thought written down, is there something else you could think in the same situation? There might be something you could think that would make you feel better, like 'Although this is hard, it's all going to turn out okay'.

1 Discuss why it is important to let people know where you are going to travel when outside your normal daily routine.
2 Explain what is meant by the expression 'trust your gut feeling' (or instinct).

1 Summarise three different ways you can 'look out for your mates or friends' when you are out together:
   a at the local shopping centre
   b on public transport
   c when playing sport.
2 Explain why it is important to make eye contact with a stranger you think is acting suspiciously, rather than avoiding eye contact altogether and looking away from them.

1 Imagine that you are going to the MCG or Rod Laver Arena to watch a cricket or tennis final. You know that there will be tens of thousands of people present and it will be crowded. You have decided to travel to the venue by train.

   Produce a table with two columns, like the one below. In the first column, list at least five things that could go wrong either travelling to and from the venue or while you are actually at the venue. For each potential 'problem', try to think of a solution that will improve your safety.

Potential problems	Possible solutions

HOW CAN I EXERCISE SAFELY?

Quiz
How can I improve my safety?

Regular physical activity is vital for good health. While there is a risk of injury with any type of physical activity, the benefits of being regularly active far outweigh the risks. Rules, regulations and guidelines relating to physical activity usually exist for a good reason: to keep you and your friends safe and to avoid injuries so that everyone can enjoy the experience.

Sometimes rules may not be directly related to a sport or activity but still need to be followed. For example, if you're inline skating, skateboarding or riding a bike, you need to pay strict attention to all traffic laws and accept recommendations regarding protective gear and warm-ups. Proper techniques and equipment also promote safety.

LISTEN TO YOUR BODY

Injuries are more likely if you ignore your body's signals of fatigue, discomfort and pain. You can avoid injury by following these simple suggestions:

- See your doctor for a full medical check-up before embarking on any new fitness program. You should also consult your doctor or physiotherapist if you have a pre-existing injury but want to start a new fitness program.

- Cross-train with other sports and exercises to reduce the risk of overtraining.
- Make sure you have at least one recovery day, preferably two, every week.
- Exercise at an appropriate intensity for your fitness level. It takes time to increase your overall level of fitness. Training too hard or too fast is a common cause of injury.
- Allow time for injuries to rest – trying to 'push through' the pain will cause more damage to soft muscle tissue and may delay healing.

iStock.com/Shutter2U

Figure 6.14 Physical activity, exercise and sport are all positive, but can be 'too much of a good thing' if done excessively.

Warning signs

If you experience any of the following symptoms, stop the activity and seek help:

- feelings of discomfort or pain
- chest pain or other pain that could indicate a heart-related issue, including pain in the neck and jaw, pain travelling down the arm or pain between the shoulder blades
- extreme breathlessness
- a rapid or irregular heartbeat during exercise.

CASE STUDY

ARE YOU ADDICTED TO EXERCISE?

Identify

Physical activity feels good and it's great for your health. It can reduce your risk of developing chronic conditions like type 2 diabetes, strengthen your bones, muscles and joints, and can even help with certain mental health conditions, such as depression. While exercise has clear benefits, it can cause problems if your love of working out crosses over into an addiction.

Dreamstime.com/Zhyk1988

Figure 6.15 Exercise isn't always good for your health.

Understand

If you constantly cancel activities with friends or family in favour of exercise – so you plan your life around your gym workouts – you might have a problem. If you exercise in spite of pain or injury, and feel obsessively guilty when you miss a session, it could be that you're addicted to exercise …

Around one in 200 people in the general population have an exercise addiction. But our new research shows among people who exercise regularly, factors including their attitudes towards exercise and perceptions of themselves mean more than one in ten could be at risk of becoming addicted … on the more extreme end of the scale, recent research among elite Australian athletes classified 34% as having an exercise addiction …

Whether exercise becomes an addiction can be related to the amount and frequency of training, appropriate nutrition, and motivation for exercising.

Over time, researchers have proposed several diagnostic criteria for primary exercise addiction. These include:

- constant preoccupation with exercise, with significant withdrawal symptoms in the absence of exercise (mood swings, irritability and insomnia)
- this preoccupation causes clinically significant distress or impairment in one's physical, social, occupational or other areas of functioning
- the above is not better accounted for by another mental disorder (such as a means of losing weight or controlling calorie intake as part of an eating disorder) …

Millions of images posted every day on social media promote a visual representation of 'perfect' bodies – pictures of muscular, 'ripped' men and slim, toned women. This 'fitspirational' trend generates unrealistic expectations, often leaving the most vulnerable with a deep sense of personal dissatisfaction … Such an environment is a fertile breeding ground for the development of exercise addiction, alongside other appearance-related disorders. An example is body dysmorphic disorder, a psychological disorder where a person becomes obsessed with imaginary defects in their appearance.

In our study, those scoring highly on exercise addiction also reported increased image anxiety and low self-esteem. Some 38.5% of overall participants were found to be at risk of body dysmorphic disorder, especially females (47%).

Our study also showed exercise addiction was a strong predictor for the use of performance and image-enhancing drugs, especially among men. Some 39.8% of respondents claimed to use a range of fitness-enhancing products, and this cohort scored three times higher on the exercise addiction scale.

While sports supplements such as protein, vitamins and amino acids are considered relatively safe, the use of prescription drugs – without medical consultation in the vast majority of cases (96%) – is more concerning. Our participants used steroids (5.9%), diuretics (4.9%), and growth hormones (1.8%). The use of sibutramine (1.1%), an appetite suppressant that has been withdrawn from most markets because it increases the risk of heart attack and stroke, was particularly alarming. Participants also used other illicit drugs, such as amphetamines (2.3%).

Source: 'Do you plan your life around your fitness schedule? You could be addicted to exercise', Katinka van de Ven and Ornella Corazza, 4 April 2019, *The Conversation*, https://theconversation.com/do-you-plan-your-life-around-your-fitness-schedule-you-could-be-addicted-to-exercise-112509

Discuss

1. Summarise three symptoms used to diagnose exercise addiction.
2. Compare the gender differences in the data noted in the article.
3. Discuss the link between exercise addiction and other forms of addiction.

INVESTIGATION

EXERCISE OBSESSION

Purpose

To investigate which individuals or groups are more likely to experience exercise obsession, along with early **signs** and **symptoms** and possible support strategies.

signs the things that can be seen when a condition is present

symptoms the feelings that accompany a condition and that are felt rather than seen

Method

1. Conduct an internet search of any articles, news items or statistics that reveal the incidence of exercise obsession. It would be ideal to compare and contrast the occurrence across different age groups, e.g. 10–20, 20–30, 30–40, as well as any gender differences.
2. Additionally, during your investigation, summarise:
   a. the main signs and symptoms of exercise obsession
   b. suggested support strategies.

Discuss

1. Is exercise obsession more likely in one particular age group than any others?
2. Does exercise obsession affect one gender more than the other, or is it evenly spread? Briefly discuss any factors that may explain any differences.
3. List the most common signs and symptoms that someone might be exercise obsessive.
4. If you thought a friend was becoming obsessive about exercise, discuss three strategies you might consider in an effort to provide them with understanding and support.

Prepare for activity – warm up

Before any exercise session, it's a good idea to gradually warm up your muscles for about 5–10 minutes to get you ready for your workout and prevent injury. The type of activity done in the warm-up should use the major muscle groups that will be used in your forthcoming activity.

Your warm-up could begin with a low-intensity activity such as a brisk walk or light run, followed by stretching that mimics the movements that are likely to occur during the activity. Stretching should start after your muscles have been warmed up, as stretching cold muscles is less effective and could lead to low-level injuries.

Cool down

Worksheet 6.6

It is also important to stretch after activity, to assist recovery. A thorough cool down can really reduce muscle soreness and stiffness. In the last five minutes of your workout, slow down gradually to a light run or brisk walk. Finish off with 5–10 minutes of stretching (emphasise the major muscle groups you have used during your activity).

Drink lots of water

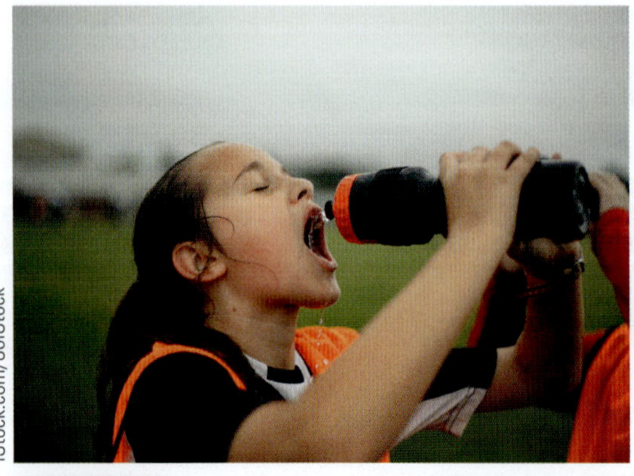

iStock.com/SolStock

Figure 6.16 Do you drink enough water during exercise?

You can lose around 1–2 litres of fluid for every hour of exercise and physical activity that you do. Dehydration happens well before you feel thirsty and may cause some serious symptoms, including increased errors, poor judgement and general tiredness, cramps, headaches, heat stress and heat stroke.

Here are some tips to avoid dehydration:

- Regularly drink fluids for several hours prior to exercise and ensure that you drink at least 500 millilitres an hour before exercise.
- Drink at least 150 millilitres every 15 minutes during exercise.
- During exercise or activities, take advantage of all breaks in play to hydrate.
- After exercise, have plenty to drink to ensure you are fully rehydrated.

9780170463096

1 How would you know that you are likely to be suffering from dehydration?
2 Describe or define 'exercise obsession' to a younger student so that they know what it is and the signs and symptoms to look out for.

1 Discuss why it is important to be adequately hydrated before starting training or competitive sport.
2 People often use the terms 'heat stress' and 'heat stroke' interchangeably. What is the difference?

1 Create a specific warm-up for a:
   a recreational runner
   b state-level netballer.
2 Investigate how 'electrolytes' found in sports drinks help athletes who are involved in activities that last for longer than two hours.

Quiz
How can I exercise safely?

HOW CAN I INCREASE MY SAFETY WHEN OUTDOORS?

Being outdoors is a great way to combine physical activity and being with friends. Even though most of the things that you do outdoors might seem harmless, there are some safety issues to think about. These include sun safety, water safety and having the right equipment.

SUN SAFETY

Exercising in hot weather puts additional strain on your body. When your body can't keep itself cool, you may suffer a heat-related illness, such as heatstroke or sunstroke. Sweating is the body's natural reaction to heat and is designed to cool your body down. When you get too hot, your **core temperature** rises, meaning sweating is no longer effective and you could develop a heat-related illness.

core temperature the temperature of the vital organs of the body (brain, heart, lungs, kidneys), measured internally

Heat illnesses

Symptoms of heat illness can include irritability, weakness, headache, feeling sick and cramps. You can avoid heat illness by following these tips:

- Drink plenty of water before, during and after exercise.
- Wear lightweight, light-coloured, loose-fitting clothes.
- Protect yourself from the sun with clothing such as long-sleeved tops, full-length pants, a hat and sunglasses.
- Exercise in the cooler parts of the day – not between 10 a.m. and 4 p.m., which is usually the hottest part of the day.
- Don't exercise as hard on hot days. Take frequent breaks and drink water or other fluids every 15–20 minutes, even if you don't feel thirsty.

UVR is B.A.D.

Ultraviolet radiation (UVR) is bad news for your skin. It's the part of sunlight that causes sunburn, skin damage and ultimately skin cancer. Every bit of UVR exposure that you receive adds up over time; this is known as cumulative exposure. There are three types of UVR:

1 UVA: causes sunburn and skin damage such as wrinkles and discolouration
2 UVB: causes sunburn and skin cancer
3 UVC: does not reach the Earth's surface but can be produced artificially, for example, by arc welding equipment.

The UV Index was developed by the World Health Organization to provide an international scale of the sun's UV strength. It measures UVR from the sun on a scale of 1 to 11+. The higher the number, the stronger the radiation and the faster your skin will be damaged and burn.

FAST FACT

UVR damages your skin quickly. Your skin gets damaged as soon as you're exposed to UVR, not just when sunburn appears.

The best way to protect your skin from UVR is to make sure you use the following five methods of sun protection every day, whenever you are outside:

1 Shade: get in it when you can.
2 Clothing: cover up as much as possible. Long sleeve and long pants, light coloured cotton fabric that reflects heat works best.
3 Hat: make it broad-brimmed.
4 Sunglasses: wraparound are best. Find a pair that meets the Australian standard AS/NZS 1067:2003.
5 Sunscreen: SPF50+. Apply it generously 20 minutes before you go outside and remember to reapply every two hours.

UV INDEX 1	UV INDEX 2	UV INDEX 3	UV INDEX 4	UV INDEX 5	UV INDEX 6	UV INDEX 7	UV INDEX 8	UV INDEX 9	UV INDEX 10	UV INDEX 11+
LOW		Moderate			High		Very high			Extreme
You can safely stay outside!		Seek shade during midday hours! Slip on a shirt, slop on sunscreen and slap on a hat!					Avoid being outside during midday hours! Make sure you seek shade! Shirt, sunscreen and hat are a must!			

Reprinted from Global Solar UV Index: a practical guide, page 8. © World Health Organization 2002

Figure 6.17 The World Health Organization's recommendations on sun protection according to the UV Index

Worksheet 6.7

FAST FACT

- All your time in the sun adds up.
- UVR has nothing to do with light or temperature, so it can't be seen or felt.
- UVR is there even when it's cloudy.
- UVR can be reflected by light onto your skin, bouncing off shiny surfaces such as water, sand, snow and concrete.

Figure 6.18 Five ways to protect yourself from skin cancer

WATER SAFETY

Australians are known for their love of the water but unfortunately, every year people drown or are injured in the ocean, swimming pools, lakes and rivers. State governments and safety authorities spend millions of dollars on campaigns to teach people how to stay safe in and around the water to reduce these, mostly preventable, drowning deaths.

Pool safety

Toddlers and young children under four years of age are especially likely to drown around the home and in home swimming pools. Constant supervision of children around water is essential:

- Supervision means constant visual contact, not occasional glances.
- Even in a supervised public pool, never take your eyes off young children swimming. If they are under five, an adult must be in the water with them.
- If you leave the pool or water area, even for a moment, take young children with you.
- All pools and spas must have a resuscitation chart nearby, where it can be read quickly.
- Don't leave paddling pools, baths, basins, sinks, buckets or troughs full of water after you have finished using them.

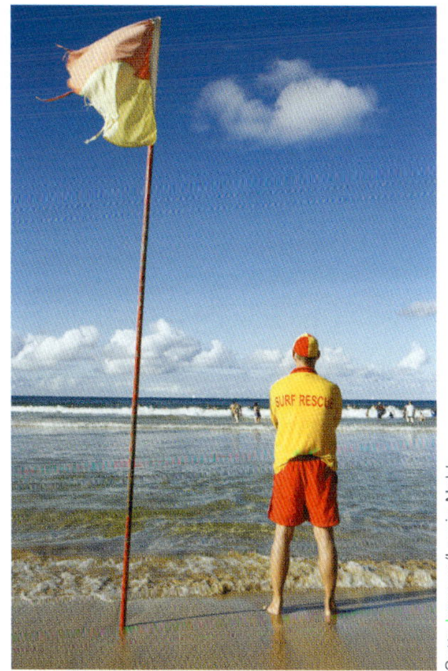

Getty Images/Laurie Noble

Figure 6.19 Patrolled beaches can still be dangerous, especially if you swim outside the flags.

Beach safety

Swim between the flags

Any beach can be dangerous. You must always swim between the red and yellow flags, which indicate the area where the beach is patrolled by lifeguards. When swimming between the red and yellow flags, always look back to the beach to check that you are still between them.

As beaches are not patrolled every day of the year, and some are never patrolled, it is important to remember these guidelines:

- Check with an adult where it's safe to swim.
- Never swim alone.
- Read and obey the water safety signs that warn people of likely risks and dangers.

Shutterstock.com/Mark and Anna Photography

Figure 6.20 Water safety signs (from left to right): Shallow Water, Boats, No Diving, No Jumping, No Swimming.

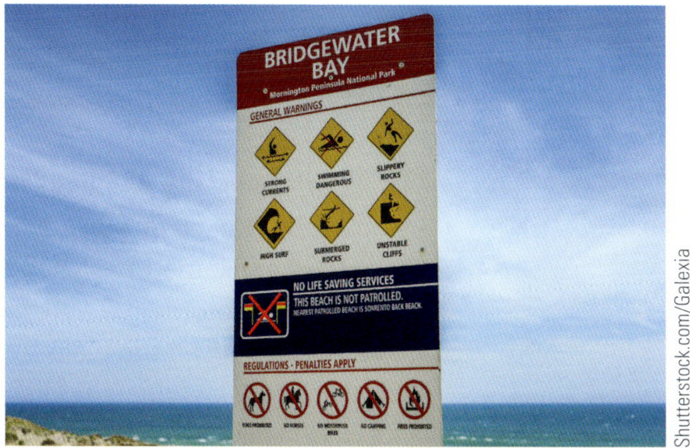

Shutterstock.com/Galexia

Figure 6.21 Danger awareness signs including warnings about strong current, dangerous swimming conditions, slippery rocks, high surf, submerged rocks, and unstable cliffs.

Weblink
Watch the video to help you spot a rip current

Rip currents

A rip is a strong water current running out to sea from a beach. Rips can easily and suddenly sweep swimmers out to sea from shallow water, sometimes several hundred metres offshore. Rips are not easy to spot, but may show some of the following signs:

- murky brown water caused by sand and seaweed being stirred up off the seabed
- foam and debris on the surface of the water, moving out to sea
- waves breaking on both sides of the rip but not inside the rip (the rip may seem calm and inviting)
- water that is darker than the surrounding water, indicating it is deeper.

What to do if you find yourself in a rip current:

- Try to stay calm and don't fight against it. Try to float with the current and attract attention for assistance.
- If you have a surfboard or bodyboard, stay on it and try to attract the attention of a lifeguard or another surfer.
- If you are a strong swimmer, you could try to escape the rip current by swimming parallel to the beach. Only ever try to swim across a rip current, never against it.

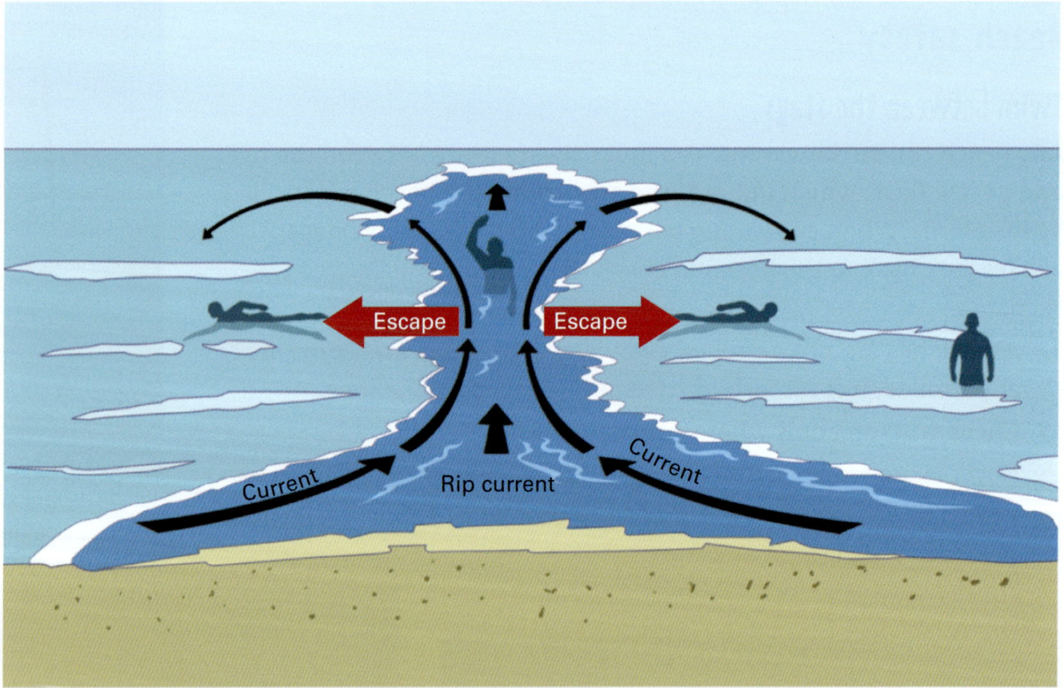

Figure 6.22 How a rip current works

Safety when surfing and bodyboarding

At a patrolled beach all surfers must surf outside the black and white quartered flags. These flags are sometimes used to create a buffer zone between the swimming area (which is between the red and yellow flags) and the board riding areas. No surfboards are allowed between the black and white quartered flags. Surfers can keep themselves and others safe by following these simple rules:

- Always let someone know where you are going.
- Never surf alone.
- Assess the conditions thoroughly – check with local authorities and the weather forecast.

- Stay aware of the conditions, as they can suddenly change.
- Use the correct equipment – leg rope and nose guard for surfing, wrist strap and fins for bodyboarding.
- If you get into trouble, stay on your surfboard or bodyboard – it will keep you afloat.
- If you are caught in a rip, stay calm and stay with your board. Try to attract attention and, if possible, paddle parallel to the beach, then catch a broken wave back to shore.
- Be aware of other people in the water and don't surf or bodyboard too close to swimmers.

CASE STUDY ⇨ WATER SAFETY CAMPAIGNS

Identify

Watch the two new water safety campaign videos using the link provided online in Nelson MindTap, and read over the article below.

Weblink
Water safety campaign videos

Understand

Migrants and international students are the targets of a new water safety video that aims to reduce drowning deaths along the Australian coastline.

The educational video has been produced by Sikh charity Turbans-4-Australia to highlight the dangers of rips and the need to swim between the flags.

The charity's founder, Amar Singh, said not enough was being done to reach the individuals and communities most at risk of drowning.

'A lot of the water safety awareness campaigns are not showing any migrant faces and I think this is key,' Mr Singh said. 'Right now, we've made it in one language, which is Punjabi, but we hope to get more funding support from state and federal governments to put it out there in multiple languages.'

Mr Singh said he would like to see the safety message broadcast widely and has lobbied the Government to include it as part of the Commonwealth visa process.

'I would like to see the video played on aeroplanes, national TV, and when people apply for a visa they get a link in their own language that says, "please, make sure you watch this",' he said …

According to the 2019 Royal Life Saving National Drowning Report, in the past 10 years, 794 people who drowned in Australian waterways were born overseas.

Discuss

1 Why is it important that water safety messages target international visitors and new arrivals to Australia?
2 Summarise the key message(s) from the water safety videos.
3 Try to obtain statistics about drownings in Australia.
   a Where do they occur most frequently (which state and which environment)?
   b Is one group of people over-represented when compared to others?

Inland water safety

Worksheet 6.8

Many people drown in Australia's rivers, lakes and dams or become paralysed after diving into shallow water. The best way to check whether it's safe to swim is to ask someone who knows the area, such as a resident, previous holiday maker, caravan park owner or park ranger.

FACE TO FACE Diving accident

Using the internet, search for a media article about a person injured while diving into shallow water. Then discuss the following as a class:

- how their life has changed since the accident
- how the accident could have been avoided
- why most accidents of this type involve young men.

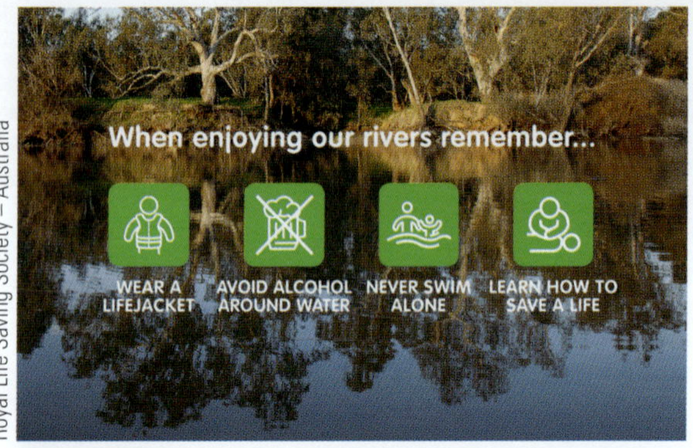

Royal Life Saving Society – Australia

Figure 6.23 Lakes and rivers can look deceptively tranquil and can present serious dangers.

Lake safety

Lakes may look calm, but are often very dangerous because of strong winds, which can create choppy conditions, and strong currents, which are likely wherever a river enters a lake. The water in lakes can also be much colder than you expect. Being suddenly immersed in cold water can cause distress and shock. If you feel cold, get out of the water immediately. Always wear a personal flotation device (PFD) such as a lifejacket when boating or doing water sports on a lake.

River safety

Rivers can be just as dangerous as lakes and the ocean. Conditions can change rapidly after heavy rain or the release of water from storage areas. What is safe one minute can be dangerous minutes later.

There are a few simple strategies for keeping yourself safe in rivers:

- Always wear a personal flotation device (PFD) such as a life jacket when boating or doing water sports on a river.
- Never swim in fast-flowing water. Check the speed first by throwing in a twig to see how fast it travels.
- If you are caught in a strong current, float on your back and travel downstream, feet first, to protect your head from hitting any objects.
- Be aware that there are likely to be objects under the water that you cannot see, such as trees, branches, rocks and discarded rubbish.
- Don't stand near the edge of overhanging river banks – they can be unstable and could crumble away.

THE RIGHT GEAR

Most sports and activities require some type of equipment; you need a board to surf, a ball to play football, a bat for cricket and so on. But you also need protective equipment, such as mouthguards, shin pads, helmets and life jackets, to reduce the risk of injury and allow you to play safely. Protective gear absorbs the impact of falls or collisions. or helps you to stay afloat.

Safety tips for equipment

Sports equipment and safety equipment can only be effective if they are used correctly and are kept in top condition. Here are some tips for making sure your equipment does the best job for you:

- Use the correct grip on bats and racquets. Holding a tennis racquet the wrong way can increase your risk of an injury to your elbow.
- Use equipment appropriate to your sport or activity and your size and age. As you grow, the equipment you use may also need to change.
- Wear the right type of shoes for your sport and replace them before they wear out. Rock climbing requires special footwear, as do a whole range of other sports, such as netball, basketball, cricket, hockey and athletics.
- Wear the same protective equipment during training that you wear during competition.
- Check your equipment regularly and replace if worn out or damaged. If you are unsure how to maintain or check your equipment, ask your coach or sporting association.

Alamy Stock Photo/david pearson

Figure 6.24 Choose the right sports shoes for the activity and the playing surface.

FACE TO FACE The right shoe

Worksheet 6.9

In pairs, discuss the following questions.

1. How are shoes made specifically for netballers different from those worn by footballers?
2. How are shoes worn by soccer players on synthetic surfaces different from those worn when playing soccer on grass?
3. How do basketball shoes increase safety for players?
4. Many students wear 'runners' to school on free dress days, but shouldn't participate in sport or physical education classes wearing these shoes. Can you give reasons for this?

INVESTIGATION

Purpose

To investigate the different parts of a sports shoe design that make it safer for specific sports.

Method

1 Choose a specific type of sports shoe (e.g. 100–400 metre athletic shoe, football/soccer boot, tennis shoe, walking shoe, basketball/netball shoe).
2 Identify as many basic parts of your chosen shoe as possible.

Discussion

1 Discuss the function/purpose of each feature you have identified. Ensure your comments are specific to the sport the shoe has been designed for.
2 Identify how each part you have discussed improves safety for the wearer.
3 Propose what might happen to a netball player wearing a basketball shoe, or a basketball player wearing a netball shoe, in terms of performance and possible safety issues?

1 Discuss why you should never swim alone using at least three sentences.
2 Describe the conditions that cause a rip to form at the beach.

1 State why you should never dive into water if you do not know the depth.
2 Consider why someone playing netball indoors should change the type of shoes they wear if they switch to playing outdoors.

1 People sometimes suffer 'wind burn' – what is this referring to? Consider if this condition actually exists, or is it simply describing something else?
2 Parts of Australia have a severely disrupted ozone layer that is said to increase UV exposure. Investigate the role of the ozone layer in protecting us from UV radiation.

Quiz
How can I increase my safety when outdoors?

HOW CAN I USE FIRST AID TO INCREASE SAFETY?

You may have heard the saying, 'failing to plan is planning to fail'. It is important to plan ahead. This includes thinking about how to reduce any risks involved in what you are doing, and considering how to respond if things go wrong.

A high percentage of patients in hospital emergency departments are young people. This is largely because they sometimes overestimate their ability, make poor decisions or are pressured by friends into doing things they otherwise wouldn't.

Having some basic first-aid knowledge and knowing how to respond in an emergency can prevent things from getting worse, and could even save someone's life.

THE DRSABCD ACTION PLAN

'DRSABCD' is an easy-to-remember action plan that will guide you through an emergency situation. It helps the first aider to assess level of consciousness and breathing in an injured person first, and then to decide on what type of basic life-support measures are needed. The letters stand for:

Danger
Response
Send for help
Airways
Breathing
Cardiopulmonary resuscitation (CPR)
Defibrillation

D — Check for **danger** Hazards/risks/safety?

R — Check for **response**

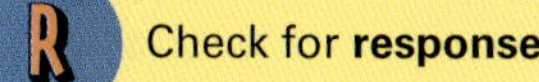

S — **Send** for help (call 000 or 112)

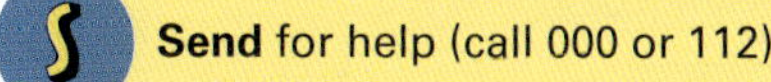

A — Open **airway**? Look for signs of life

B — Check for breathing (if no breathing, start CPR)

C — Give 30 chest **compressions**, 2 breaths (5 cycles in 2 min)

D — Attach **defibrillation** as soon as available and follow prompts

Figure 6.25 DRSABCD life-support flow chart

D = danger

Always ensure that the area is safe for you, the injured person and any bystanders before approaching an accident or emergency. If you can, remove the source of the risk or danger, or remove people from it. Only do this if there is no risk to you. For example, if you sense that an argument might be getting out of control, you might try to calm the people down. If that is not possible, move away from the dangerous situation.

R = response

If it is safe to approach, first assess whether the person is conscious or unconscious. Consciousness refers to the person's awareness of and response to their surroundings.

To assess an injured person, use verbal and touch techniques. Give a simple command such as 'open your eyes, squeeze my hand' (verbal), and squeeze both shoulders firmly (touch). If you are dealing with a child or infant, firmly rub their breastbone instead of grasping their shoulders. Never shake a small child. A person who fails to respond to these techniques is unconscious. Your action plan depends on the level of consciousness:

- **If the person is conscious**, assess for and manage other injuries or illnesses and continue with the DRSABCD action plan.

FAST FACT

An unconscious person is not aware of their immediate surroundings and cannot protect themselves from further danger. The people who are first to reach them become their most important life support.

➪ **If the person is unconscious**, roll them onto their back and call 000 or 112 (on a mobile phone) and assess their airways.

S = send for help

Call 000 or 112 for an ambulance or, if possible, ask another person to make the call while you proceed with checking the airways and breathing, and assessing the need to start CPR.

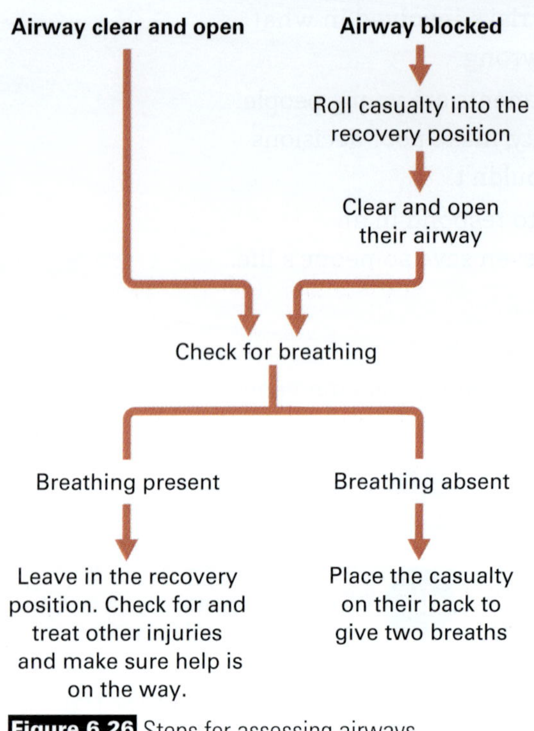

Figure 6.26 Steps for assessing airways

A = airways

Recovery position

If the injured person's air passages are not clear (they may contain food, vomit, blood or other fluids), the person must be turned into the recovery position so that you can clear the airways. These instructions show you how to put the injured person into the recovery position.

> **FAST FACT**
>
> Dial 112 on any mobile phone to contact an emergency service. This connects you to the same call service as 000. You do not need to use a SIM card, PIN or phone credit to activate the mobile, but you must be in an area that has phone service coverage. Dialling 112 from a fixed-line telephone in Australia will not connect you to emergency services.

Table 6.1 Recovery positions

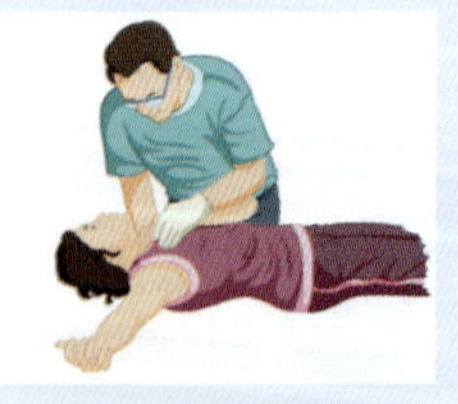 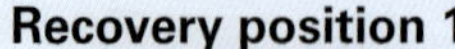 **Recovery position 1**	**1** Kneel beside the injured person (the patient). **2** Place the patient's arm, the one furthest away from you, straight out at a 90-degree angle to their body. **3** Place the patient's other arm across their chest. **4** Bend the patient's knee that is closest to you.
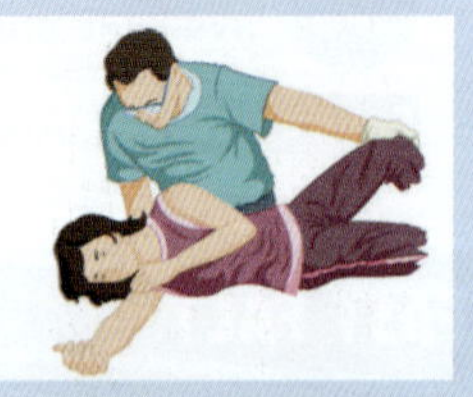 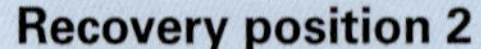 **Recovery position 2**	**5** Place one hand on the patient's bent knee and the other hand on their shoulder. **6** Gently roll the patient away from you. **7** Ensure that the patient's bent knee touches the ground. **8** Place the patient's upper arm across their lower arm.
 **Recovery position 3**	**9** Ensure the patient's head is tilted back and their face is turned slightly downward. This will allow fluids to drain from the mouth. **10** If possible, put a rubber glove on your hand, then put two fingers in their mouth and sweep the mouth clear of any foreign material, if required. Once the airway is clear and open, assess for signs of breathing.

9780170463096

Clearing the airways

If an unconscious person is lying on their back, they may not be able to breathe properly because their tongue can block the airways. This is potentially life threatening. Gently tilting a person's head back and lifting their chin up at the same time allows their tongue to move out of the airways, allowing them to breathe. Very gently move the head backwards, especially if there is the possibility of a neck, back or spinal injury. If the person has fallen, or their head has been injured, this is extremely important.

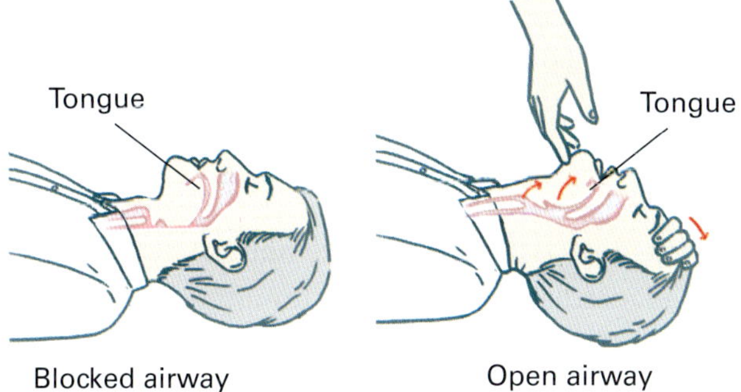

Figure 6.27 An unconscious person lying on their back may have their airways blocked by their tongue (left). Place your fingers under the chin and lift it upwards. This lifts the tongue from the back of the throat and opens the airways (right).

> **FAST FACT**
> Tilt an adult's or child's head fully back to further open the airways. Babies' heads should not be tilted backwards, as this may cause further blockage to the airway.

B = breathing

Oxygen is vital for life. Lack of oxygen over an extended period will cause the heart and brain to stop working. To find out whether the injured person is breathing, you should look, listen and feel for signs of breathing:

- **Look** for movement of the lower chest for 10 seconds.
- **Listen** for the sound of regular breathing.
- **Feel** for air escaping from the mouth or nose with your cheek.

If the person is not breathing or not breathing normally, begin cardiopulmonary resuscitation (CPR).

C = cardiopulmonary resuscitation (CPR)

CPR is recommended for a person who is unconscious and not breathing, or not breathing normally. The purpose of CPR is to keep oxygen and blood circulating around the body until emergency medical help arrives. CPR is a combination of mouth-to-mouth breathing, which forces air into the lungs, and chest compressions, which keep blood circulating around the body.

Table 6.2 What life support you should give, and when

Condition	Breathing?	Basic life support (BLS) requirements for an adult
Conscious	Yes	Place the injured person in a comfortable position, assess and manage their other injuries, and closely watch to ensure that they continue breathing.
Unconscious	Yes	Turn the injured person into the recovery position (see Table 6.1), call for help, keep their airways open and closely watch that they continue breathing. Assess and manage their other injuries.
Unconscious	No	With the injured person on their back, give two breaths. Check for signs of life.
No signs of life	No	Cardiopulmonary resuscitation (CPR) is needed. Start CPR with 30 compressions and two breaths. Work towards giving 100 compressions in one minute.

Table 6.3 Recommended CPR for adults, children and infants

	CPR	Method of compressions	Depth of compressions
Adults and children 9–17	30 compressions and two breaths	Two-handed pressure over middle of chest	One-third of the depth of the chest
Child (1–8 years)	30 compressions and two breaths	Two-handed pressure over middle of chest	One-third of the depth of the chest
Infants (0–1 year)	30 compressions and two breaths	Two-finger pressure over middle of chest	One-third of the depth of the chest

Note: This is a guide only. Aim to give 100 compressions each minute.

Weblink Watch the video on how to give CPR to an adult

D = defibrillation

WB Worksheet 6.10

An automated external defibrillator (AED) is a device used to restore normal heart rhythm to a patient whose heart has stopped (cardiac arrest). An automated external defibrillator is applied outside the body. It automatically analyses the patient's heart rhythm and advises the first aider whether or not a shock is needed to restore a normal heartbeat.

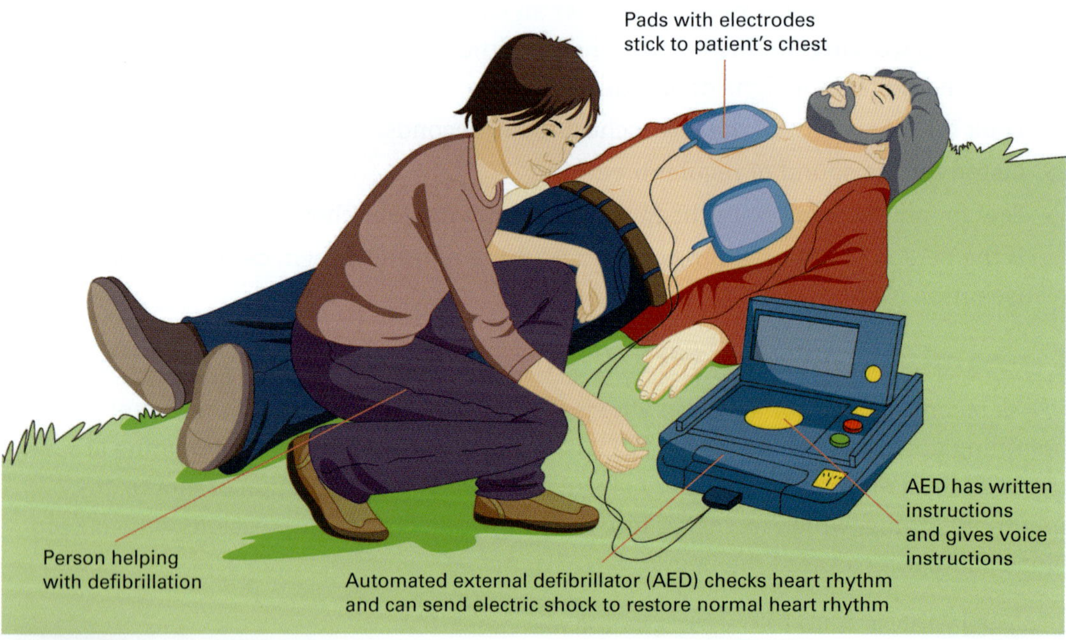

Figure 6.28 Using an AED is simple – follow the voice prompts and you could save someone's life!

The AED is very simple to use. The first aider follows a set of voice prompts with visual guides. Don't be afraid, and always remember that it's better to do something than nothing in these situations.

Once you have used DRSABCD and assessed that a person is conscious (they have a heartbeat and are breathing), their other injuries or conditions (e.g. bleeding) may need your attention.

SPORT INJURIES

ligament fibrous tissue that connects two bones, especially at a joint (knee, ankle, etc.)

If you play sport, you may have had a common sports injury such as a sprain, strain or knock. A sprain happens when you overstretch or tear a **ligament**. The joint (e.g. an ankle) is affected, but there is no dislocation or break in the bone. Symptoms include rapid pain, swelling, bruising and a warm feeling at the injured site. A strain is an injury to the muscles or **tendons** and has similar symptoms to a sprain. Strains tend to happen through overtraining or overuse, often when muscles 'overstretch'.

tendon a structure that connects muscle to bone

The treatment for sprains, strains and other sports injuries depends on the injury and on the amount of damage. Mild injuries may only need first aid. Some sporting injuries and sprains may need assessment and treatment from a doctor or paramedic. Some sprains need a cast or rigid protection to assist with recovery. Serious sprains and strains may require surgery, if the body is so damaged it is unlikely to repair itself.

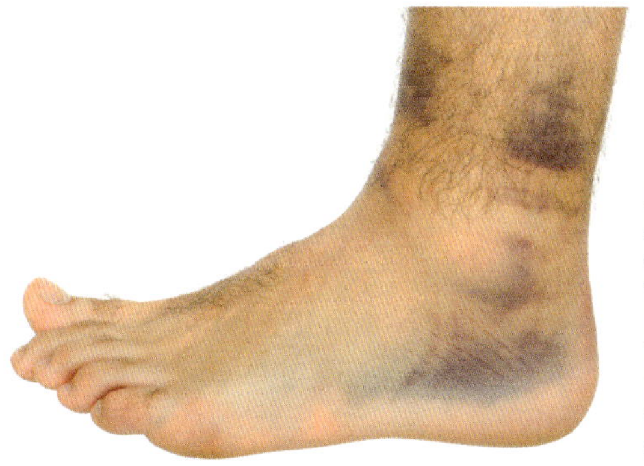
Shutterstock.com/Luis Santos

Figure 6.29 An ankle sprain can cause strain on the Achilles tendon.

If the injury is not serious, stop what you are doing and move to a safe place (or ensure that where you are is safe). Follow the **RICER** steps for a quick recovery from minor sprains, strains and other sporting injuries.

Worksheet 6.11

Table 6.4 RICER

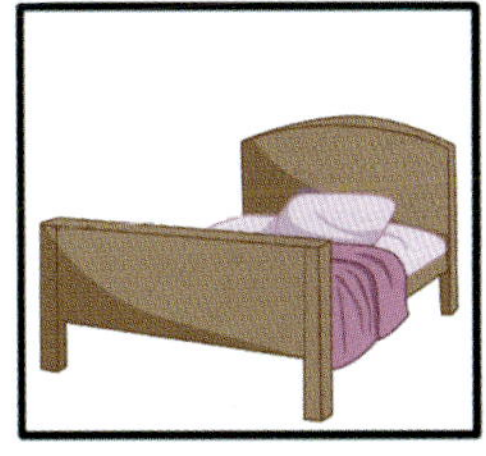		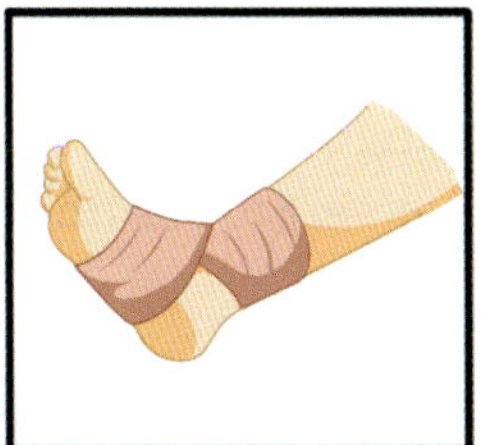	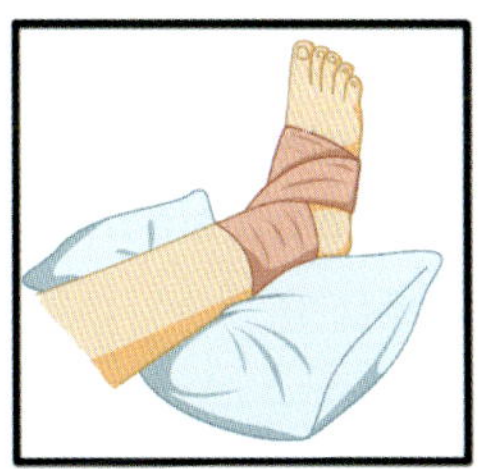	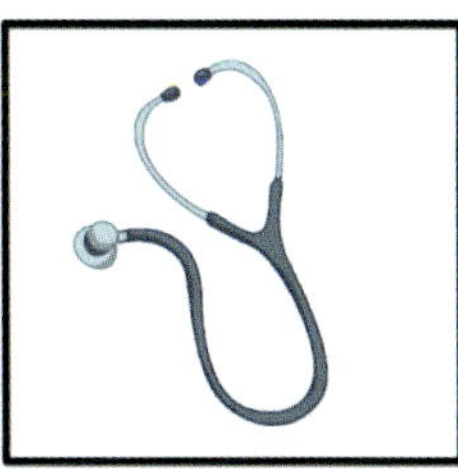
R = rest	**I = ice**	**C = compression**	**E = elevation**	**R = referral**
Rest reduces further damage. Move as little as possible to avoid more injury. Don't put any weight on the injured part of the body.	Apply a cold pack or ice to the injury for 20 minutes every two hours. Continue this treatment for the first 48 to 72 hours. Ice cools the tissue and reduces pain, swelling and bleeding. Place a cold pack wrapped in a towel onto the injured area. Do not apply cold packs directly to the skin. Extra care must be taken with people who are sensitive to cold (such as children) or who have blood circulation problems.	Wrap with a flexible crepe or elastic bandage, covering the injured area and the areas above and below. Compression reduces bleeding and swelling. Ensure the bandage is not too tight.	Elevate the injured area to stop bleeding and swelling. Place the injured area on a pillow for comfort and support.	Refer the injured person to a qualified professional such as a doctor or physiotherapist for precise diagnosis, ongoing care and treatment. A full recovery is then more likely.

FAST FACT

Even though RICER is easy to remember and appears in most first-aid books and manuals, more and more sports trainers and physicians are applying compression first, and then ice. The advantage of this approach is that direct contact with the skin is avoided, and when the ice is removed, the compression limits blood flow and minimises further damage.

Getting better

If you've been injured and you try to come back to sport and physical activity too soon, you risk reinjuring yourself, maybe more seriously than before. Don't let anyone, including your parents, friends, teachers or even coaches, pressure you into returning to training or competition before you have fully recovered. Your doctor, physiotherapist or sports trainer will give you specific advice on when you should return to your sport or activity.

concussion a brain injury that occurs following a knock to the head. It can alter brain function and cause unconsciousness.

hypersensitive extremely sensitive

trigger an event that sets off another event

Taking time to heal is particularly important if you've had a **concussion**. Lots of athletes try to return to sport or activities too quickly after getting a concussion; some footballers even try to return to the field after only a short time out because they think they're fit to keep playing. But jumping back into the game too soon puts a player at greater risk of suffering another concussion, as the brain can still be healing. Repeat concussions can lead to dangerous and sometimes delayed brain injury. Always get clearance from a medical expert to resume sport or physical activity if you've had a concussion.

ASTHMA

iStockphoto/davidf

Figure 6.30 Asthma puffer with spacer

People with asthma have **hypersensitive** airways. When exposed to certain **triggers**, such as dust, pollen or exercise, the airways become narrow, making it harder to breathe out. An asthma attack can develop over a few minutes or a few days. The symptoms of a mild asthma attack include chest tightness, coughing, wheezing and shortness of breath. These symptoms can quickly worsen to include gasping for breath, inability to speak and blue colouring around the lips. There may be little or no improvement after using inhaler medication. If a person is having an asthma attack, you should commence asthma first aid.

FAST FACTS

- More than 2.7 million Australians, or just over 11 per cent (about one in 10 adults and one in nine children) have asthma.
- More boys than girls have asthma, but it is more common in women than men aged over 15.
- Asthma is more common among First Nations Australians than other Australians.

Asthma first aid

The following four steps are known as the 4×4 first-aid plan and are an easy way to remember what to do if someone is having an asthma attack.

Step 1 Sit the person upright, be calm and reassuring. Do not leave the person alone.

Step 2 Without delay, shake a blue reliever puffer, using a spacer if available. The spacer increases the amount of medication inhaled into the lungs. Give four separate puffs, one puff at a time. Ask the person to take four breaths from the spacer after each puff.

Step 3 Wait four minutes. If the person still cannot breathe normally, give four more puffs.

Step 4 If there is still no improvement, or you are concerned at any time, call 000 or 112 immediately. Tell the operator that the person is having an asthma attack. Keep giving four puffs every four minutes while you wait for emergency assistance.

CASE STUDY: AIR POLLUTION AND ASTHMA

Identify

Asthma is the 9th leading contributor to the overall burden of disease in Australia, and is the leading burden of disease in children aged 5 to 9.

Understand

Air pollution is now recognised in the Australian Government's 10-year National Preventive Health Strategy, with other long-standing health risks such as smoking and obesity.

CEO of Asthma Australia, Michele Goldman said they were buoyed to see their recommendation to include air quality and other environmental health risks reflected in the Strategy.

'Much like unhealthy food and water, breathing polluted air directly affects someone's health, such as increasing risk of getting asthma or cancer,' Ms Goldman said.

The following is a summary of some of the findings resulting from the 2019/2020 Black Summer Bushfires:

1 PEOPLE WITH ASTHMA EXPERIENCE SERIOUS HEALTH IMPACTS FROM BUSHFIRE SMOKE

94% of people with asthma reported symptoms during the Bushfire Crisis

People with asthma were 4 times more likely to attend Emergency Departments or be hospitalised and 7 times more likely to report steroid use

2 CHILDREN AND YOUNG PEOPLE WITH ASTHMA WERE MOST IMPACTED BY BUSHFIRE SMOKE

2x They were twice as likely to report attending the Emergency Department or being hospitalised because of asthma symptoms

3 PERIODS OF BUSHFIRE SMOKE EXPOSURE INCREASES ANXIETY AND DEPRESSION IN PEOPLE WITH ASTHMA

"I have been on edge. Unable to exercise which increased my anxiety."

"I sometimes feel useless, seeing how other people seem to be able to function normally when I'm struggling to do basic tasks."

4 SUSTAINED EXPOSURE TO BUSHFIRE SMOKE REDUCED PARTICIPATION IN EVERYDAY ACTIVITIES, PARTICULARLY FOR PEOPLE WITH ASTHMA

People with asthma were significantly more impacted physically, financially and socially, compared to people without asthma

Two thirds (66%) of people with asthma reported they had reduced capacity to participate in daily activities

5 PEOPLE WITH ASTHMA ARE TWICE AS LIKELY TO EXPERIENCE FINANCIAL STRESS DUE TO BUSHFIRE SMOKE

25% of people with asthma reported experiencing financial stress

13% of people without asthma reported experiencing financial stress

10% of people with asthma reported lost salary during the Bushfire Crisis

6 CURRENT PUBLIC HEALTH MEASURES AGAINST BUSHFIRE SMOKE DO NOT PROTECT PEOPLE FROM EXPERIENCING ASTHMA

People with asthma were more likely to report taking action to reduce their exposure to bushfire smoke including:

- staying inside with doors and windows shut
- using an air conditioner with the recycled setting
- using a facemask
- using a HEPA air cleaner

Source: Asthma Australia, Bushfire Smoke Impact Survey 2019 –2022, https://asthma.org.au/what-we-do/advocacy/smoke-survey-results/

Figure 6.31 Six important facts about asthma

Discuss

1 Discuss how poor air quality in Australia can restrict day to day participation in activities for Australian children and how this impacts their families.

2 List three measures the Australian government can take to improve participation rates in physical activity for children who have asthma whilst they are in school.

3 Investigate any apps that may exist to provide information about air quality. Rate your top three air quality apps in terms of information provided and user friendliness.

BLEEDING

Bleeding is usually associated with cuts and abrasions, but injuries to the body can also result in bleeding inside the body (internal bleeding). This can range from minor (bruising) to massive (life-threatening bleeds). The purpose of first aid for severe external or internal bleeding is to slow the loss of blood until emergency medical help arrives. Always use gloves, if available, when treating someone who is bleeding.

Minor bleeding

tetanus a serious disease caused by bacteria that enter the body through a cut or wound

Small cuts and abrasions that are not bleeding much can be managed fairly easily. Press down with your hand for about 30 seconds to stop the bleeding, then clean the wound with water or saline solution, if available. If you cannot clean the wound properly, seek further medical help. Cover the wound with a clean dressing such as a bandaid or a gauze pad and bandage. Be aware that deep, narrow cuts (e.g. from stepping on a nail, or a cat bite) may cause **tetanus**. If you are not sure if your tetanus immunisation is current, see your doctor.

Alamy Stock Photo/Angela Hampton Picture Library

Figure 6.32 Always tilt the head forward to reduce backflow of blood from a nosebleed.

Nosebleeds

Bleeding from the nose is not usually severe, unless it is associated with a head injury such as being hit with a hockey stick, being punched or falling heavily on your head. First-aid suggestions for nosebleeds include the following:

- Sit the person upright and ask them to tilt their head forward (resist the temptation to tilt the head back to slow or stop the flow of blood, as this sends it down the nasal passages and into the roof of the mouth, and may cause breathing difficulty).
- Use your thumb and forefinger to pinch the nostrils shut and hold for at least 10 minutes.
- Release the hold gently and check for bleeding. If bleeding continues, pinch the nostrils shut for another 10 minutes.
- Get medical help if bleeding continues beyond 20 minutes.

Newspix/Michael Klein

Figure 6.33 Major blood loss is often associated with shock.

Severe external bleeding

Even a small cut can result in severe external bleeding, depending on where it is on the body. Severe bleeding can lead to shock, a serious, life-threatening condition where the injured person no longer has enough blood circulating around their body. To manage severe external bleeding:

- Check for danger before approaching the injured person and, if possible, send someone else to call for an ambulance.
- Lay the person down with the injured area above the level of their heart (if possible).
- Ask the person to apply direct pressure to the wound with their hand or hands to stem the blood flow. If they can't do this, apply direct pressure yourself.
- If possible, pull the edges of the wound together before applying a dressing or pad. Secure it firmly with a bandage.

- Do not remove any object embedded in the wound. Apply pressure around the object.
- Do not remove initial dressings, even if they become saturated. Add fresh padding over the top and secure with a bandage.

Internal bleeding

The most common type of visible internal bleeding is a bruise, which occurs when blood from damaged blood vessels leaks into the surrounding skin. Some types of injury can cause visible bleeding from an opening such as the mouth and ears, but some internal injuries can cause bleeding that remains inside the body; for example, within the skull or abdominal cavity.

Listen carefully to what the person tells you about their injury – where they felt the impact, for example. The signs and symptoms that suggest concealed internal bleeding depend on where the bleeding is inside the body, but may include:

- pain at the injured site
- swollen, tight abdomen
- nausea and vomiting
- pale, clammy, sweaty skin
- breathlessness
- unconsciousness.

First aid cannot manage or treat any kind of internal bleeding, but you can help by treating or preventing shock and calling for medical assistance:

- Check for danger before approaching the person
- If possible, send someone else to call for an ambulance.
- Check levels of consciousness.
- Lay the person down on their back if conscious.
- Cover them with a blanket or something to keep them warm.
- If possible, raise the person's legs above the level of their heart.
- Don't give the person anything to eat or drink.
- Offer reassurance. Manage any other injuries, if possible.
- If the person becomes unconscious, place them on their side. Check breathing frequently and be ready to begin cardiopulmonary resuscitation (CPR) if necessary.

Fairfax Syndication/Robert Pearce

Figure 6.34 Suspected internal bleeding requires urgent medical attention.

> **FAST FACT**
>
> Most sports have a 'blood rule', which requires any person who is bleeding to be removed from the field, both so they can be treated and to protect others from any transfer of blood. If any blood spills on the court, field or playing surface, this should be cleaned up in order to minimise the risk of cross-contamination and infection.

Protect yourself and others

It's important to protect yourself and others when treating any form of bleeding. Any break in the skin will allow blood and other fluids to be lost, which could contaminate first aiders. It will also allow germs to enter the body. If the wound is minor, the aim of the first aider is to prevent infection. Severe wounds may be very daunting to deal with, but the aim is to prevent further blood loss and minimise shock.

Any open wound presents a risk of infection. It is important to maintain good hygiene procedures to prevent cross-infection between you and the injured person:

- If possible, wash your hands with soap and water before and especially after administering first aid. Dry your hands thoroughly before putting on gloves.
- First-aid kits contain gloves. Always put on gloves beforehand if available. If not, improvise and try to provide a barrier between yourself and the casualty's blood (if practical and possible).
- Do not cough or sneeze over the wound.

Quiz
How can I use first aid to increase safety?

1 If someone grazes their knees or elbows playing baseball or softball, summarise the steps that should be taken to provide first aid.
2 Discuss why the blood rule should be applied strictly during any type of sport or recreational activities.
3 Why should compression be applied before ice for a rolled ankle?

1 Outline why a spacer is better for asthma sufferers than using just a puffer.
2 If you have a nosebleed, why is it a bad idea to tilt your head back to slow the rate of blood loss?
3 Discuss why people who regularly fight are placing their brains at high risk of a delayed brain injury.

1 Contrast how the activities performed by a paramedic are very different from those provided by a first aider.
2 A 'corkie' is a familiar name for an injury that often occurs in the upper thigh and results in internal bleeding. Why is internal bleeding potentially more dangerous than external bleeding?
3 If you are at a party and see someone get hit in the head, but they seem to be alright, how should you care for them?

CHAPTER 6 REVIEW

1 Identify your own personal network of trusted adults who you can turn to if you are not feeling safe.

2 In most cases if you don't feel right in a setting or situation, it probably isn't safe and requires a quick evaluation. List three signs that you might not be safe.

3 When you feel that your safety could be at risk and things could go wrong for you or your friends, list four strategies you might put into place to improve your safety.

4 If you are experiencing emotional, physical or sexual abuse, this requires the intervention of a trusted person in your support network. It's not your fault! Who are the people that make up your support network?

5 Bullying occurs when a person, or group, uses their power over another person or group to threaten them or force them to do something. If bystanders take no action, how does this empower the bully further?

6 Discuss how social networking can be a great way to stay in contact with friends and family. This can be very important for socially or physically isolated individuals, such as those with a physical disability or students living in rural or remote areas.

7 Social networking needs to be used appropriately – it's easy to forget who you are communicating with online. Before posting anything online, think about who will be able to access or see it. What are three considerations you need to make before posting something online?

8 In most cases, content that is posted online stays there forever. Discuss what having a digital footprint means.

9 Around water, what should you do to check that it's okay to swim?

10 How should you treat someone with a nosebleed?

11 Name at least two signs of internal bleeding.

THE GREAT OUTDOORS

7

IN THIS CHAPTER

You will have the opportunity to explore ways in which you can use the outdoor environment where you live and go to school. This might be to play a game on the way home in the local park or to take a more structured approach, challenging yourself to get involved in outdoor physical activities such as surfing or mountain biking. The focus will be on participating in activities that challenge you individually and/or as a member of a team. It will also remind you of the importance of teamwork and looking after the environment in which you live. You will develop skills in various settings and the confidence to build on these skills so that you can challenge yourself further physically. Working in a team or group may take you out of your comfort zone and challenge the way you interact with your peers.

Shutterstock.com/Nicetoseeya

By the end of this chapter, you should be able to:

- understand what outdoor recreation is and the benefits of being active in the outdoors
- understand minimal impact in outdoor recreation activities and be willing to apply this when participating in activities outdoors
- use basic skills to participate in a variety of outdoor and challenge activities such as Ultimate Frisbee, Ultimate disc golf and various initiative and team challenges
- read a map, plan and navigate a route through various environments using existing routes or tracks
- understand what orienteering is and develop those skills
- participate in a range of initiative games, in pairs or teams
- create team challenges and find solutions to challenges
- evaluate how you performed in each activity.

WHAT IS OUTDOOR RECREATION?

Quiz
Pre-chapter

Before you start, take the pre-chapter quiz to find out how much you already know.

Video
Case study: What are the benefits of being active outdoors?

Outdoor environments can be grouped into four key areas:

- aquatic environments, including lakes, rivers/creeks and beaches/sea
- bush environments
- alpine/mountain environments
- urban/park environments.

iStock.com/dstephens

Alamy Stock Photo/Design Pics Inc

Figure 7.1 Outdoor recreation opportunities

FACE TO FACE Class discussion

Worksheet 7.1

What does outdoor activity mean to you? Provide examples of where you have participated or could participate in outdoor activities. Your responses will largely depend on where you live. People who live near the coast (the majority of the population of Australia) will experience a different range of activities from those in inland areas. They will also depend on how you interact with your local area and your experiences further afield, such as places you might have been for holidays.

UP AND MOVING

Posters

In pairs or small groups, using a 'Y chart' graphic organiser, explore what outdoor activity means to you. Think about what it looks like, feels like and sounds like. Present your information on a poster to be displayed around the room. Examine each poster and identify differences in how people, even in the same area, view the great outdoors. What are the similarities and differences between the views that are displayed?

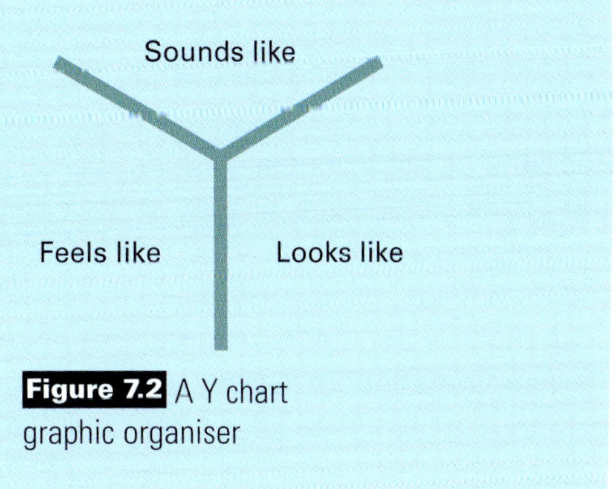

Figure 7.2 A Y chart graphic organiser

Scaffold
Y chart

subalpine the regions at the foot of mountains

CASE STUDY THE AUSTRALIAN ALPS

Identify

Australia's main alpine and **subalpine** areas are the Snowy Mountains in New South Wales, the Bogong High Plains in Victoria, and central and southwestern Tasmania. They occur above about 1400–1500 m on the mainland, and 700–1000 m in Tasmania. Although Australia's mountains are relatively low by global standards (Mt Kosciuszko, the continent's highest peak, rises only 2228 m above sea level), there is true treeless, alpine vegetation above the climatic treeline ...

iStock.com/Svibrav

Figure 7.3 Mount Kosciuszko in the Snowy Mountains, New South Wales, is Australia's highest mountain.

Understand

The alpine climate is cold, wet, snowy and windy, with a short growing season. The soils are highly organic and can hold tremendous amounts of water. Alpine plants are short: mostly tussock-forming snow grasses, rosette-forming herbs such as snow daisies, and ground-hugging shrubs ... The Australian Alps are hugely important for conservation, water production and recreation. Most alpine areas are within national parks and are home to many unique plants and animals ...

Major rivers – such as the Murray, the Murrumbidgee and the Snowy – begin in the Alps. Water from alpine catchments is worth A$9.6 billion a year to the Australian economy. Millions of people visit every year to camp, walk, ski, ride and take in the scenery. The Alps are one of Tourism Australia's 'National Landscapes' and the local tourism industry is worth hundreds of millions of dollars annually.

... the alps face multiple threats, including global warming, invasive species, disturbances such as fire, increasing pressure from human recreation, and unsound ideas about how to manage the high country. The climate has already changed. Since 1979, average temperatures during the growing season on the Bogong High Plains have risen by 0.4C, while precipitation has decreased by 6%. Since 1954, the depth and duration of the snowpack in the Kosciuszko region have declined.

Rising temperatures are a serious problem because the Australian Alps are relatively low mountains and the alpine species, already at their distributional limits, have nowhere else to go. Woody vegetation may increase – the treeline may rise and shrubs are likely to expand into grasslands and herb fields, which may make the landscape more prone to fire. Mainland alpine ecosystems can regenerate after large fires. But Tasmania's alpine vegetation is extremely fire-sensitive, and more frequent fire is likely to be detrimental to all alpine ecosystems.

… feral animals and plants are a clear threat and will become more difficult to manage in the future without concerted action now. Horse and deer numbers are increasing with alarming speed. These animals are occupying habitats well above the treeline. Many alien plant species have invaded the alps over the past half-century, a trend likely to be exacerbated by climate warming …

There is cause for hope, however. The Australian Alps are on the National Heritage List, which is protected by federal law. There is also still time. The world is acting on climate change. Some species may adapt genetically, while some likely changes to vegetation may happen slowly. Scientists and land managers are working together to anticipate and manage change in the alps.

Source: 'EcoCheck: Australia's Alps are cool, but the heat is on', by Dick Williams and James Camac, *The Conversation*, 29 April 2016, https://theconversation.com/ecocheck-australias-alps-are-cool-but-the-heat-is-on-56997

Discuss

1. Describe some features of an alpine environment.
2. List three ways human recreation can possibly threaten the Australian Alps.
3. How does being on the National Heritage List aim to protect the Australian Alps?
4. Research if there are any other outdoor environments in Australia that are on the National Heritage List.

WHAT RECREATIONAL ACTIVITIES CAN I DO IN NATURAL/OUTDOOR SETTINGS?

The word 'recreation' was first used in the 14th century. It meant 'refreshment or curing of a sick person'. In the past, people spent their time going to work, completing daily chores, eating and sleeping. There was very little free time left to relax or do anything else. 'Free time', traditionally called 'leisure time', is time that is free from obligations (such as paid and unpaid work) and doesn't include tasks that are necessary for existing, such as sleeping and eating.

People still work long hours but technology has provided more leisure time. For example, travelling using a car, train or bus has reduced the time taken to get to work or go shopping. As society has evolved, so have the number of activities that people can choose to do in their leisure time. People can use this time to play sport, watch TV, read a book, participate in a hobby or play games on the internet. The list of leisure-time activities is endless, as are the reasons why people choose to do these activities. They range from wanting to use up excess energy by playing sport or physical training, to wanting to relax and unwind from a stressful day.

Recreational activities are activities that people choose to do in their leisure time. They can be active and inactive, indoor and outdoor, healthy and unhealthy. Generally, recreation activities are non-competitive, but some more adventurous recreational activities have a competitive element to them.

As well as traditional recreation activities, there are some more extreme or adventurous activities that are typically more demanding physically, mentally and technically because of the more extreme environments they occur in and/or because of the length or difficulty of the event. Some examples are ice climbing, glacier climbing, ultramarathons, Ironman triathlons and extreme obstacle events such as 'Tough Mudder'.

Getty Images/Robert Perry – PA Images

Figure 7.4 Tough Mudder

INVESTIGATION  TOUGH MUDDER

Purpose

To investigate the motivation of people who enter the 5 kilometre or 16 kilometre 'Tough Mudder' outdoor obstacle course and to learn about possible benefits obtained by participating in this team challenge.

Method

1 Conduct online research to find testimonials and personal accounts/reflections from people who have competed in, or are about to compete in, Tough Mudder Australia.
2 Try to find personal reflections and examples of the benefits people discuss as a result of taking part in this challenge.

Discussion

1 Discuss two common motivations behind people entering Tough Mudder.
2 Would you consider doing the 5K challenge? Why or why not?
3 List at least three benefits people who complete Tough Mudder say they experience.
4 If you entered the 5K challenge, what do you believe would be the most memorable part of the experience?
5 Tough Mudder organisers often choose remote areas to set up their courses and use a large amount of constructed obstacles in preference to naturally occurring ones such as trees, rocks and other vegetation. Discuss two different reasons why this may be their preference, or a condition placed upon them by the land owner(s).

FACE TO FACE Being active

For each of the following outdoor environments, think about which 'active' activities are usually associated with the environment. List as many as you can.

- Aquatic environments, including lakes, rivers/creeks and beaches/sea
- Bush environments
- Alpine/mountain environments
- Urban/park environments, for example, '**parkrun**'

parkrun free, weekly 5-kilometre timed runs held in parks around the world

BENEFITS OF BEING OUTDOORS

Being involved in active outdoor recreation or adventure activities has many benefits. These are mainly for the individual, but society benefits as well. If people have a passion for the outdoors, they are more likely to look after it and **advocate** for it. These benefits can be physical, social, emotional and spiritual:

advocate to support, speak out about or recommend something

- physical: improvements to physical health
- social: relationships with family and friends or peers
- emotional: such as self-esteem and mental wellbeing
- spiritual: awareness of being part of a bigger system (universe).

Individual benefits

Individual benefits of outdoor recreation include:

- connection with other people and communities
- enhanced capacity to nurture self
- positive impacts on cardiovascular health, cancer prevention, diabetes and mental health
- increased skills (physical, problem solving, self-awareness).

Community/social benefits

Participation in outdoor recreation and adventure recreation contributes to many positive outcomes in the community, including:

- improved quality of life
- healthier families
- enhanced health outcomes
- improved physical, mental, social, community and environmental health and wellbeing.

Worksheet 7.2

Environmental benefits

Two benefits to the environment of outdoor recreation are increased individual connections to nature and an enhanced awareness of the natural environment.

CASE STUDY  SPENDING TIME IN NATURE CAN IMPROVE YOUR MENTAL HEALTH

Identify

A growing amount of research has suggested spending time in nature can improve your mental health. Whether it is a quick stroll through your local park or going for a dip in the ocean, studies suggest it can have a restorative effect on our minds.

Understand

Australia's national parks are worth $145 billion to the economy from the improved mental health of people who visit them, according to research.

A team from Griffith University conducted surveys of nearly 20,000 people who used national parks in Queensland and Victoria, and were able to extrapolate the figure from there.

Professor Ralph Buckley, who headed the research, said: "When we go to a national park and feel better because of it, this number measures that value."

About 10 per cent of Australia's GDP is spent on mental health issues, and the researchers estimate this could be 7.5 per cent higher if there were no national parks. National parks are believed to offset $29 billion in health costs in Queensland, while the team extrapolated the global figure to $8.7 trillion.

Dr Ali Chauvenet, one of the co-authors of the study, said: "Protecting national parks is important not just because of traditional ideas such as biodiversity and protecting the environment, but because people's wellbeing and mental health would be affected if these were not there," she said.

Professor Buckley said he hoped the findings would give a good indication of the inherent value of national parks, which he feared was becoming overlooked.

"If you encourage people to enter parks and you pay the parks service to run those parks then the payback in terms of improved mental health and productivity is much, much bigger than anything you would get out of tourism development."

"It's possible that park visits could then become a routine part of the healthcare system, prescribed by doctors and funded by insurers," Dr Chauvenet said.

iStock.com/Blue Planet Studio

Figure 7.5 A new study has found national parks are worth billions in mental health benefits to the Australian economy.

Source: 'EcoCheck: Australia's Alps are cool, but the heat is on', by Dick Williams and James Camac, The Conversation, 29 April 2016, https://theconversation.com/ecocheck-australias-alps-are-cool-but-the-heat-is-on-56997

Discuss

1 Discuss how walking through a national park or track can improve someone's mood.

2 State what you think this means: "The payback in terms of productivity is much bigger".

3 a Mindfulness requires people to disconnect. How does surfing, or participation in other outdoor pursuits, allow people to disconnect?

b Identify an example of an outdoor activity that allows you to be mindful. Specifically, discuss how it allows you to disconnect from school, part-time work, family, etc.

WELLBEING CHECK IN

BEING MINDFUL IN NATURE

Identify

With the right viewpoint, you can discover lots of things you haven't noticed before. You might even discover something about yourself.

This activity has been developed in collaboration with Dr David Bakker of MoodMission

Understand

Mindfulness is a way of focusing the mind and letting go of distractions or unhelpful thoughts. We can do anything mindfully to increase our appreciation of it by paying closer attention. If we get distracted from what we're mindfully focusing on, we just put the distraction to one side and gently refocus. Research has shown that spending time in nature has positive mental health effects. But it can be hard to get these benefits if we're distracted from our surroundings.

Practise

1 Think of a familiar spot where you can spend some time outdoors. It could be in a garden, at a park, in the bush or somewhere else surrounded by trees and plants.
2 Go to that place and leave distractions behind. That means put your phone away, no headphones, etc.
3 While you are there, try to pay special attention to your surroundings. Look at the trees and count the branches. Notice their shapes. Notice how many colours you can see in each plant. Compare these colours with neighbouring plants.
4 You will probably get a bit distracted while doing this. You might have distracting thoughts about yourself, other people, the past, the future, etc. That's okay, and totally normal. When you get distracted, put the distraction to one side and then refocus on your surroundings. This will help to let go of any anxieties or negativity.

Reflect

What did you notice while you were outside? Did you discover something about how your mind works? A lot of people learn something about themselves when they try to focus on something other than themselves.

INVESTIGATION

OUTDOOR RECREATION IN AUSTRALIA

Purpose

To investigate the top five to eight types of recreational activities in which Australians participate, in natural/outdoor settings.

Method

Students should undertake their own research by searching sites such as the Australian Bureau of Statistics (ABS), the Australian Institute of Health and Welfare (AIHW), the Outdoor Council of Australia (OCA) or the Australian Sports Commission. Try to make your search specific when entering key words on the sites' search page(s).

Discussion

It is likely that the different websites will each have slightly different information, depending on when they conducted their research, but there should be some similarity in the top five to eight outdoor activities Australians participate in.

1 Provide a list of the top five to eight outdoor/recreational activities Australians participate in.
2 Describe any gender differences you may have discovered during your research.
3 Some outdoor pursuits rank equally regardless of the state people are in, while others differ from state to state. Discuss any state-to-state differences and explain why the participation rates might differ.
4 Discuss the role outdoor education programs run by schools play in encouraging people to participate in recreation/outdoor leisure activities once they leave school.

PROMOTING MINIMAL IMPACT

Natural environments provide an excellent way for people to learn about themselves, the world they live in and how humans and nature interact. The places you choose to go outdoors to camp, walk, ride, climb, paddle, fly and sail are very special. The uniqueness of the place adds to the outdoor activity experience. As more people discover the pleasures of outdoor recreation, it becomes more apparent that the bush, beaches and waterways need care and protection to ensure that they are available to many more generations to enjoy.

Everyone participating in outdoor recreation has a responsibility to follow a **minimal impact** code of practice. By observing a few simple rules, everyone can make a difference and your special outdoor places will stay special.

minimal impact reducing the potential damage to the environment

Seven principles of minimal impact

Minimal impact can be promoted by following some basic steps. 'Leave no trace Australia' recommends seven principles to follow when using the outdoors.

1 Plan ahead and prepare – take time before your trip to be clear about your intentions and prepare carefully for all aspects of the trip.
2 Travel and camp on durable surfaces – whenever possible, be aware of the damage you can cause to plants and nature. Use established paths, tracks and campsites.
3 Dispose of waste properly – 'pack it in, pack it out' is a useful saying. Any rubbish that you carry into the bush must also be taken out with you.
4 Leave what you find – people visit natural areas for many reasons and it is vital that these areas remain undisturbed.
5 Minimise campfire impacts – ensure campfires are only used in designated areas, if they are to be used at all.
6 Respect wildlife – your visit will have some sort of impact on wildlife, so keeping this to a minimum will contribute to peaceful coexistence.
7 Be considerate of your hosts and other visitors – be aware of other people using the natural environment and the traditional owners of the land.

CASE STUDY ⇨ DANGEROUS LITTERING

Identify

Queensland police say a discarded cigarette likely sparked the bushfire which destroyed 11 homes and the historic Binna Burra Lodge in the Gold Coast hinterland in September 2019.

Understand

Officers said two local teenagers – aged 17 and 19 – had been questioned about the incident and detectives had determined the fire was an accident.

'A prosecution will not be commenced against those persons ... they are afforded privacy just like anyone else in their position,' a QPS spokesperson said. Last week, police stated they would not reveal what sparked the blaze as they feared those responsible could be vilified in the small, tight-knit community. But after a backlash from locals, authorities have now released more details.

Binna Burra Lodge chairman Steve Noakes welcomed the police decision to be more transparent with the community.

'It's nice to know the actual cause of it,' he said. 'Maybe it helps to get the message out about how super careful we all have to be at this time, right across Australia with this terrible bush fire season we're having.' ...

'Such a small simple mistake can have such severe impacts on people and their lives,' he said.

'People really have to take so much care not to act recklessly with anything that can cause a little tiny fire that can grow into a large devastating bushfire.' ...

Queensland introduced laws for dangerous littering in 2011, for litter that is likely to cause harm to a person, property or the environment.

It includes throwing a lit cigarette butt onto dry grass in high fire danger conditions, and individuals can be fined $533.

Discuss

1. Which one of the leave no trace 'laws' does littering address?
2. Why is it important that people extinguish campfires before going to sleep in the bush or moving on from a campsite where campfires have been lit?
3. Why do you believe the 17- and 19-year olds were not charged by police, but may have been fined?

CASE STUDY ⇨ LEAVE NO TRACE – MINIMISE IMPACT WHEN VISITING PARKS AND FORESTS

Identify

We all love to visit Queensland's parks and forests, and we want to leave them for our children, and theirs, to enjoy. How can we make sure we leave no trace?

9780170463096

Planning

- Find out about any specific regulations for the area you will visit.
- Prepare for extreme weather, hazards and emergency.
- Ensure you are aware of the minimal impact guidelines.
- Pack all equipment you will need, such as plastic bags for carrying out waste.

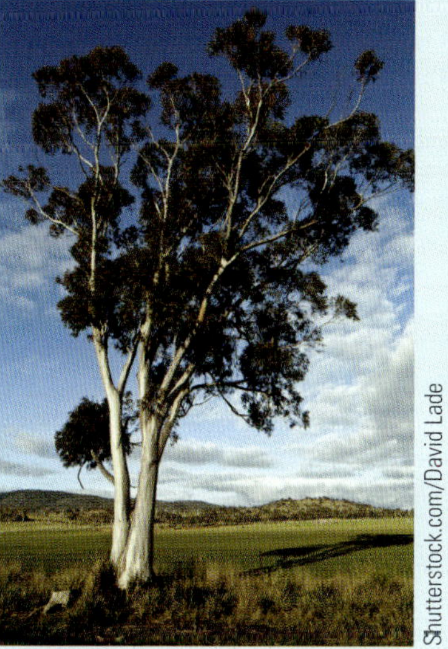

Shutterstock.com/David Lade

Figure 7.6 Parks and Forests QLD has guidelines to minimise impact.

On arrival

A park ranger may tell you about minimal impact and safety messages, and about the features of the park.

- Stay on established tracks at all times.
- Observe and listen.
- Leave things where they are.
- Remember: healthy animals only eat their natural food; don't feed any animals.
- All native plants, animals, rocks, historical and cultural remains are protected by law in national parks. Ensure that they remain undisturbed.

Tread lightly

Staying on the track minimises erosion, limits spreading of diseases and weeds, and keeps you from getting lost.

- Stay on the track at all times. Even if they are rough or muddy, don't widen tracks or take short cuts. Using the sides of a track or cutting corners increases damage, erosion and visual scarring. Walking through muddy sections keeps the impact to a narrow area.
- Keep groups small. Smaller groups have a lower impact.
- On rougher tracks, stay on rocks and hard ground wherever possible. Avoid stepping on any plants.
- Cutting new tracks is illegal.
- Do not mark tracks with stones, tape or other materials. This is unsightly and can confuse other park users.
- Avoid the spread of soil diseases by cleaning boots, tyres and camping equipment before entering a new area.

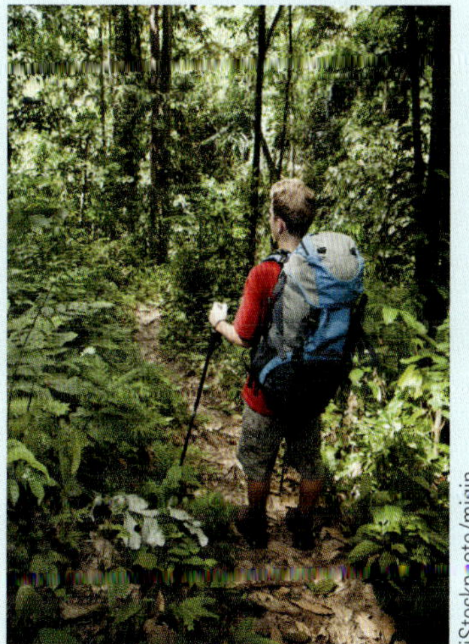

iStockphoto/migin

Figure 7.7 Stick to the path

Rubbish

You may find bins in some parks. Bins full or overflowing with rubbish attract animals looking for food. Animals can get sick eating rubbish, and the rubbish may be spread throughout the park. To avoid this, it is best to take your rubbish home with you, even if there is a bin provided.

- Pack to minimise rubbish. Don't take potential rubbish such as bottles, cans and excess wrapping.
- Take a plastic bag or container for your rubbish.
- Carry out all rubbish including food scraps and lolly wrappers.
- Pick up rubbish you may find along the way.
- Make sure animals can't get to your rubbish bag – plastic bags full of rubbish are a target for animals looking for food. When camping, take care, especially at night.

Human wastes

Human wastes can contaminate waterways and cause disease.

- Use toilet facilities where they exist, even if you don't really need to go.

In areas without toilets:

- Take a hand trowel and choose a spot at least 100 metres away from campsites and creeks or rivers.
- Use a different area each time. This spreads the impact over a wider area.
- Dig a hole 15 centimetres deep. Bury all faecal waste (excrement) and paper, mixing it with soil to help decomposition and to discourage animals.
- Carry out sanitary items.

Happy camping!

Alamy Stock Photo/Sigrun Eriksen

Figure 7.8 Take rubbish home with you.

Discuss

1. Jonathan is going camping in a national park. He is hiking to get there, and notices that he could take a shortcut off the path, directly over a small hill, instead of following the path around. What should he do?
2. Sarah loves peacock feathers and finds two that have been shed by a peacock on the side of a walking track. Can she take them home without breaking the rules set down by Parks and Forests QLD?
3. Nguyen rushes ahead of the rest of his classmates and comes to a section where the path breaks off in three directions. He decides to write a message in the soil so when his classmates catch up, they will know which way he has gone. Would you do the same? Briefly discuss.
4. You are going on a Year 8 camp to an outdoor environment. You will be spending some of the camp in tents and the remainder in huts. During the camp, you will be participating in activities that make use of the natural environment, including bushwalking, water activities and an obstacle/challenge course that is situated among a group of native trees.
   Imagine that you have not received any information about 'leaving no trace' and minimising your impact on the natural environment. Write a brief account of what 'doing the wrong thing' would look like while on camp in relation to the seven key principles.

Worksheet 7.3

FACE TO FACE Leave no trace

After reading the minimal impact guidelines for schools created by Parks and Forests QLD, critically evaluate the information. In pairs, discuss whether you think it covers all the principles of 'Leave no trace Australia' (page 277).

MOUNTAIN BIKING

sustainability maintaining/conserving the natural environment and not depleting the natural resources

To ensure that the outdoors and natural areas are used in a **sustainable** way, a number of organisations and activity groups have created their own minimal impact guidelines and codes of conduct. An example for mountain biking is provided.

Mountain bikers code of conduct

Respect the landscape

Respect your local trail builders and be a good steward of the physical environment. Keep singletrack single by staying on the trail. Practise Leave No Trace principles.

9780170463096

Carry out your rubbish. Do not ride muddy trails because it causes rutting, widening and maintenance headaches. Ride through standing water, not around it. Ride (or walk) technical features, not around them.

Share the trail

On multi-use trails mountain bikers give way to horses and foot traffic. There are some regional and local differences on single-use trails so make sure you check trail signage.

Ride open, authorised trails

Building unauthorised singletrack or adding unauthorised trail features are detrimental to our access. Poorly built features could also seriously injure other trail users. Get involved and help your local club or trail care alliance maintain and construct trails.

Ride in control

Speed, inattentiveness and rudeness are the primary sources of trail conflict among user groups. If you need to pass, slow down, ring a bell or verbally announce yourself, and wait until the other trail user is out of the path. Use extra caution around horses, which are unpredictable. Be extra aware when riding trails with poor sight lines and blind corners, and make sure you can hear what's going on around you

Plan ahead

Be prepared and self-sufficient. Every mountain biker should carry what they need for the ride they're undertaking, and know how to fix a flat tyre or make minor repairs. Download a GPS trail app such as TrailForks on your phone for navigation or carry a map in unfamiliar locations. Ride with a partner or share your riding plan with someone if you're heading out solo.

Mind the animals

When it comes to wildlife, live and let live. Never frighten animals. Respect them and their environment.

Adapted from Responsible Riding, IMBA Rules of the Trails. Reproduced with permission from Mountain Bike Australia.

Weblink
Watch the video to improve your mountain biking skills.

FACE TO FACE Promoting sustainable recreation

The Mountain Bike Australia Code of Conduct can easily be applied to other recreational settings, with the key words of 'respect', 'share', 'control', 'plan' and 'mind' providing great foundations within the school or local community environment. For example, let's look at a school environment.

RESPECT the schoolyard and do not do anything to damage buildings, furniture, outdoor facilities or plants. For example, do not walk through garden beds.

SHARE all parts of the school with other students. Do not stop other students from using parts of the playing fields or gym during lunchtimes.

CONTROL the language you use and be aware of people around you. Do not run in corridors or locker areas, and use language that is both respectful and understood by all. Deliberately speaking in languages other than English to prevent others understanding you should be avoided.

PLAN AHEAD for classes and other aspects of school life. Pack your lunch and sports clothes yourself; don't rely on your parents to do this for you. Make sure you know what subjects you have each day and bring the required books and equipment for each.

MIND animals that may be present on your school grounds – birds, dogs, native animals, etc. – and do not deliberately frighten or hurt them.

1 In pairs or small groups, design a way of promoting an understanding of sustainable outdoor recreation in your local area. Focus on an activity that is popular in the area or that you are interested in. This might involve collaboration with local conservation groups, council and schools.
2 Look at your local area in terms of recreational use.
   a Is there land that could be used differently or more effectively as an area for recreational activities?
   b Is there any land in the local area used for recreation that is of cultural significance to the local First Nations Peoples?

FACE TO FACE Local strategies

1 Using Google Maps, locate your house and make a list of all the outdoor recreation options available to people in your local area. These might include local parks, bike or walking tracks, state forests, sporting fields, national parks, rivers, creeks, dams or lakes.
2 From this list, identify how these resources could be used by the local population to be active. For example, a local dam could be used for water sports; a local park offers various recreational opportunities. Identify any connections that First Nations Peoples have with these local recreational areas.
3 Select one local area that you believe has the potential to be used by the local population. List a number of recreational uses for this area. For example, part of a park could be used as a BMX track. Select one of the options from your list and create a map that shows the key features that you would like to see in that area.
4 Before presenting this to the class, ask a classmate to evaluate your strategy and provide suggestions and feedback. You will then need to evaluate their comments and justify why you agree or disagree.

Worksheet 7.4

GEOCACHING

Geocaching is an outdoor 'treasure hunting' activity that was first established in 2000. Participants use a Global Positioning System (GPS) receiver or mobile device and other navigational techniques to hide and seek containers, called 'geocaches' or 'caches', anywhere in the world.

A typical cache is a small waterproof container containing a logbook in which the geocacher enters the date they found it and signs it with their code name. After signing the log, the geocacher must replace the cache exactly where it was found. Larger plastic storage containers or ammunition boxes can also contain items for trading, usually small toys or trinkets.

There is a lot of information on the internet about geocaching. Watch some videos to get a good idea of what it's all about.

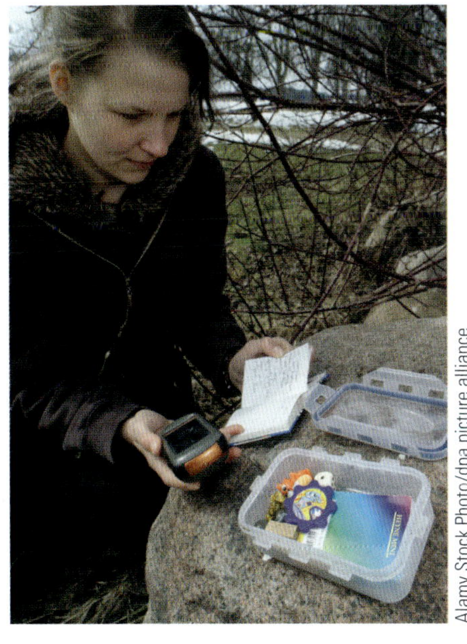

Alamy Stock Photo/dpa picture alliance

Figure 7.9 A typical geocache

Worksheet 7.5

Worksheet 7.6

1 Discuss why it is important to 'leave no trace' when using outdoor environments.
2 List three strategies campers should always use in outdoor environments to minimise the risk of them contributing to bushfires.
3 When camping, fishing or engaging in any other outdoor pursuit, state why it is important to let someone else, other than those you may be with, know your plans.
4 Think of some memorable times you've had outdoors. (Do not include sporting activities). Can you describe what you gained from the experience? It might have been a fun day because everyone was happy, or you were challenged because you had to walk up some steep hills, or you saw some amazing views. Sometimes, the benefits that can come from experiences like this are overlooked – you might have to dig deep to identify them.

   Complete a table like the one provided, identifying benefits for six activities.

Outdoor experience/ activity	Physical benefits	Social benefits	Emotional benefits	Spiritual benefits
e.g. Visited the beach				

Weblink
Watch the video to find out more about geocaching.

Scaffold
Memorable activities

1 Discuss a time when you have been in an outdoor environment and it has made you feel good. Try to provide at least one example of an individual experience and one example of an experience when you were with friends.
2 Have you ever come across litter/rubbish when walking or playing outdoors and picked it up and put it in the bin? Discuss why this is something more people should do, even though it's not their litter.
3 Consider why it is important to put in place laws that stop developers and industries from purchasing large areas and using these for housing or primary industry, such as farming or mining? Why or why not?

1 Consider what the following places do to encourage more people to use the outdoor environments in their own time. Ensure that your examples are specific, for example: *Install shade sails between the canteen and resource centre*, rather than just stating *install shade sails*.
   a Your school
   b The local park
   c An outdoor swimming pool
2 Most of us use recycling bins at school and home, but overseas countries have stopped taking our recyclables, causing problems for local and state authorities. Investigate alternatives for this 'waste' other than just dumping it into landfill/tips.

Quiz
What is outdoor recreation?

WHAT TYPE OF MOVEMENT CHALLENGES CAN I COMPLETE?

Movement challenges do not need to be complex. Games that require little equipment are often the best. Chapter 8 has some more information about games and equipment.

Weblink
Watch the video to introduce yourself to Ultimate Frisbee

ULTIMATE FRISBEE

Ultimate Frisbee is a game played with simple rules and minimal equipment. It is a fast, free-flowing game that combines elements of netball, soccer, gridiron and touch football. It is a considerable challenge to place an accurate pass to a partner. Building your skills and learning to use the frisbee in different ways requires patience and practice. It's also a great workout, with a lot of running and jumping.

Rules of Ultimate Frisbee

- Two teams of seven players each play on a rectangular field with two end zones.
- The objective is for the team with the frisbee to pass it up the field without dropping it and catch it in an end zone to score a point. The other team tries to intercept the frisbee or knock it down.
- The frisbee may be moved in any direction by passing it to a teammate. Once a player has the frisbee, they are not allowed to run with it and they have 10 seconds to throw it.
- The game is self-refereed. If a player committing the foul disagrees with the foul call, the play is redone.
- Players can substitute after a score and during an injury timeout.

Newspix/Simon Chillingworth

Figure 7.10 Ultimate Frisbee in action

9780170463096

Frisbee grips

A frisbee is a light plastic disc with a lip. It is designed to fly aerodynamically when thrown with rotation and can be caught by hand.

Backhand grip: use for a backhand throw. Fingers are curled under the frisbee's rim and the thumb is placed on top of the frisbee.

Forehand grip: use for a forehand throw. The index and middle fingers are extended under the frisbee. The ring finger and little finger support the outside of the frisbee. The thumb is on top of the frisbee.

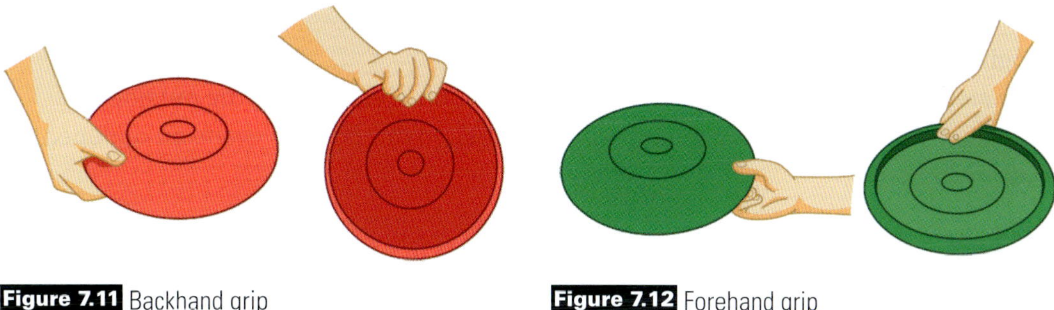

Figure 7.11 Backhand grip

Figure 7.12 Forehand grip

Throws

Backhand throw: the back of the hand faces the target, as in a tennis backhand. This is the most common throw, and the most powerful.

Forehand throw: the frisbee is thrown on the same side of the body as the throwing arm, like a tennis forehand.

Figure 7.13 Backhand throw

Figure 7.14 Forehand throw

Figure 7.15 Hammer throw

Hammer throw: the frisbee flies upside down. Use a forehand grip and throw from above the head facing the target, like a tennis serve.

Roller throw: the frisbee hits the ground then rolls. This can be either a backhand or forehand throw, and the inside edge of the frisbee should hit the ground before the outside edge. (Not a legal throw in Ultimate Frisbee.)

Skip (bounce) throw: the frisbee hits the ground then bounces up and keeps flying. It can be backhand or forehand, and the outside edge of the frisbee should hit the ground before the inside edge. (Not a legal throw in Ultimate Frisbee.)

Fake throw: when a player pretends to pass to create space and deceive the opponent.

Footwork: pivoting

Pivoting is a footwork movement to change direction, where a player keeps one foot still and steps with the other.

Figure 7.16 Pivoting

Attacking skills

Leading: sprinting strongly to the frisbee, either directly forward or diagonally to the free space (away from the defender or opponent).

Dodging: evading an opponent. Moving a few steps away from the intended catching position, then placing the outside foot strongly on the ground and pushing off in the desired direction to evade an opponent or receive a pass.

Defending

Guarding a player who may or may not have the frisbee. One-on-one defending techniques include defending in front, from the side or from behind.

Figure 7.17 Leading **Figure 7.18** Dodging **Figure 7.19** Defending

Catching

Pancake catch: the frisbee is caught with one hand on the bottom and one hand on the top. This is the easiest method of catching.

Two-handed catch: the frisbee is caught in two hands that are side by side. If the frisbee is above shoulder height, the player's fingers will be on top of the frisbee (and thumbs underneath), otherwise their fingers will be underneath the frisbee (and their thumbs on top).

Figure 7.20 Pancake catch **Figure 7.21** Two-handed catch – above shoulder **Figure 7.22** Two-handed catch – below shoulder

One-handed catch: the frisbee is caught with one hand. If it is above shoulder height, the player's fingers will be on top of the frisbee (and their thumbs down), otherwise their fingers will be underneath the frisbee (and their thumbs on top).

Figure 7.23 One-handed catch

Weblink
Ultimate Australia

Ultimate disc golf

Disc golf is an individual sport that combines the rules of golf with throwing a flying disc (golf disc). While the game is similar to golf, the fact that the disc flies through the air makes it an entirely different challenge. A range of discs are used, from putters to midranges or drivers, or even discs that go straight, left or right. Players can change discs for each throw to increase the fun. Like other outdoor sports, there is also the challenge of coping with changing weather conditions, such as wind and rain.

Disc golf has a number of advantages over golf, including:

- it is easy to learn
- it is suitable for people aged 10 to 95 years, and for some disabled players
- it is inexpensive – you only need one disc, which costs less than $20
- most courses are in public parks, so it is free to play
- it is environmentally friendly – no watering of greens or fairways needed.

Weblink
Watch the video showing the rules of disc golf.

Disc golf challenge

Aim

To design and participate in your own disc golf course

Equipment for each team

- Five plastic hula hoops
- Five markers/domes of the same colour or shape
- Five flexi poles (useful but not essential)
- One frisbee for each player

Method

1. The class is divided into teams of four or five.
2. Each team has a different starting point and finishing point.
3. As a team, decide on the distance to each 'hole' (metres), and where each hole will be placed in relation to obstacles. (For example, you may decide to place a hole behind a tree or the goal posts.)
4. A 'hole' is marked by a hula hoop.
5. Design a score card with the names of your team members, number of holes, distance between holes (metres) and a column for the number of throws.
6. Designate a scorer for each hole so that everyone takes on the responsibility of scoring.
7. Decide who will throw first and play the course, recording the number of throws it takes for each hole.

8 A hole is completed when the frisbee lands in the hoop. (Hoops can be attached to a fence or tree so that they are not all lying on the ground).

9 At the end of each hole, record who was first, second, etc.

10 At the last hole, the winner is the player who has taken the fewest throws for each hole.

When this course has been played once, there are several options:

a Each team can swap and play another team's course to see if they can beat the score.

b Team challenge: each team completes each course. The scores are then added up and the team with the fewest total throws is the winner.

Discussion

Scaffold
Reflection table

In pairs, reflect and evaluate on your participation in the disc golf activity. Use a table like the one provided to assist in this process.

Planning and participation	What was successful and why	How I could improve

1 What strategies were most successful, and why?

2 What responsibilities or roles were fulfilled as part of playing disc golf?

3 What benefits have you gained from this experience? Can you identify similar benefits to other activities, such as physical, social and emotional?

MOVEMENT CHALLENGES WITH A TWIST – USING YOUR SENSES!

At any moment throughout the day your brain is flooded with a billion bits of information that it simply cannot process. In order to cope, the brain selects the most important information and discards the rest. The next challenge involves heightening your awareness of certain things that are usually taken for granted, such as your sight. Research has shown that when one or more of your senses are removed, one or more of your other senses works harder to compensate for the loss. In the next challenge your sight will be removed. How will you cope? Will your other senses compensate?

Hug a tree challenge

Aim

To see how well you can use your other senses when you are blindfolded

Method

To begin, you need to get into pairs. One partner will be blindfolded. This person is guided to an object that is 20–30 metres away. If completing this activity outdoors, trees make ideal objects because they are unique in texture and shape. However, any object is fine as long as it is distinctive.

The non-blindfolded person leads their partner to the object. Try to make the journey as difficult as possible, twisting and turning so that your partner becomes disoriented. The return journey should be equally confusing for them.

Once introduced to the object, the person has 20–30 seconds to familiarise themselves with it. They must use their sense of touch and smell as well as their hearing. The experience will be even better if the object has a peculiar shape or surface that will help the blindfolded person identify it.

Once your partner has explored the object, bring them back to the start point, ensuring they are still blindfolded. Remove the blindfold and ask them to identify and locate the object with their eyes open. The sighted person should not provide any clues until the object has been found. Swap roles and repeat the process. If possible, time how long it takes for each partner to locate the object.

When both of you have completed the task twice, discuss what factors helped the blindfolded students to identify the secret object. What parts of the object were hardest to identify?

Variations

Repeat the activity in groups of three, where two students are blindfolded and joined together so they have to work as one. This will create interesting discussion and variations of distances and direction!

Communication in movement challenges

Communication is about giving and receiving messages. It can be both verbal (spoken) and non-verbal. Some examples of non-verbal communication, often called 'body language', are eye contact, facial expression and posture.

Worksheet 7.7

Being able to communicate effectively is important. Communication in a team environment is essential to the success of the team. Successful communication occurs when everybody understands each other, can express their own opinions and ideas, and everyone listens and is open to what others have to say. Simple and clear messages are key.

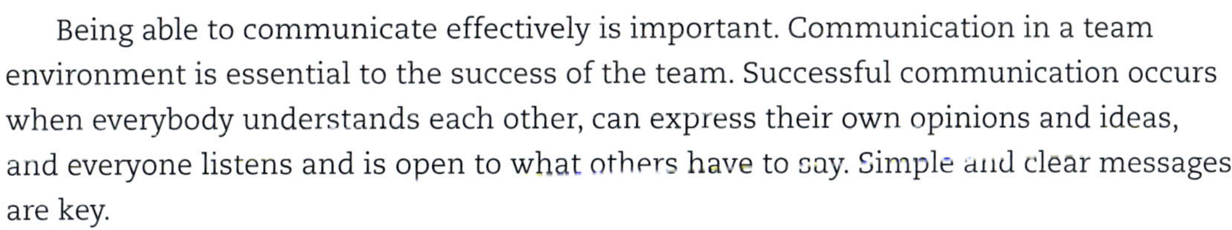

INVESTIGATION

HOW CAN I COMMUNICATE EFFECTIVELY?

Purpose

To investigate effective and ineffective communication examples relevant to physical activity, sport and the workplace.

Method

1. Conduct an internet search of articles, research or clips that show both effective and ineffective communication practices.
2. Summarise the top three practices in each case, as well as addressing the discussion points below.

Discussion

1 Provide a summary of body language cues that indicate someone is interested and listening to what is being said by a peer/teammate.
2 Describe how you can tell if someone is not interested or engaged with what is being communicated to their group. For example, a coach delivering a message at half time or a supervisor talking to a team of workers.
3 When communicating information to another person, provide ways you can check to ensure that they have understood what has been communicated.

Worksheet 7.8

There are many principles of effective communication, including speaking slowly and clearly so that people can understand you. Have you ever been in a noisy classroom when there is so much noise that you cannot even hear yourself think? However, it is possible to have a conversation with one person and ignore all the other noise. The next activity puts that theory to the test.

Hog call challenge

Aim

To effectively communicate with your partner so that you can find each other when you are both blindfolded.

Method

1 Find a partner.
2 Each pair must agree on two words they are going to use to communicate (for example, orange + juice or peanut + butter). Agree which person will say which of the two words.
3 When the class is ready, each pair announces their two words and partners move to opposite ends of the gym or playing field. Each puts on a blindfold.
4 Focus your attention on the two words that you and your partner agreed to.
5 Once you are in the start position you will begin to shuffle away from your original position. You now need to assume the 'bumpers up' position, which means walking with your hands up in front of you in case you bump into somebody or something.
6 From the starting position, you call out the agreed word, then listen and walk towards the matching word until you find your partner, who will also be listening for your agreed word. So if one partner was 'peanut', they would be calling this out, and the other would be calling 'butter', until each partner finds the other.
7 Once you find each other you can take off your blindfolds and watch the other pairs trying to find each other. (A more difficult variation is to use animal noises!)

Discussion

What did you learn from this process about effective communication?

Source: Reprinted with permission from playmeo. To access more interactive group activities, go to www.playmeo.com.

DEVELOPING NAVIGATION SKILLS

Figure 7.24 Mountain biking following a track requires navigation skills.

At some time in the future you may explore the great outdoors by bushwalking, cycling, canoeing, kayaking, horseriding or cross-country skiing. Any trip into an unknown environment requires successful **navigation** from the starting point to the destination and, if necessary, a return route. This becomes more challenging in mountainous or remote regions.

navigation the process of finding a way to get from one place to another

Where you have grown up may have an impact on your navigation skills in certain areas. For example, students from the bush may find it challenging to navigate across the city using various methods of transport. Similarly, people from the city will need to learn to navigate in the countryside, such as in national parks and state forests, where they may camp, bushwalk and use other forms of recreation. In both city and bush environments it is important to be able to use a map and read directions.

BASIC MAP-READING SKILLS

A map is simply a diagram of an area of land that shows all the important features and landmarks, such as roads, valleys, hills and buildings. All these features are placed at the correct location and the correct 'scale distance' apart. Maps are used to indicate the direction and distance from one place to another. The four cardinal points of the compass are used to indicate direction. These are north, south, east and west. The half cardinal points are used to more accurately indicate a direction. These are north-east (NE), south-east (SE), north west (NW) and south-west (SW). The top of the map is always the north, the bottom the south, the right-hand edge the east and the left-hand edge the west. Whenever two places are considered, the first and most important aspect is the direction 'from' and 'to' those places.

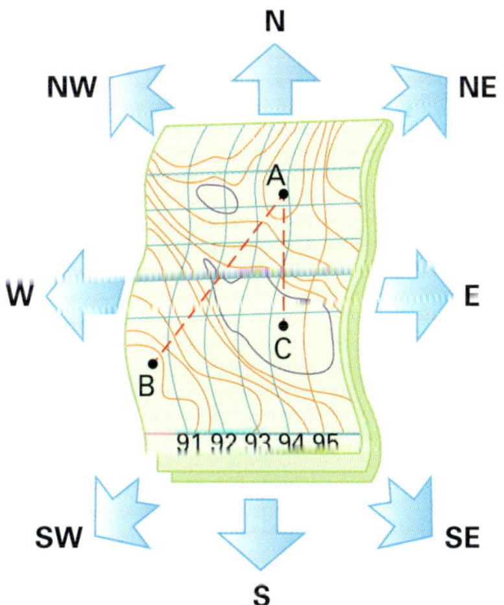

Figure 7.25 Directions on a map

Scale

Maps are usually much smaller than the area they represent; the size of the reduction is known as the scale of the map. Map scale is an important factor in deciding what type of map to use. The larger the scale of the map, the more detail it has, so the easier it is to navigate. However, the larger the scale, the less area it can cover in one sheet of paper. If the scale of the map is too large you will be continually 'walking off' the edge of the map.

Scale is normally expressed as a ratio. A scale of 1:25 000, which is a common scale for a map, means that 1 centimetre on the map represents 25 000 centimetres

(250 metres, or a quarter of a kilometre) on the ground; 1:50 000 means 1 centimetre on the map represents 50 000 centimetres (500 metres, or half a kilometre) on the ground (smaller scale). Generally, the larger the scale, the easier it is to navigate because there is more detail. As a general rule, the faster you travel, the smaller the scale of the map. For example, walkers will use 1:25 000 and 1:50 000 scale maps, but a cyclist might use a larger one because they are likely to cover more distance.

Shutterstock.com/Globe Turner

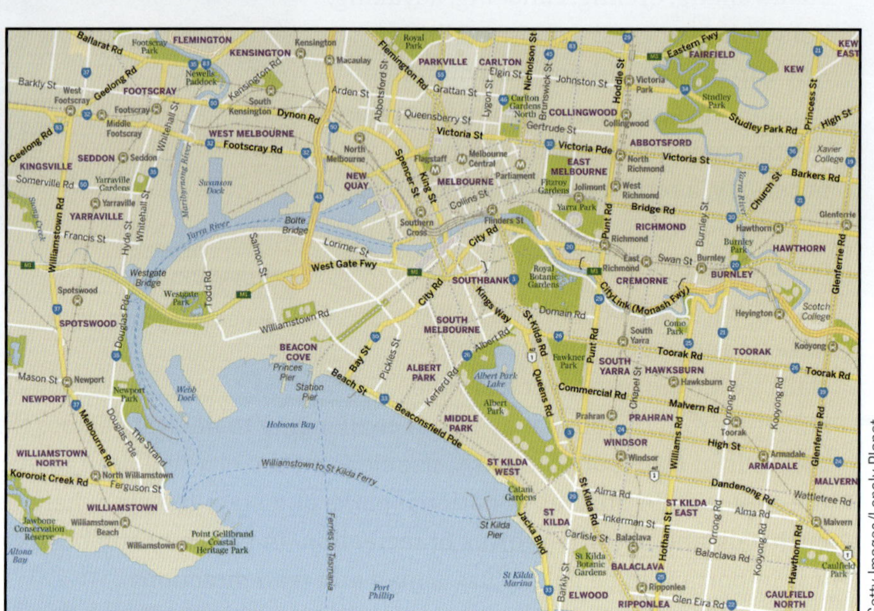

Getty Images/Lonely Planet

Figure 7.26 Comparison of scales: the map of Australia on the left is at a small scale compared to the map of Melbourne on the right, which is at a larger scale.

Distance

It is vital to be able to measure the distance between two places on the map quickly and accurately. The usual method is to use a piece of string and lay it along the proposed route from the start point to the end. Then lay the string straight along the scale rule on the margin of the map, ensuring that one end starts at zero on the scale. Alternatively, you can use the straight edge of a piece of paper, marking the paper at each sharp turn and then laying it along the scale line and reading off the distance. Distance can also be measured using Google maps or a similar program.

Time

When you plan your next walk, paddle or cycle, in addition to identifying the route, noting the landmarks and path or road junctions will enable you to pinpoint your positions. You also need to measure the distance between those junctions when you change direction or take another path or road.

A normal walking pace along roads or footpaths might be around 5 kilometres per hour, meaning it will take about 10–12 minutes to walk a kilometre. Before you begin, make sure you are travelling in the correct direction (identify landmarks to pinpoint your position), then note the time and complete the next section of your journey. Keep on repeating this procedure until you reach your destination. Compare the actual time that it took to walk each section with your estimated time. You can now work out the average time that it takes you to walk a kilometre or 500 metres. This will help you better estimate how long your next walk will take, depending on the terrain.

Measuring distance by travelling time is effective for a cyclist, canoeist, rower, horserider or even a cross-country skier. You just need to work out the average time it takes you to complete a kilometre.

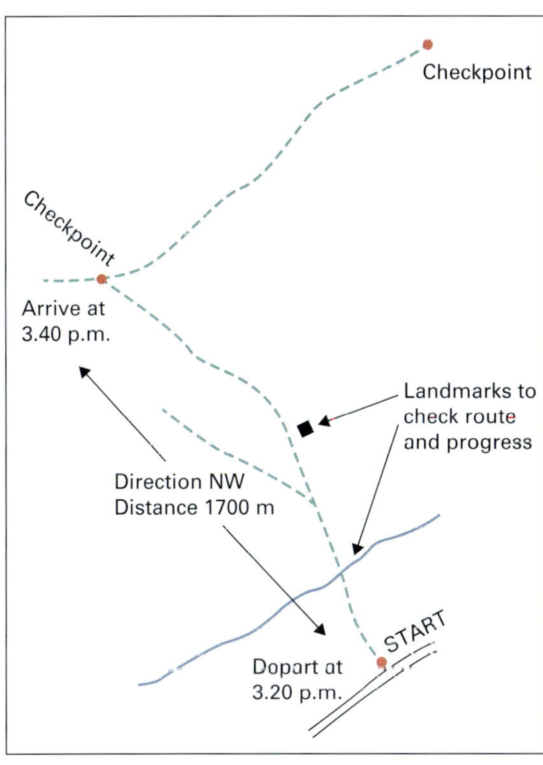

Time required to travel 1700 metres is 20 min

Time required to travel 1000 metres is $20 \times \frac{1000}{1700}$

$= 11.76$ min or

≈ 12 min per kilometre

$= 6$ min per 500 metres

$= 1.2$ min per 100 metres

1. Set map and locate position
2. Recheck start position
3. Note direction (NW)
4. Note distance 1.7 km/1700 m
5. Note landmarks to track position and progress
6. Note time
7. Travel using set map to point direction when required

Figure 7.27 Distance and time

Orientation

Using a road or river

Use a line feature such as a road, path, ridge, river or anything that has a line or direction on your map. Holding your map horizontally, turn it around until the feature on the map is parallel with the feature on the ground. Your map should now be correctly set (oriented) and all the other landmarks and features should also be in the correct direction on the map. This is useful when you are not entirely sure where you are but you have a rough idea of the area you are in. It is still possible to have the map entirely the wrong way around, however; check that the landmarks are on the correct side of the features you have identified.

Using a landmark

Another method is to use a landmark or feature. To use this method, you must know where you are. Find your position on the map, look around the landscape and find a feature that is marked on the map, e.g. a farmhouse. Lay a pencil or straight edge through your position on the map and then, holding the map horizontally, turn the whole map around until you can sight along the straight edge from your position on the map, through the spot feature on the map to the spot feature on the ground.

Knowing where you are at any time is important – pinpointing your position using features on the map and the ground allows you to track your position, whether you are on a bike, walking or canoeing. This skill can be practised on a bus or train or in a car, as long as you have a map of the area you are travelling through. The only difference is that it will happen a lot more quickly in a train, bus or car.

Grid references

Maps are generally covered with lines running east to west called 'eastings' and lines running north to south called 'northings'. On 1:25 000 and 1:50 000 scale maps, the

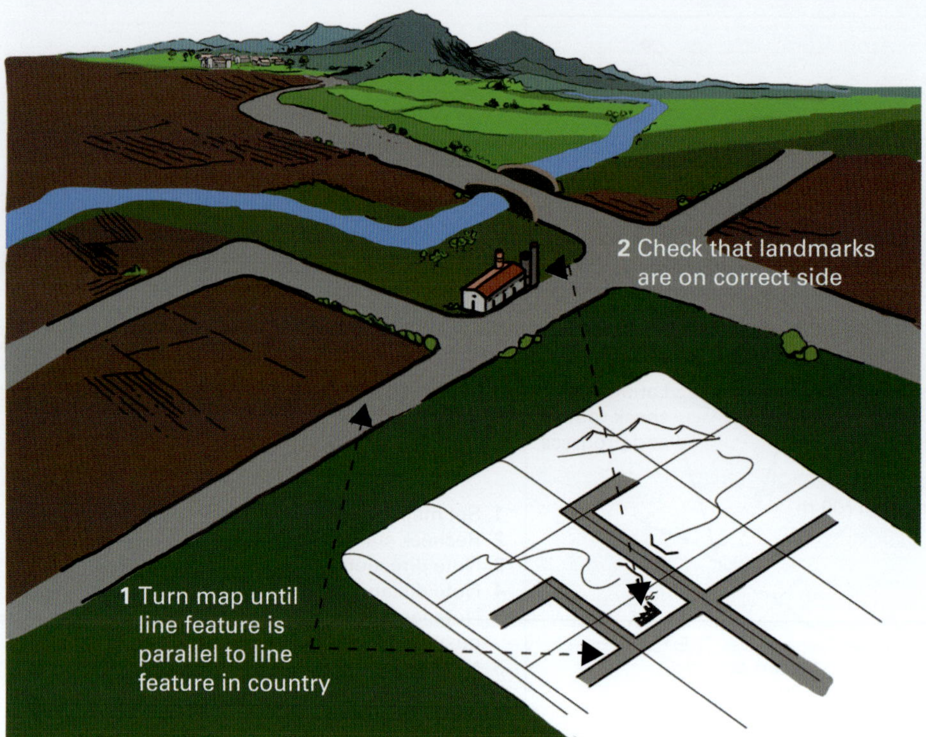

Figure 7.28 Orienting the map using a line feature

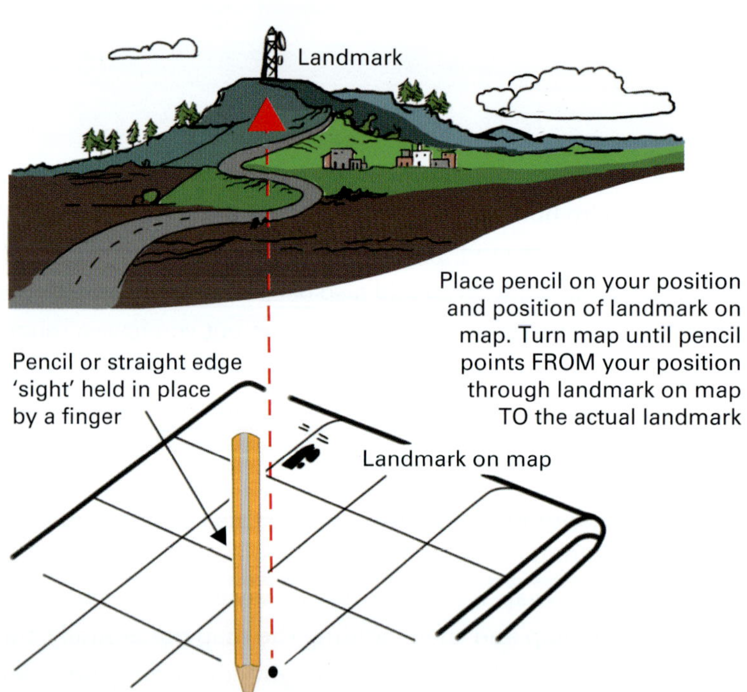

Figure 7.29 Orienting the map by landmark. Sometimes it is easier to orient the map from a hill or higher point. You must know where you are or your approximate position.

distance between the grid lines always represents 1 kilometre of actual country. Each line is identified by two figures ranging from 00 to 99. In a six-figure grid reference, the first three figures indicate how far to the east the place is while the second three figures show how far north it is. The first two figures in each group of three are read from the map, while the third figure is measured. If you remember that the letter E comes before the letter N in the alphabet, you will have no difficulty remembering that eastings come before northings. Using a six-figure grid reference, you can work out where a place is located to within 100 metres. Practise until you are comfortable with reading grid references.

Figure 7.30 Using a vantage point

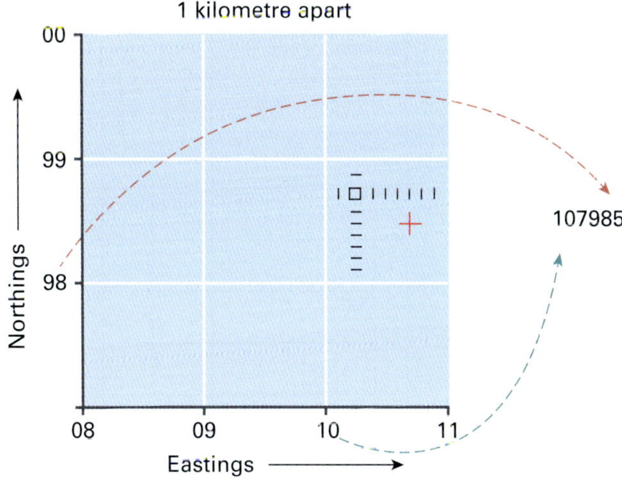

Figure 7.31 Using a grid reference

UP AND MOVING Map-reading challenge

1. Find a basic map or a plan or sketch of your school.
2. Rule some grid lines on top of it at a scale of 1:5000 (1 centimetre to 50 metres, or 20 centimetres to every kilometre). The scale can be modified depending on the size of the school grounds.
3. Each group will be given five or six identifiable markers.
4. Write the grid references on your map of where you placed these markers. Once all groups have placed their markers out on the course and marked them on the map, groups will then swap courses. The aim will be to find all the markers using the six-figure grid references provided on the map.
5. Initially these markers could be large and bright, but as players become more confident reading the map they can be a bit smaller and more difficult to find. This will challenge groups to develop their observation and reading of the map references.

ROUTE CARDS

The basic skills of navigation can easily be applied around a town or its surrounding areas. However, if you go into the country or the bush, being more prepared about the route you will take is an important part of ensuring your safety. This is a necessary part of preparation for schools and other groups before they venture into more challenging locations. Route cards are one method of preparing your route.

A route card helps to break up the journey into sections or 'legs' on the basis of direction. Whenever there is a major change in the direction at a path junction, you start another leg. The start of each leg can be regarded as a 'checkpoint'. Each section on a route card allows you to plan your route carefully. It allows for the route to be checked and followed by someone else. It also allows you and your group to constantly check your position, and it means that if you do get lost because you missed a path or track, you should easily be able to backtrack to the last checkpoint.

ROUTE CARD (Use one per day)				Names of group members						Name of group or unit
Day of the week	Date	Day of venture 1st, 2nd, etc.								Address
									Setting out time	Ph:
Leg	Place with grid ref Start	General direction or bearing	Distance in km	Height climbed in metres	Time allowed for leg	Time for stops or meals	Total time for leg	Estimated time of arrival (ETA)	Details of route to be followed	Escape to:
1	TO									
2	TO									
3	TO			DO						DO
4	TO									
5	TO			NOT						NOT
6	TO									
7	TO			USE						USE
8	TO									
	Totals								Supervisor's name, location, ph. number	

Figure 7.32 A sample route card for use in planning a route in more demanding environments

Scaffold
Route card

iStockphoto/7000

Figure 7.33 Tracking a planned route is the same whether you travel by water or on land.

Once you have everything prepared for your trip you can follow these instructions to use the route card successfully, no matter what method you are using to get from point A to point B.

- Before you start, locate your position on the map and orient the map, or vice versa. Recheck your position to make sure it is correct. Note and check the landmarks or features that will enable you to pinpoint your position as far as your first checkpoint. If this is not possible, note a feature before that checkpoint, such as a bend in the river that the path is following.
- Note the direction you are travelling in, using a compass or the compass on your smartphone.
- Estimate the time required to reach the next checkpoint.
- Record the time and then depart, using your map to point the direction.
- Track your position on the map using landmarks.
- Try to avoid stopping, but use your map all the time to anticipate the detail and features on the ground. You will find paths and entrances, and trails that were not on the map, which may confuse you. This is why you need an idea of how long it will take to get to the first checkpoint and the direction you will be travelling in.
- At every major change of direction, look back and take in the scene from the opposite point of view from which you are heading.
- Before setting off on the next leg, repeat the same procedure as you did at the start of the journey.

9780170463096

This process is essential to successful navigation. It is much easier to track your position constantly than to have to sort yourself out if you get lost. Following a planned route is the same whether you are mountain biking, canoeing or on a horse. If you are in a national park, there is plenty of information available about where and what you can do and key information about safety.

1 Discuss why movement challenges just as important as mental challenges for students to participate in at school.
2 Identify two ways to communicate effectively and two ways to communicate ineffectively.

1 Think of a movement challenge you have found difficult.
   a Discuss some of the challenges it presented.
   b Discuss why you believe it is important to keep trying to get better at the movement challenge, rather than giving up.
   c What have you done to improve at a particular challenge in the past?
2 In every class, someone will be better at a particular challenge than others in the class. Suggest ways these classmates can continue to improve their skills, while improving the skills and hopefully experiences of their classmates as well.

1 With a classmate, modify an existing sport that is played outdoors and design a new outdoor challenge that your class can participate in. You will need to write down the aims of the game, a few specific rules (try to keep it simple), the size of the playing field and any equipment that may be needed.

WHAT ARE SOME ORIENTEERING CHALLENGES I CAN PARTICIPATE IN?

Quiz
What type of movement challenges can I complete?

Weblink
Watch the video to introduce yourself to orienteering.

Orienteering is a good way to work on your navigation and map-reading skills. Begin with small areas and, as your skills improve, you can then progress to more difficult terrain.

Orienteering is a combination of outdoor adventure with map reading and navigation. It involves navigating though the bush, parks or streets with a specially produced map and orienteering compass. The aim is to locate checkpoints (controls) on various features along the way, such as a boulder, track junction, bench seat or street lamp. Controls are generally identified by distinctive orange and white flags.

The skill in orienteering lies in finding the best route between controls. It is a stimulating mental challenge as well as a fun physical activity, and it doesn't need much equipment. Each event may offer several courses, with differing lengths and levels of difficulty, to cater for a variety of skill levels.

A typical course might have plenty of control sites, and may change direction several times. It is important to have handrails for participants to follow between

consecutive control sites. Roads, paths, fences, edges of playing fields and drains make suitable handrails. The start and finish should be located in a spot that allows for good supervision.

Figure 7.35 shows the basic features or legends that are found on orienteering maps. These are different from normal maps, but the map-reading skills required are the same.

Figure 7.34 An orienteering course map from Westerfolds Park

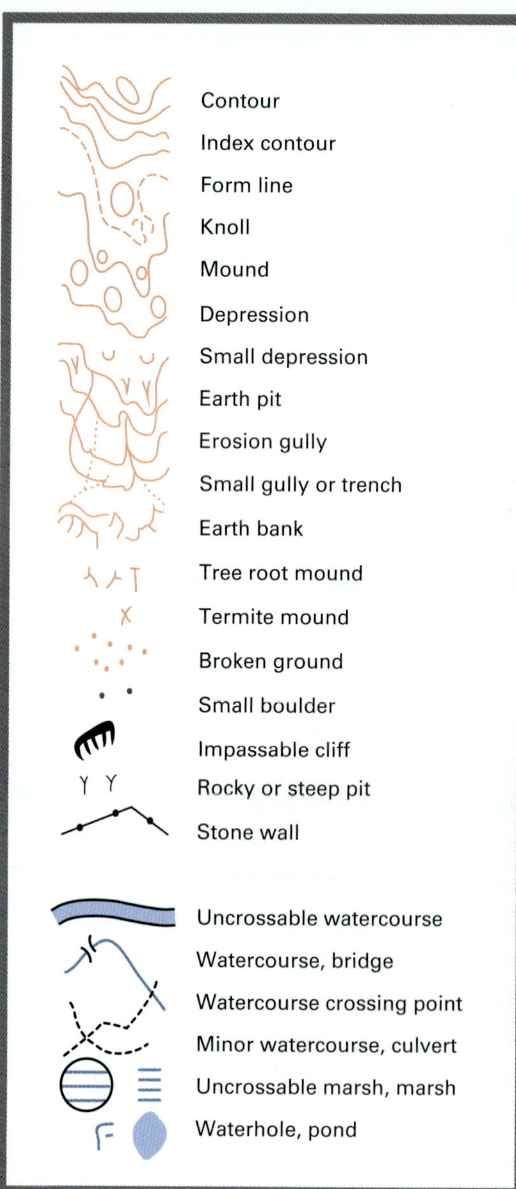

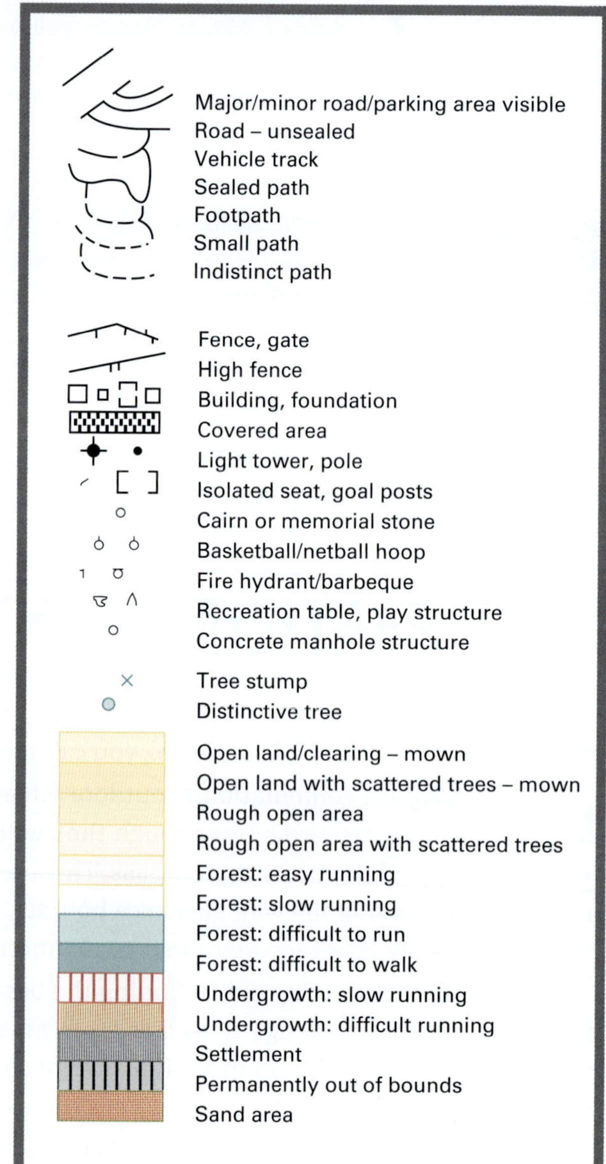

Figure 7.35 Sample legends from orienteering maps

FACE TO FACE Become an orienteering commentator

This activity is designed to introduce you and a partner to the concepts of orienteering while also improving literacy skills, using the map provided.

Follow the course on the map below with your finger as your partner makes up a story about an orienteering run and becomes a 'commentator'. Start at '1'; your partner then describes how to get to '2' by calling out the different features on the course. Try to imagine other people around on the course, either competing against you or simply enjoying the environment with their family or friends. Swap roles every few paragraphs and make sure you keep up with the action.

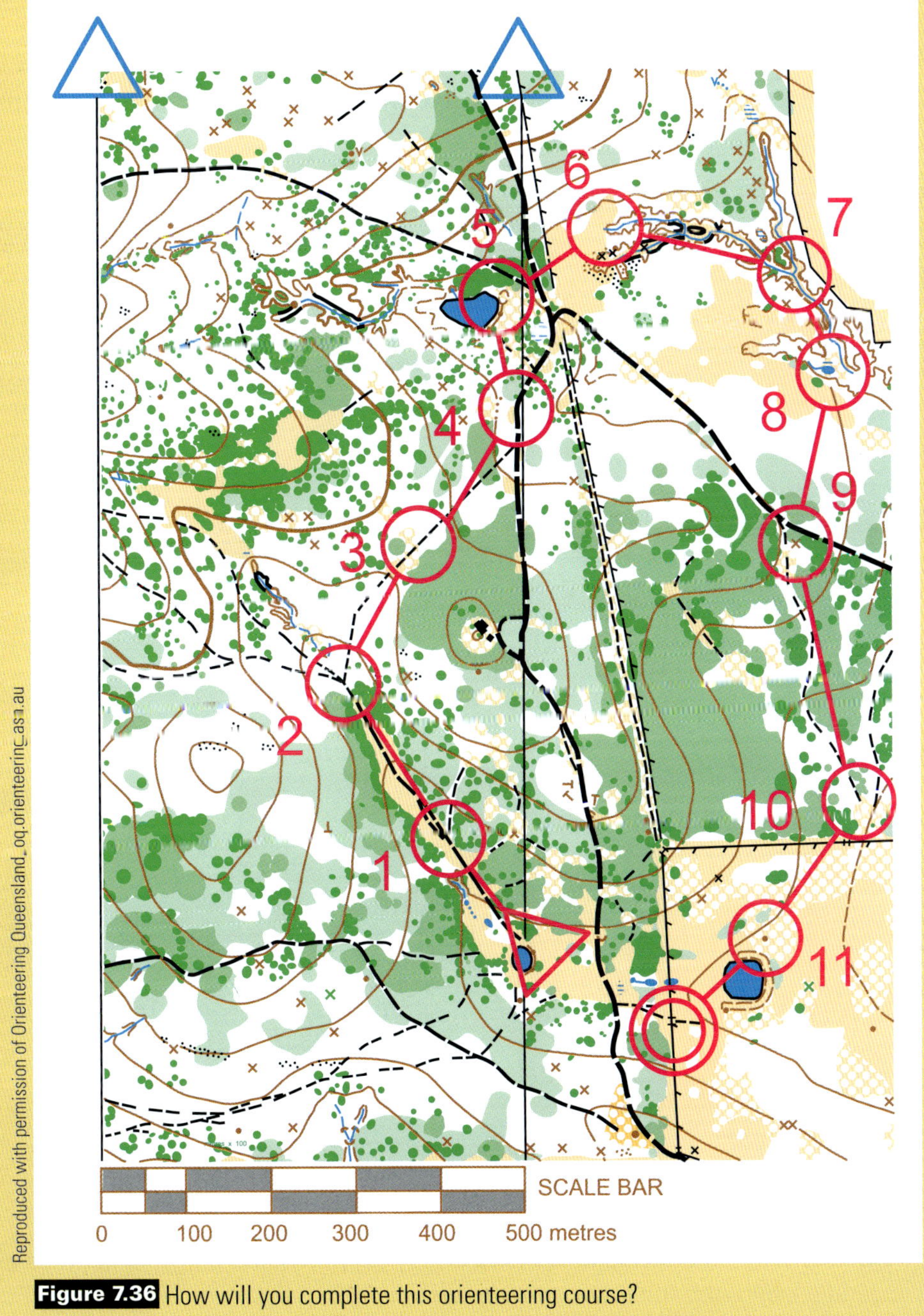

Figure 7.36 How will you complete this orienteering course?

The clue descriptions

Start at clearing
1 (41) Track junction
2 (42) Track junction
3 (43) Saddle
4 (44) Track bend
5 (45) Pond, deep, NE edge
6 (46) N watercourse, W end
7 (47) Watercourse junction
8 (48) Pond, shallow, S edge
Follow pink streamers
9 (49) Termite mound, N side
10 (50) Track, SE end
11 (51) Mound, NE side
140 metres to the finish

The next activity contains some simple orienteering suggestions that can be followed as a class with your teacher. They do require some preparation (e.g. maps, controls and time placing these out).

Worksheet 7.9
Worksheet 7.10

Star relay

A small area of the school grounds or local park may be used for a first orienteering activity. The markers (controls) can be placed in this area. Depending on the size of the controls and the terrain, it may not be necessary to mark them on a map. As the area becomes larger, more skills (physical and mental) are required to find the controls.

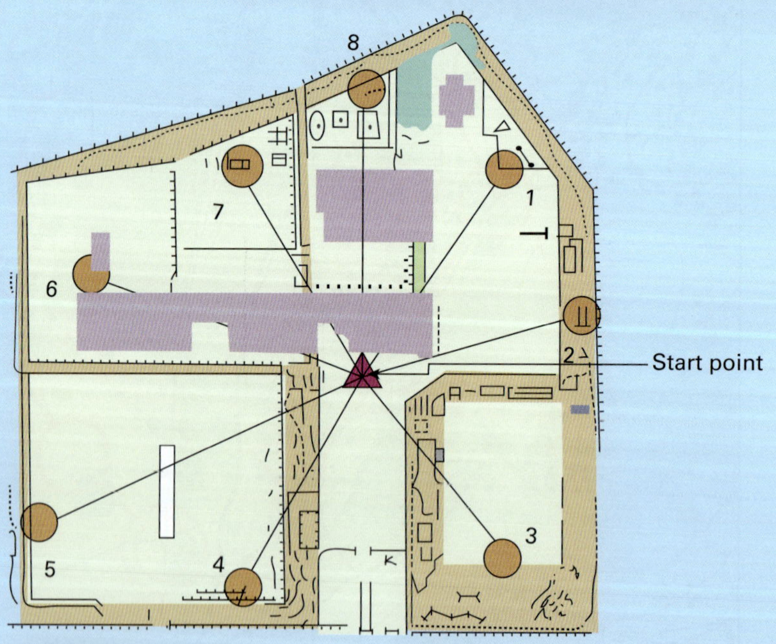

Figure 7.37 A star relay for orienteering

Equipment required

- **Map**: showing the boundaries, fences, buildings, tracks and paths, large trees and vegetation, and ALL the other small features (posts, poles, bins, seats, tables, etc.)

9780170463096

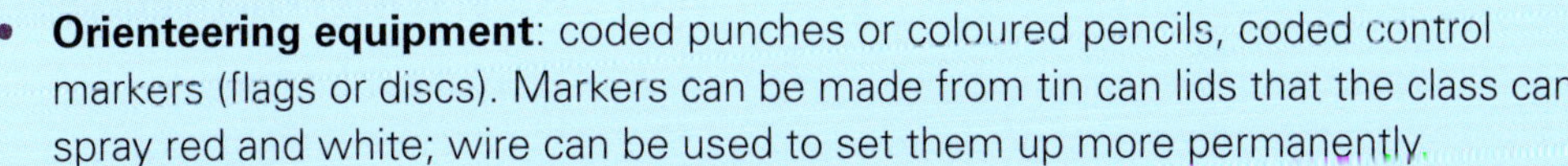

- **Orienteering equipment**: coded punches or coloured pencils, coded control markers (flags or discs). Markers can be made from tin can lids that the class can spray red and white; wire can be used to set them up more permanently.
- **Control cards**: to record the coded answers. These can be created on the computer and printed out.

Worksheet 7.11

What to do

- Place eight controls out into your designated area and mark the map as shown in the example illustration.
- Number 1 has code #1 START at the triangle.
- Work in pairs; each person has a map and they share a control card.
- Student A navigates to #1, punches the card and returns to the start, hands the card to Student B who then navigates to #2, and so on until all controls have been visited.
- To avoid overcrowding at #1 control, different students start at different numbers, and then continue in a clockwise direction until all the controls have been visited.

Shutterstock.com/Rawpixel.com

Figure 7.38 Finding the next control point in orienteering

Summary of key orienteering skills

1. Orient the map: one of the key aspects of orienteering is to ensure that you relate the map to your surroundings (see page 293). Maps always have a north arrow.
2. Identify and recognise the features: as you use maps more often, this skill will become more natural.
3. Make decisions about which way to go. Frequently, this will depend on what is in front of you; sometimes there will be a track that may be quicker, or you could avoid a steep hill.
4. Use 'handrails' (line features such as fences or the sides of buildings) to lead you towards the control. As the maps become more detailed, this will assist you in reaching the controls quickly.
5. Use 'attack points' (identifiable features close to the control) to lead you to the control.

Create your own orienteering course

In pairs, create your own orienteering course with a combination of different challenges for the other students in your class. Different challenges might mean smaller controls; controls placed in more difficult positions; or a time challenge to see which team can complete the course in a specified time. You will need to design the challenge, then trial the course by running it to see if it can be achieved in the allocated time.

FACE TO FACE How did I go?

In pairs or individually, reflect and evaluate on your participation in orienteering activities, especially the activity in which you designed your own course. By evaluating how you performed, you can learn how to improve various aspects of teamwork, planning and performing.

These questions may help your evaluation.

1 What responsibilities or roles did you take on? For example, who designed the course or created the controls? Who decided the best strategy to take to complete the course?

2 Think about this process and describe how your role or a combination of roles led to your team being successful.

3 How can you transfer some of the skills that you learnt in orienteering to other learning situations?

1 Discuss why it is important to use the 'leave no trace' principles while orienteering.

2 List at least two differences between orienteering and geocaching.

1 When working in teams during orienteering challenges it is important to have leaders, but equally important to consider everyone's opinion. What can happen if only one person's ideas or opinions are considered?

1 Research and list the contact details of two orienteering clubs or organisations you could contact to find out more about orienteering challenges in your local area or region.

Quiz
What type of orienteering challenges can I compete in?

HOW CAN I PARTICIPATE IN INITIATIVE GAMES?

Initiative games have been created to develop skills of decision-making, taking charge, carrying out a plan and leadership. In these games you must act to solve a problem or a particular challenge. These activities usually involve movement, some sense of risk or an unknown outcome. They require many skills from participants to complete the task, resulting in better cooperation and interaction between the players.

After an initiative game it is important to reflect on how you performed in the activity and what you learnt from the experience. Often the teacher will give you open-ended questions designed to draw out your learning for a specific focus, such as cooperation or communication. These questions will get you thinking deeply and making judgements about the process.

WELLBEING CHECK IN

This activity has been developed in collaboration with Dr David Bakker of MoodMission

Identify

The best leaders are able to encourage the people around them to deal with challenges. Are you a good leader to yourself?

Understand

Sometimes we get caught in the trap of 'negative self-talk'. Negative self-talk is when you say or think bad things about yourself. Although everyone might do it from time to time, it can cause problems if it happens too much. Sometimes people think that being hard on themselves will help motivate them. This isn't completely true though; research suggests that the best way to motivate yourself or others is through positive encouragement.

Practise

1 Think of an upcoming challenge you have.
2 Now think to yourself, 'If I had to encourage someone else to deal with that challenge, what would I say to them?' Write down some of your ideas.
3 Take a moment to consider whether any of your friends are facing the same challenge or a similar one. Is there anything you could say to them that would be encouraging? Write down some of your ideas.

Reflect

The next time you notice any negative self-talk, what could you do for or say to yourself to combat it? Is there anything you could do to remind yourself to use the strategies you come up with?

Stepping stones

Two of the most difficult skills to master are discipline and focus. Most people are good at these in short bursts of time, but generally it takes a lot of practice to use these skills over a long period of time. Before you start this next challenge, think about some of the important attributes of being successful as a group or team. Briefly discuss how these can have an impact on a team when they are challenged!

Aims

To promote teamwork, communication and trust; to encourage creativity; to develop coordination and focus.

Method

You will be placed in groups of about five or six behind a line that represents the safety zone. Your challenge is to move to the other safety zone, about 15–20 metres away. You can only move across the 'forbidden area' between safety zones by using a 20–30 centimetre plastic disk known as a stepping stone.

Rules

- Every player must be in contact with the stepping stones at all times.
- If a stone is left untouched while your team is crossing the forbidden area, it will be taken away from your team.
- If someone touches the forbidden area, then the whole team must start again.
- If someone touches the forbidden area and happens to be the only person touching a stepping stone at that time, then that stepping stone will also be removed.
- You have a set time to complete the challenge. (Depending on the time available, two or more teams will be competing at the same time.)
- If the stepping stones are higher than the floor, be careful not to fall off as you move from one step to another.
- Your team has approximately five minutes to plan their strategy. Good luck!

Discussion

Spread out in pairs or groups of three from the group that you were in. You will need to respond to the following question with your partner:

1. What was your group best at in the last challenge? Now change partners and share your thoughts again with another member of your group. As a class you can volunteer your thoughts based on these discussions.
2. What does your group need to improve on for the stepping stones challenge?

   Discuss this again and then repeat as above. Finally, the whole class can discuss their thoughts about this question, based on the discussions with their groups.
3. Before you started this challenge you discussed what skills are needed for successful teamwork. Identify the skills you used in completing the stepping stones challenge.

   Repeat your discussion as above.
4. Propose variations to increase the difficulty of the challenge. For example, would a shorter time make a difference? Or if each team has to carry something with them (for example, a medicine ball)?

Source: Reprinted with permission from playmeo. To access more interactive group activities, go to www.playmeo.com.

UP AND MOVING Madagascar

Aim

To build communication and teamwork.

Method

You are involved in a new company that has decided to relocate its main office. Because your company has done so well it requires new staff. The office relocation will involve you standing on a tarp and turning the tarp over while standing on it. Your new staff are represented by markers/domes spread around the tarp. You will have to collect these as part of the challenge.

Rules

- Your team must all be standing on the tarp to begin. Then, with no one touching the floor, the tarp has to be turned over completely to the other side.
- If any member of your team touches the floor, the team will have to start again. Each member of your team will be given a stepping stone.
- Team sizes will vary according to the size of the tarp (a 2 × 3-metre tarp will have 8 to 15 people).
- Once you have turned your tarp over, you must collect all the new employees (domes or markers) by using only the stepping stones in your team. The group must be in contact with every stepping stone at all times; any 'untouched' stepping stones will be removed.
- If someone happens to touch the floor, they must return to the tarp and they lose their stepping stone.
- Stepping stones can be used on the floor when the group is attempting to flip the tarp over.

Source: Reprinted with permission from playmeo. To access more interactive group activities, go to www.playmeo.com

String maze challenge

Aim

To build communication and teamwork.

Method

Find a space in the gym or outdoors for the string maze to be set up. You can use trees, furniture, door handles, netball/basketball posts, etc. Construct the maze by zigzagging the string through the space, wrapping it around branches or furniture at different heights and sides. You will need at least 30 metres of string (available from craft/hardware retailers) to set up the challenge.

Rules

You must get from one end/side of the maze to the other without touching the string.

Variations

Work in pairs; one is blindfolded and the other guides them through by using effective communication, making sure they are careful to avoid touching the string.

Set a time limit and note how far through the course different groups get before time is up, or time each person/pair through the course and deduct one point for each 'string touch'. The winner is the fastest with the least point deductions.

Source: Reprinted with permission from playmeo. To access more interactive group activities, go to www.playmeo.com.

INVESTIGATION

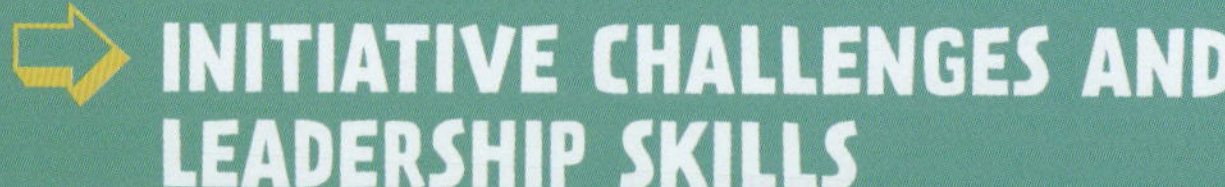

INITIATIVE CHALLENGES AND LEADERSHIP SKILLS

Purpose

To investigate initiatives/challenges that are different to those listed, and to be able to take the rest of the class through the challenge (with help from the teacher).

Method

1 Search the internet for suitable initiatives or challenges you believe you can take the rest of the class through (safety needs to be carefully considered).
2 An activity may need to be modified so it can be undertaken in a certain school setting. For example, the equipment, space or rules may need to be varied to enable the activity to 'work' in specific school settings.
3 Make sure your activity is accessible to everyone in the class.

Discussion

1 Write a clear outline of what the initiative or challenge involves – remember that images can convey concepts very effectively.
2 Clearly state any rules that need to be respected while undertaking the challenge.
3 Consider how you will minimise impact on the environment as a result of the challenge being undertaken.
4 At the completion of the initiative/challenge, reflect on how effectively you communicated the task requirements to your peers. If you could do the activity again, what would you do differently, and why?

UP AND MOVING

Traffic light debrief

Now that you have attempted this challenge it is time to see how and what your group experienced from the challenge. Each group is given three coloured cones, which represent the following:

- Red = things you would like to stop happening in the group
- Yellow = things that you are not sure about
- Green = things that you would like keep going

As a group, think about your performance in the previous activity and consider one thing about how the group worked together. Stand next to the coloured cone that best represents your thoughts about the challenge. Allow everyone in the group to give their explanation of why they are standing there.

Source: Reprinted with permission from playmeo. To access more interactive group activities, go to www.playmeo.com.

Quiz
How can I participate in initiative games?

CHAPTER 7 REVIEW

1 Describe what the 'great outdoors' means to you.

2 What are the different types of outdoor recreation?

3 Create a list of at least three benefits of being outdoors.

4 What are the main principles of 'Leave no trace Australia' in relation to use of the great outdoors?

5 When camping or visiting a national park, what are some ways that you can ensure your impact is minimal?

6 Why is it important to have a route planned before you undertake a journey in more physically challenging or remote country?

7 What is orienteering? How does orienteering help with navigating and using a map?

8 Discuss how initiative games can be used to improve confidence, teamwork and communication.

9 What have you learnt about yourself when you have been blindfolded or had to complete a certain challenge?

10 What are some important attributes for a team to be successful?

11 What have you learnt about your experiences in a team-based challenge?

12 Identify an example of an outdoor activity that allows you to be mindful.

PLAYING THE GAME AND BEING A GOOD SPORT

IN THIS CHAPTER

You will learn about different types of sport and games – how to play them, how to create them and how to get the most out of them.

'Sport is very important for building character because when you're involved in sport your individual character comes out, your determination, your ability to be part of the team and the acceptance of the collective effort is extremely important in developing your country.'
Nelson Mandela (2002)

shutterstock.com/Nicetoseeya

By the end of this chapter, you should be able to:

- reflect on different types of games
- identify the characteristics of an enthusiastic and competent sportsperson
- propose games to suit a wide range of different abilities
- understand the significance of games in different cultures
- apply movement skills to solve strategic and tactical problems in different games and sports.

HOW DO YOU BECOME A COMPETENT, LITERATE AND ENTHUSIASTIC PARTICIPANT OF SPORT AND PHYSICAL ACTIVITY?

Quiz
Pre-chapter

Before you start, take the pre-chapter quiz to find out how much you already know.

Video
Case study: What makes an enthusiastic, literate and competent sportsperson?

One of the goals of health and physical education in Australia is to provide you with the tools and information that you need in order to become a competent, literate and enthusiastic sportsperson.

Alamy Stock Photo/Gary Mitchell, GMP Media

Figure 8.1 Australian, Canadian and English long jumpers come together to celebrate their medals at the Gold Coast 2018 Commonwealth Games.

THE ROLES OF SPORT

If you ask yourself, 'What is sport?', you will probably come up with a different answer to your classmates. Sport means different things to different people. Far from just being the games played at lunchtime or on the weekend, or the football, basketball or cricket matches that are on TV, sport has a much wider meaning when its role is examined beyond your local community or internationally.

United Nations (UN) an international organisation founded after World War II. Its aim is to maintain peace and security and to improve living standards and human rights around the world

The **United Nations** views sport as a way of promoting peace because it disregards the borders of countries, religion, wealth and social class. Sport can be a way to strengthen social bonds and promote friendship, peace, acceptance, tolerance and justice.

There is more to physical education than learning how to play different sports and learning new skills. 'Games' are learnt so players can become skilled and thinking participants in a sport. However, while sports may share similar strategies, tactics and skills, each sport carries its own unique culture, customs and traditions in order to promote the ideals set out by the United Nations.

The modern Olympic Games are a good example of the United Nations ideals in practice. Pierre de Coubertin, the 'father' of the modern Olympic Games, probably understood these concepts better than most. He once said, 'May joy and good fellowship reign, and in this manner, may the Olympic torch pursue its way through ages, increasing friendly understanding among nations, for the good of a humanity always more enthusiastic, more courageous and more pure.'

CASE STUDY  PIERRE DE COUBERTIN, 1863–1937

Identify

Pierre de Frédy, Baron de Coubertin was born into a noble family in Paris, France. He was a keen player of many sports, including boxing, fencing, horse riding and rowing. He believed that through physical education and organised sport, people could improve not only their physical strength, but also their character and social skills.

Alamy Stock Photo/Sueddeutsche Zeitung Photo

Figure 8.2 Pierre de Coubertin – 'father' of the modern Olympic movement

Understand

Coubertin was an admirer of ancient Greece and became interested in reviving the ancient Olympic Games. He believed that sporting competition among **amateur** athletes from all around the world would help to promote peace and understanding across different cultures. He also believed that the benefits of competitive sport were in the competition itself, not in the winning.

In 1894, he organised what would become the first meeting of the International Olympic Committee. It was agreed at this meeting that an international games would be held every four years, with the first to take place in Athens, Greece, in 1896. He served as president of the International Olympic Committee for 29 years and was responsible for developing the Olympic ideals that are still in place today.

amateur an athlete who is not paid to compete

In 2019, the Australian Olympic Committee announced a new Australian Olympic Change-Maker program. This has evolved from the Pierre de Coubertin Award, which was established to recognise secondary school students who embodied Olympic values in Australia.

The Australian Olympic Change-Maker program recognises and rewards secondary school students who are demonstrating the Olympic spirit through leadership and driving positive change in their communities. This can take many forms – from minor to major, and from the ordinary to the extraordinary, such as leading teams, coaching juniors, supporting seniors, making a difference at a sports club, championing a national cause or effecting change on the world stage.

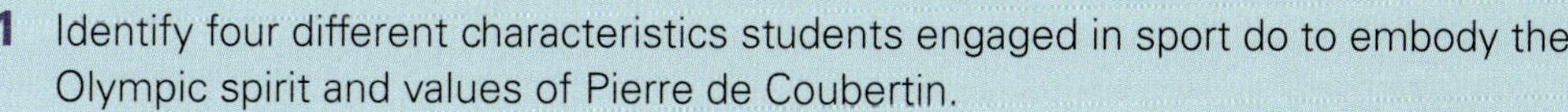

Discuss

1 Identify four different characteristics students engaged in sport do to embody the Olympic spirit and values of Pierre de Coubertin.
2 What specific things do these students do that embody the Olympic spirit?

Worksheet 8.1

What are competent sportspeople?

Competent sportspeople are people who have a wide range of movement skills and a broad understanding of a game, which enable them to participate in the sport in a meaningful way. They understand and can carry out a range of tactics and strategies that are appropriate for their game of choice, and they are knowledgeable game players.

What are literate sportspeople?

Literate sportspeople understand and appreciate the rules, rituals and traditions of sport. They can tell the difference between good and bad sporting practices, and they are willing to act on that knowledge to improve the practice of sport.

Worksheet 8.2

What are enthusiastic sportspeople?

Enthusiastic sportspeople participate in sport as part of a physically active lifestyle. They act in ways that preserve, protect and enhance their sport culture and help make playing sport more accessible to all people within their society.

Getty Images/Daniel Pockett; Getty Images/Michael Dodge – CA

Figure 8.3 Ash Barty is an extremely competent sportsperson, who has played both elite tennis and cricket.

Newspix/Brendan Radke

Figure 8.4 Competition is just one part of becoming a competent, literate and enthusiastic sportsperson.

WHAT WOULD MY ROLE MODEL DO?

This activity has been developed in collaboration with Dr David Bakker of MoodMission

Identify

Sometimes we get caught up in the moment and act in a way that we later regret. Maybe instead we could aspire to act like someone we look up to.

Understand

It's easier to see things from your perspective than anyone else's. After all, you're you! But this means that it's also harder to see solutions to problems that aren't in your current view. One way of coming up with creative, successful solutions to problems is to look to others for inspiration. It helps if you admire the person you look to.

Practise

1. Who are your role models? They could be sports stars, family members, friends, celebrities or even fictional characters. If you're struggling to think of any, try and think of people who value similar things to you and who you admire for their achievements. Write down two or three role models.
2. Think of a problem you've been experiencing recently. It can be as big or small as you like. Examples might include running late, having trouble staying organised or not getting enough sleep.

3 Now think about this: If one of your role models was in your situation, how would they fix it? Try and picture them in your exact situation. Would they talk to their friends? Get help from someone else? Make a committed plan to fix the problem? Write down what you think they might do.

4 How could you replicate what your role model would do? Now's your chance to make your own plan, inspired by the best.

Reflect

Did thinking about the problem from this different angle give you new solutions? Or did it help in some other way? Maybe you got a renewed sense of motivation to do something about the problem?

HOW TO BECOME A COMPETENT, LITERATE AND ENTHUSIASTIC SPORTSPERSON

Becoming a competent, literate and enthusiastic sportsperson involves understanding and participating in three key features of physical education:

- healthy competition
- positive sporting behaviour
- different sporting roles.

Getty Images/Richard Heathcote

Figure 8.5 Close competitions add to the excitement of sport.

Healthy competition

Healthy competition means that everybody will participate in a wide variety of different sports and games. There are no activities where anyone for any reason is excluded and, if necessary, everybody participates, regardless of their gender or sporting preferences. Every team will have a diverse mix of players who can each contribute to the team in a meaningful way and celebrate victories and losses together. Healthy competition means that no single team of players will ever dominate in every sport. Many teachers and coaches will ensure that competitions are close contests, which will add to the excitement of playing.

Positive sporting behaviour

Fair play and integrity are very important behaviours to learn and practise. This means that you, your classmates and your teacher need to be constantly practising and reinforcing positive sporting behaviour during your physical education class.

Alamy Stock Photo/Chris Putnam

Figure 8.6 Positive sporting behaviour should always be demonstrated.

FACE TO FACE Positive attitudes

Make a list of the behaviours demonstrated by people who have a positive sporting attitude.

Different sporting roles

You will be expected to participate in a range of sporting roles. Sport doesn't just happen! Lots of different people do a range of different jobs to allow you and others to participate in a sporting event. Very few people simply arrive at a sporting venue and begin to play. Teams need to train and practise, learn from coaches, arrange events to raise money for the club, design club uniforms and celebrate successes, both on and off the field. As well as being a player in your team, you may also be the captain, coach, trainer, manager, cheerleader, scorekeeper, journalist, equipment manager or another one of the important roles that need to be filled during a sporting contest.

ROLES

What roles do you need to fulfil in order to become a competent, literate and enthusiastic sportsperson? There are three types of roles you will be expected to carry out during your physical education classes:

- player roles
- duty team roles
- team roles.

Shutterstock.com/Dean Clarke

Figure 8.7 A student refereeing a game

Player roles

The first and most important role is that of the player. Everybody in your class is expected to actively take on this role and make a significant contribution to the team and the competition. To do this successfully, everybody needs to make an effort in learning new skills and techniques, playing hard and fair, supporting teammates, respecting the decisions of game officials and respecting opponents, both on and off the field.

Duty team roles

Certain roles have to be performed so that a sport can function properly. These roles are usually carried out during the competition phase of a sporting event. It is important that you learn how to carry out some of these roles. Most sporting competitions require at least a referee/umpire and a scorekeeper. Additional duty roles may include assistant referee, touch judge, boundary umpire, judges and so on. It is the duty team's responsibility to manage fair participation during the sporting competition.

Team roles

These are the non-playing roles that are necessary for teams to function. All the teams within your class will have students taking on roles other than player. Some of these roles may include coach, manager, trainer or fitness leader. It is important that you have a turn at a variety of these roles during your physical education lessons.

Table 8.1 The different roles and responsibilities you may be asked to carry out during your physical education lessons

Roles	Responsibilities
Player roles	
Player	➩ Try your best to learn new skills, apply those skills to solve problems in the games (tactics), and reflect upon what works and doesn't work (strategy) ➩ Play hard and fair ➩ Support your teammates ➩ Respect the decisions of officials ➩ Respect your opponents
Duty team roles	
Referee/umpire	➩ Make rule decisions during a contest ➩ Ensure the contest flows in a fair and just way
Scorekeeper	➩ Record and compile scoring sheets ➩ Keep an ongoing account of the status of competition (e.g. when points were scored/saved and by whom) ➩ Communicate scoring records to responsible team role holders (e.g. manager, etc.) and your teacher
Team roles	
Coach	➩ Provide team leadership and motivation for team members ➩ Conduct and coordinate team practice and training sessions ➩ Facilitate skills, tactics and strategy for the team ➩ Decide on team line-ups and substitutions to ensure play is equitable and fair ➩ Work with your teacher to ensure all team members are learning and contributing to the team in a meaningful way
Captain	➩ Represent and lead the team on the field ➩ Provide leadership and direction to the team on the field ➩ Assist and encourage teammates ➩ Ensure teammates maintain high levels of sportsmanship and respect for officials and their opponents
Manager	➩ Control the administrative functions of the team ➩ Complete and submit all necessary forms for the team ➩ Assist team members to carry out their necessary duty team and team roles where appropriate
Fitness leader	➩ Lead team warm-up, cool down and stretching sessions ➩ Conduct fitness sessions during team training ➩ Monitor team fitness improvements and gaps during the competition
Trainer	➩ Know the common injuries associated with a sport ➩ Monitor first-aid kits and ensure they are adequately stocked ➩ Notify the teacher of any injuries during training or competition and keep records of these ➩ Assist the teacher in the event of a sporting injury
Journalist	➩ Compile records and statistics and publicise them ➩ Contribute to regular school sport/PE reports ➩ Write a match report after every match ➩ Submit reports where appropriate

BECOMING A THINKING AND COMPETENT PLAYER

The most important role that you will be asked to carry out during your physical education lessons is that of a player. What does it take to be a thinking and competent player of games and sports?

Games and sports come about when people choose to construct an activity that is governed by a set of rules or laws. These rules outline why and how an activity should proceed or be played. Imagine you are standing with some friends in the playground of your school, in an open area with a ball. Ask yourself, what gets a game started with that ball? What happens to determine what you do with that ball? How many players on a team? How can you move the ball? How do you score points? And so on. These types of rules are known as fundamental rules, and they have a major influence on what a game or sport looks like and how it is played.

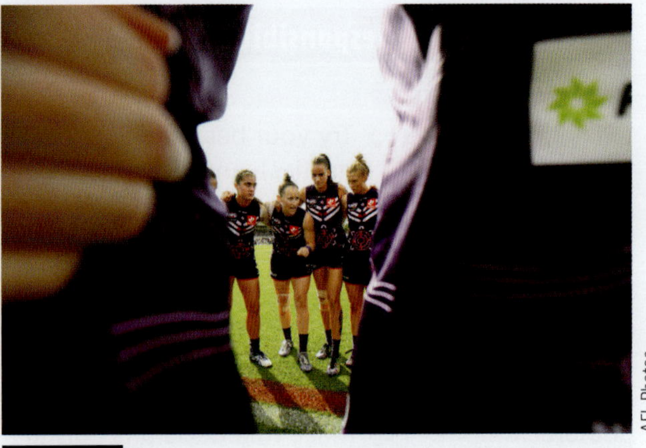

AFL Photos

Figure 8.8 Can you spot the team captain within this photo?

FACE TO FACE Schoolyard play

Think of a game you play with your friends in the schoolyard. In small groups, discuss the following questions:

1 How does the game get started?
2 How many players can play?
3 How do you score points?
4 What special rules exist?

Fairfax Syndication/Jon Reid

Figure 8.9 Schoolyard football often has its own unique rules.

Fundamental rules

There are four types of rules that apply to games. These are called the 'fundamental rules of games', and they can be broken down as follows:

- **Scoring rules** relate to the skills involved in scoring points.
- **Players' rights rules** relate to the rights of players and fairness in scoring points (i.e. taking turns, serving, batting rotations, kick-offs, etc.).

- **Freedom of action rules** relate to the way that players can interact with the ball or object, which gives the game a specific character (i.e. you can't run with the ball, you have only six touches before you lose possession, you must hit the ball within a specified area).
- **Physical engagement rules** relate to ensuring the rules of scoring, players' rights and freedom of action are respected through fair play and appropriate conduct (i.e. no tackling from behind, no contact can be made with the ball carrier, tackling must be below the shoulder and above the knee).

Rules also apply to the time and space of the game. They set out how points are scored, what skills are needed (how to play the game) and how to be a successful player.

Worksheet 8.3

When you change the rules of a game, the team strategy, tactics and the skills needed to achieve success are also affected.

CASE STUDY ⇨ QUIDDITCH

Identify

Quidditch is no longer limited to the pages of Harry Potter. The sport was first brought to the 'Muggle' world in 2005 in the US. Quidditch is increasing in popularity in Australia. Players even have to ride a broomstick!

Quidditch Australia/Ajantha Abey Quidditch Photography

Figure 8.10 The Australian national quidditch team at 'brooms up', the start of the game.

Understand

There are four positions in quidditch. The keeper, stands guard over the goals and generally leads and directs play on pitch. The chasers, three on pitch at one time, are the goal scorers of the game, and move the quaffle ball around the pitch. Two beaters are always on pitch. Using a dodgeball, beaters throw the ball at the other team to 'beat' them. If the opposing player is hit with the dodgeball, they must 'dismount' from their broom and return to their own goal hoops. One seeker from each team appears on pitch 18 minutes into the game. They must chase a 'snitch' around. Unlike the books and films, in real-life quidditch, the snitch is a person. The seekers must grab a small ball attached to the snitch, which ends the game.

Quidditch is considered one of the first gender-neutral sports. Only four individuals of any one gender can be on pitch at any time. If a team is found to be 'breaking gender', they receive a blue card and one player (the captain) is sent to a time-out box.

Quidditch is a full-contact sport, similar to rugby, with tackling allowed. Like any other full contact sport, this can result in injuries, and players train to both tackle and be tackled. However, not everyone can be tackled. You can only engage in physical contact with players of the same position as yourself. Only beaters can tackle other beaters; only chasers can tackle other chasers. Likewise, you cannot interact with balls owned by other positions. This means only beaters can interact with the bludgers; and only chasers and keepers can interact with the quaffles.

Discuss

1. From the above description of the game, identify one of each of the following rules:
   a. Scoring rules
   b. Players' rights rules
   c. Freedom of action rules
   d. Physical engagement rules
2. What impact do you think being a gender-neutral sport will have on:
   a. injury rate
   b. social inclusion?

   Justify your responses.
3. Try to recall other made-up sports from books and movies or TV. Pick one, and design a way to make it playable in real life.

HOW DO COMPETENT PLAYERS THINK?

Have you ever noticed how many sports seem similar? Or that the skills and tactics you use in schoolyard games are also present in many other games and sports? This is because most sports and schoolyard games share similar skills and similar tactical objectives.

Alamy Stock Photo/supershoot

Figure 8.11 Do I 'sweep' or 'block'?

Tactical objectives

Tactical objectives are the thinking skills that go with every skill you use in a game. They are the decision-making and problem-solving skills you use to help you achieve the objective in any game, which is usually to win the game or score points.

Tactical decisions

The tactical decisions you have to make during game play are usually related to either what you need to do or how you are going to do it. When you are considering what to do in different game situations, you need to be able to recognise different 'cues' and predict possible outcomes of your decisions. For example, there is no value in attacking a space near the goal if you can't see the cues or can't predict the chance of scoring or losing possession of the ball.

Figure 8.12 Two tactical decisions: moving into a space with opposition (left) or with no opposition (right)

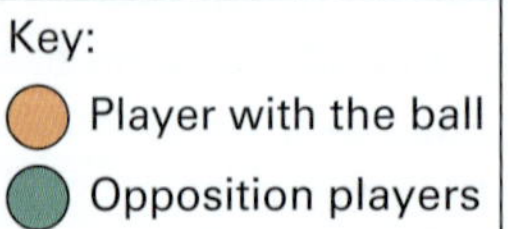

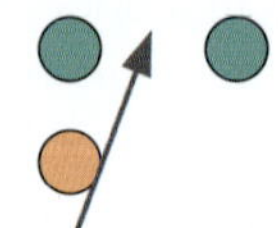

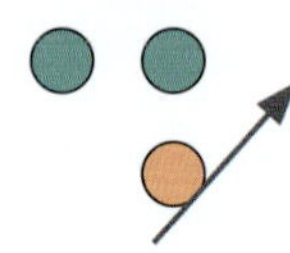

9780170463096

When you are making a decision about how you are going to achieve your desired outcome, you first decide on the best way to achieve it, then you choose the skills and movements you need. For example, where a large space is available but time is limited, a quick action may be best; when you have more time but accuracy is vital, some patience and control may be preferable. Such situations often arise near the scoring zones in games such as football, netball, basketball and hockey.

Figure 8.13 Do I shoot or pass?

Newspix/Steve Tanner

1 Make a list of team roles that your teammates might be able to fill. For any new roles (those not included in Table 8.1), list the responsibilities of the role. Ensure you consider the roles of the team coach and captain.
2 As a team coach, what skills will you need in order to be able to respond to the individual needs of your players?
3 As a team captain, what skills will you need to ensure that conflicts are effectively resolved, both within your team and also during matches?

1 Summarise five of the most important rules you can think of that apply to your favourite sport or game.
2 Next to each of the rules in your list, identify which of the four categories of fundamental rules they apply to (scoring rules, players' rights rules, freedom of action rules, physical engagement rules).

1 Investigate what SEPEP is, and identify its key goals in relation to physical education in Australia.

WHAT ARE THE KEY FEATURES OF A PHYSICAL EDUCATION PROGRAM?

There are several features of a physical education program that will allow you to become a competent, literate and enthusiastic sportsperson. These are closely linked to how sport and physical activity are conducted in community settings. These features include:

- having seasons
- having team membership
- participating in formal competition
- maintaining accurate records
- participating in crowning events
- participating in festivities.

In addition to these features, a competent, literate and enthusiastic sportsperson is able to modify existing games and even create their own. In order to be able to modify and create games, you need to understand how they are classified and what makes different types of games unique or similar.

THE GAMES CLASSIFICATION SYSTEM

The games classification system categorises games of similar intent. Games of similar intent are those that share similar tactical decisions. Specifically, the games classification system classifies most sports and games as being one of the following types:

- target
- net/wall
- striking/fielding
- territorial.

This system helps you recognise that skills and tactics learnt from one particular sport can be transferred into other games of similar intent. It also removes any doubts you may have about a particular sport (for example, 'that's a boys' sport' or 'I'm not good at netball'), instead allowing your knowledge and problem-solving to be shared across many different sports.

The games classification system also allows you to examine different sports in order of tactical complexity (from the simplest tactical problems to solve to the most difficult). Target games are the simplest in terms of tactical complexity, followed by net/wall games, then striking/fielding games. Territorial games are the most tactically difficult. The tactics you learn in the simpler games also appear in the more complex ones. In other words, there are tactics you will acquire in target games that will be used in all other game types. For example, many net/wall, striking/fielding and territorial games involve hitting a target of some type (for example, open space, a boundary line or a goal) in order to score points.

Worksheet 8.4

The next section briefly examines the four types of games and how they are further classified according to the difficulty of the tactics.

Target games

projectile any object that can be propelled into open space, e.g. ball, frisbee, stick, boomerang

The aim or tactical intent of target games is to place a ball or other **projectile** near, in or on a target to achieve the best possible score. These games encourage and develop a high degree of precision in the skills of hand–eye coordination, breathing, and concentration on a specific target. All target games can be classified as either 'unopposed' or 'opposed target games.

Unopposed target games

In unopposed target games, the shot you play is not affected by that of your opponent. In other words, you will always play the best shot you possibly can in order to score the most points or win the game. Examples of unopposed target games are golf and archery.

Opposed target games

In opposed target games, your opponent's shot will affect the decisions you make regarding your shot and will limit the shots you are able to play. Examples of opposed target games are lawn bowls and snooker.

Shutterstock.com/Patrick Foto

iStock.com/vgajic

Alamy Stock Photo/Kevin Wheal

Alamy Stock Photo/Janine Wiedel Photolibrary

Figure 8.14 a-d Golf and archery are both unopposed target games; lawn bowls and pool are both opposed target games.

Movement skills in target games

Target games typically require you to master two types of movement skill. The first is stability, or balance. The second is object control skill, which is the skill required to send an object away from the body, such as throwing or striking.

FACE TO FACE Opposed or not?

With a partner, identify each of the following target games as being either opposed or unopposed:

- snooker
- darts
- bocce
- ten-pin bowling
- quoits
- marbles.

UP AND MOVING — Film me!

Using your smartphone or another video recording device, ask a friend to record you playing a target game during class. Review the footage and answer the following questions.

1 How balanced was my body during my shot?
2 What could I do with my body to improve my balance?
3 How effective was I in sending the object away from my body?
4 What could I do to more successfully send the object away from my body?

FACE TO FACE — Tactical decisions in target games

As a class, discuss some common tactical decisions you need to make when playing target games.

1 Which shot is most likely to score me the most points? Use examples from both opposed and unopposed target games.
2 Which target do I shoot at to score the most points? Use examples from both opposed and unopposed target games.
3 When do I play a shot for points or a shot to protect my points? Use examples from opposed target games.
4 How do I shoot at an obstructed target? Use examples from opposed target games.

UP AND MOVING — Blindfold challenge

Working in pairs, play one of the target games you have studied. One person plays the game blindfolded and the other person gives instructions and feedback for each shot their partner plays. Start with playing an unopposed target game before moving to opposed target games.

Consider the types of feedback that you need to give your partner. Specifically, your feedback should:

- improve your partner's skill execution
- inform your partner where their shot falls in relation to the target
- inform your partner of obstacles or opponent shots.

Net/wall games

Worksheet 8.5

The aim, or tactical intent, of net/wall games is to send a ball or projectile into an opponent's court so it cannot be played or returned by your opponent. Open space on the opponent's court is your target. You need to be able to place a ball or projectile into the open space of your opponent's court while covering and defending as much of the open space as you can in your own court.

Most net/wall games are played on a 'court', but the surface of the court can be made of nearly anything, including concrete, clay, wood or even sand.

iStock.com/.shock

Alamy Stock Photo/Jürgen Hasenkopf

Getty Images/alacatr

Alamy Stock Photo/Hikupic

Figure 8.15 Tennis is typically played on hardcourt or clay surfaces, but occasionally (and rarely) on grass. Volleyball is typically played on hard surfaces like wood or soft surfaces like sand.

Movement skills in net/wall games

Net/wall games require you to master three types of movement skill.

1 **Locomotor skills**
   » Locomotor movement is when you move from one place to another. Locomotor skills include rolling, balancing, sliding, jogging, running, leaping, jumping, hopping, dodging, galloping, skipping, floating and moving the body through water to safety.

Source: VCAA

2 **Non-locomotor/Stability skills**
   » Moving on the spot without any change in location. Non-locomotor skills include twisting (the rotation of a selected body part around its long axis), bending (moving a joint), swaying (fluidly and gradually shifting the centre of gravity from one body part to another), stretching (moving body parts away from the centre of gravity), turning (rotating the body along the long axis) and swinging (rhythmical, smooth motion of a body part resembling a pendulum).

3 **Manipulative/Object control skills**
   » Movement skills that require an ability to handle an object or piece of equipment with control, such as kicking, striking, dribbling or catching a ball.

CASE STUDY 

SEPAK TAKRAW – AN ANCIENT ASIAN GAME FINDS A HOME IN AUSTRALIA

Identify

The ancient east Asian sport of Sepak Takraw, also known as Chinlone or Caneball, is helping refugees from Myanmar feel more at home in Australia.

Understand

The game is a national sport in many South-East Asian countries and is similar to volleyball but players use their feet instead of their hands to pass a ball made of handwoven **rattan**.

123RF.com/Sippakorn Yamkasikorn

Figure 8.16 A game of Sepak Takraw in progress

rattan fibre from a palm tree that is used to make a variety of products, including furniture, baskets and balls for Sepak Takraw

Many refugees from the Karen ethnic group that fled the war-torn country of Myanmar now play the game regularly in Australia, thanks to competitions run around the country.

The origins of caneball are unclear as it is played in many countries across South-East Asia, including in Thailand and Malaysia where it is called sepak takraw. In Myanmar, the game is thought to be about 1500 years old, and was originally more like a dance then a competitive sport.

When [Takraw player] Eagle came to Australia in 2008, after spending more than 13 years in a refugee camp in Thailand, he wanted to use the game to help young Karen people.

He set up Australia's first National Caneball Competition through his work at Multicultural Youth Services in Canberra.

In Bendigo, central Victoria, Nay Chee Aung, who works as a caseworker at Bendigo Community Health Services, recently organised the second annual Victorian competition. More than 120 people participated from around the state, including those from Werribee, Ringwood, and Geelong.

'My main goal when I started the tournament was to bring the community together,' Mr Aung said. '[I wanted] to reduce the social isolation and increase participation, which is really good for the community and also lets other communities know that this is the game that we play.'

Eagle is also encouraging women to become involved, despite it being a male-only sport in Myanmar and this year, six female teams played in the Canberra competition.

'Here in Australia women have equal rights, so I want women to feel important to our community,' he said.

Discuss

1 Conduct online research to find out where the game of Sepak Takraw is considered a national sport. Then search for a video clip of some professionals playing Sepak Takraw.
2 Like quidditch, Sepak Takraw is fighting gender norms in Australia. How is it doing this?
3 Based on the case study, why do people play sport?

Handball

1 During recess and lunchtime in most schools in Australia, you will find students (and some very keen teachers) playing handball. Make a list of all the rules you can think of that relate to playing handball at your school. For example:
- What happens if the ball hits a line?
- How many times is the ball allowed to bounce in each square?
- Who serves?
- What rules govern the serve and a fair return?

2 Using your smartphone or another video recording device, ask a friend to record you playing handball during class or recess. Then swap roles. Review the footage and answer the following questions individually.
- **a** How effective was I in covering the court space during the game?
- **b** What could I do with my body to improve my covering of the court space?
- **c** When did I look uncomfortable playing a shot?
- **d** What could I do with my body to be more comfortable in those situations?
- **e** How effective was I in sending the object away from my body?
- **f** What could I do to improve the way I send the object away from my body?

INVESTIGATION

Purpose

To explore the tactical decisions made in net/wall games.

Method

1 Play a game of tennis, or watch a video of a match from the Australian Open.
2 Analyse how yourself and others make tactical decisions during the game by answering the questions below.

Discussion

1 What position on the court gives me the most opportunity to return the ball:
- **a** on a serve?
- **b** during a rally?

2 What shot should I play to move my opponent out of their court space?
3 When should I play a shot into the back of my opponent's court?
4 When should I play a shot into the front of my opponent's court?
5 When should I play a shot to the left or right of my opponent's court?
6 What are the advantages and disadvantages of playing a lob shot?
7 What are the advantages and disadvantages of playing a drop shot?
8 What are the advantages and disadvantages of playing a smash/spike?
9 What are the advantages and disadvantages of playing ground strokes to the left and right edges of my opponent's court?

Worksheet 8.6

Striking/fielding games

The aim or tactical intent of striking/fielding games is to score more points than the opposition over a specified period of play (usually called an innings). Striking/fielding games typically involve two distinct phases of play. There is the offensive phase (when your team is batting) and the defensive phase (when your team is fielding).

Figure 8.17 Offence (batting) team
Shutterstock.com/sirtravelalot

Scoring

offence the phase in a game when you are attacking the opposition

When your team is in **offence** (batting), this is usually the only time during the game when you can score points against your opponent. You can usually score points (runs) in striking/fielding games using either of the following methods:

- running between or around a designated area, for example, running between the wickets in cricket or around the bases in softball
- hitting the ball into a designated area of the ground, for example, hitting the ball into the boundary in cricket to score four runs or hitting the ball over the fence between the foul lines to score a run in baseball.

defence the phase in a game when you are resisting an opposition's attacking phase

When your team is in **defence** (fielding), this is the only time you can prevent your opponents from scoring points (runs). You can usually prevent points being scored against your team in several ways, such as:

Figure 8.18 Defence (fielding) team
Alamy Stock Photo/Bruce Leighty – Sports Images

- bowling/pitching a player out, such as hitting the stumps in cricket or pitching three strikes in softball
- catching the ball on the full (i.e. before it touches the ground for the first time) when a batter has hit it into the field
- preventing a player from reaching a scoring zone by tagging them or hitting a target with the ball, such as tagging a player between bases in baseball, or hitting the stumps or throwing to first base before the runner gets there in cricket and softball, respectively
- occupying sufficient open space to prevent a batter from running to score points.

Movement skills in striking/fielding games

The movement skills involved in striking/fielding games are complex and varied but may include:

- **locomotor skills** such as running, walking and jumping
- **manipulative/object control skills** such as hitting a ball, stopping a ball along the ground or in the air (catching), throwing to a target and bowling or pitching.

Striking/fielding games encourage participation in various positions (strikers, runners, fielders, bowlers) by all players throughout the game.

Striking/fielding games: tactical decisions

When batting, there are several types of tactical decisions that you are likely to face:

- What shot do you need to play in order to score the maximum number of points?
- When are you best positioned to play such a shot?
- What shots are safest to play based on the fielding and delivery you face?
- When do you choose to run/steal points?

When fielding, you are likely to face different types of tactical decisions:

- What fielding structure allows the team to cover the most amount of space?
- In which position do you need to stand in order to reduce the number of points this particular batter can score?
- What type of ball delivery is likely to get this batter out?
- What type of ball delivery is likely to reduce the number of points this batter can score?

Territorial (invasion) games

There are two types of territorial games, but both have a similar aim or tactical intent.

Worksheet 8.7

First, there are 'goal' territorial games where points are mainly scored by putting a ball into a goal or net. Examples of these include soccer, lacrosse, water polo, Gaelic football, basketball, netball and hockey.

Second, there are **end zone** territorial games, where points are mainly scored by taking a ball into a designated scoring zone. Examples of these include rugby, gridiron, and touch football.

end zone a section at either end of a sporting field where a player or ball/projectile must cross in order to score points

Movement skills in territorial games

All territorial games, such as soccer, ice hockey, basketball, rugby and netball, involve many locomotor skills. These include running, stopping, turning, jumping and stability skills such as guarding.

Object control skills are also important in territorial games. They include sending away (kicking or throwing), receiving (catching and trapping) and retaining (dribbling and carrying).

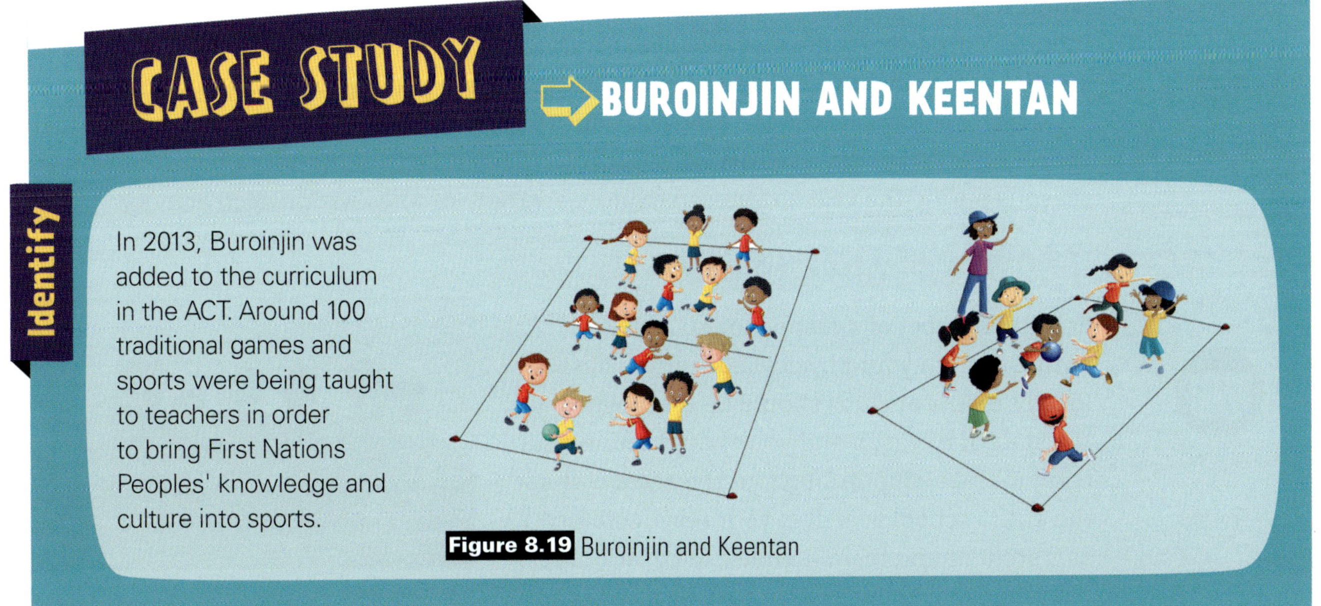

Figure 8.19 Buroinjin and Keentan

Understand

Buroinjin shares some similarities with sports like handball and netball. A central component of the game is passing to each other, meaning everyone is included in the game. The ball was originally made with kangaroo skin and stuffed with grass. Buroinjin developed from a game played by the Kabi Kabi people of southern Queensland.

Discuss

1 Do a web search for the First Nations Australian games of either Keentan or Buroinjin to find out more information.
2 Read about the game you chose. Identify the locomotor, stability and object control skills needed to play this game.
3 Now make a list of as many sports as you can think of that have similar skills and tactical decisions to the First Nations Peoples' sport you researched.
4 If you have time, play a game of Keentan or Buroinjin.

Tactics in territorial games

Territorial games generally involve a lot of physical activity and provide opportunities to develop a variety of skills, including locomotor movements, disposing of the ball, foot and hand–eye coordination and challenging other players for possession of the ball. Players must try to keep possession of the ball for as long as possible in order to attack open space and increase their scoring opportunities. When they are not in possession, they seek to block their opponent's open space and try to regain possession of the ball. Territorial games require players to develop communication skills with team members, and group decision-making is vital for success.

HOW TO DESIGN A SEASON

Worksheet 8.8

Scaffold
Use the planning table scaffold in Nelson MindTap to plan your next season of physical education with your teacher and class.

Most sporting competitions are based around seasonal participation. That is, sports are predominantly played during either the winter or summer months. In Australia, most territorial games, such as football and netball, are played during the winter months, with the striking and fielding sports, such as cricket and softball, played in the summer months. Some sports are played all year round, but are often broken up into several smaller competitive seasons so players can commit to playing for a set period of time. The idea of seasons is used in physical education to focus on learning about a particular sport or game type for a set period of time. In many cases, it is better to have only a few seasons during the year so you can explore a variety of roles within each activity.

TEAM MEMBERSHIP

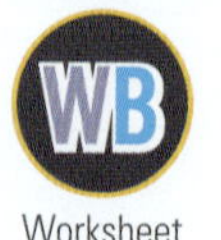

Worksheet 8.9

Worksheet 8.10

Being a member of a team should create enthusiasm in your physical education classes. Ideally, teams or clubs should be established at the start of a season and you should retain your team membership throughout the season, if not the entire school year. It is your responsibility as a student and team member to contribute to planning, practising and competing as a team. Having a long-term team membership will

iStock.com/pixdeluxe

Figure 8.20 Being part of a team has social meaning for everyone.

enable you to benefit from the **social meaning** of your sporting experiences, as well as from the personal growth that often comes from positive sport experiences.

social meaning knowledge and learning that comes from contact with other people

FORMAL COMPETITION

A typical sporting season has a schedule of formal competition and combined practice/training sessions. The team membership and formal competition together provide the opportunity for planning and goal setting.

INVESTIGATION

HOW CAN TEAMS BE PICKED TO ENSURE BALANCED AND HEALTHY COMPETITION?

Purpose

Pierre de Coubertin said, 'The most important thing is not winning but taking part; the essential thing in life is not conquering but fighting well.' Picking teams and making sure that balanced and healthy competition takes place is an important part of successful learning; winning is not as important. If team selections are done well, early competitive experiences can inspire you and can drive your personal sporting journey. If team selections are done poorly, your competitive experiences can end the sporting journey before it has even begun.

Method

1 There are a few things you and your teacher can do to ensure teams are balanced, and to increase the likelihood of a healthy competition.
   - Don't pick captains who then pick their teams. People usually pick their friends or the talented athletes first. Nobody ever wants to be picked last, and they're unlikely to do their best if they feel that nobody really wanted them on the team in the first place.
   - Don't randomly assign students by number. This doesn't usually work. Physical education classes are rarely large enough for balanced teams to result from this process.
2 In your PE class, investigate three other ways you could select teams.
3 In groups, identify the advantages and disadvantages of each method.

Discussion

1 How could a teacher break a class into evenly matched groups for any particular activity?
2 List three ways in which students could break themselves into evenly matched groups for an activity.
3 Which method do you personally think is the best way to form teams? Explain why.

RECORDS

It is important to keep accurate records of training sessions and competition matches. Records can include statistics on shots for goal, points scored, times, blocks, steals, assists and so on. Keeping accurate records provides feedback for individual players, your team and the class as a whole. Records also help your teacher assess your progress as a competent, literate and enthusiastic sportsperson. These records and statistics can also

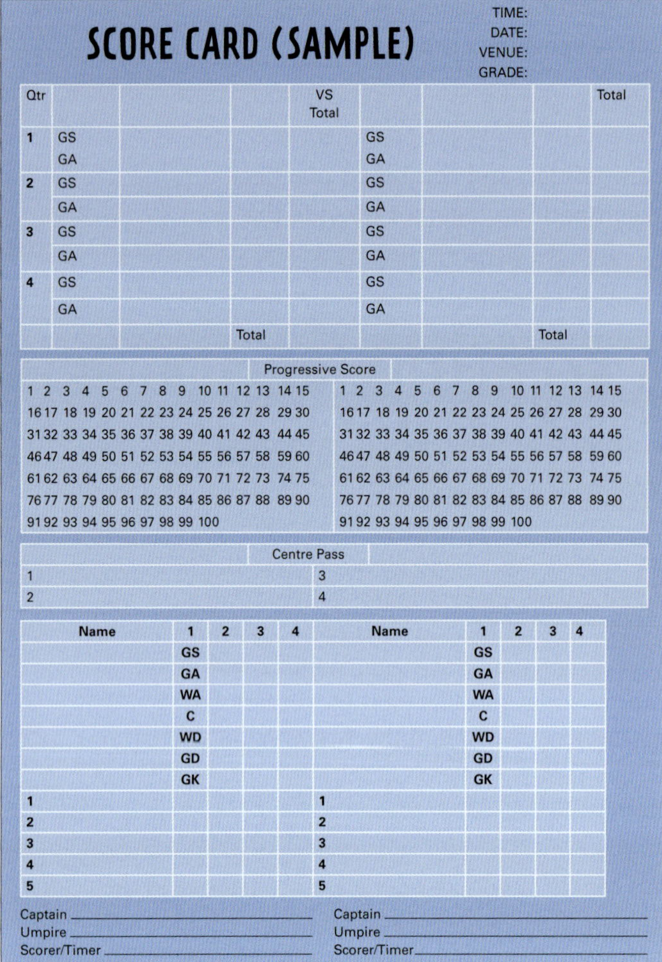

SCORE CARD (SAMPLE)

TIME:
DATE:
VENUE:
GRADE:

Qtr				VS Total				Total
1	GS				GS			
	GA				GA			
2	GS				GS			
	GA				GA			
3	GS				GS			
	GA				GA			
4	GS				GS			
	GA				GA			
			Total				Total	

Progressive Score

1 2 3 4 5 6 7 8 9 10 11 12 13 14 15	1 2 3 4 5 6 7 8 9 10 11 12 13 14 15
16 17 18 19 20 21 22 23 24 25 26 27 28 29 30	16 17 18 19 20 21 22 23 24 25 26 27 28 29 30
31 32 33 34 35 36 37 38 39 40 41 42 43 44 45	31 32 33 34 35 36 37 38 39 40 41 42 43 44 45
46 47 48 49 50 51 52 53 54 55 56 57 58 59 60	46 47 48 49 50 51 52 53 54 55 56 57 58 59 60
61 62 63 64 65 66 67 68 69 70 71 72 73 74 75	61 62 63 64 65 66 67 68 69 70 71 72 73 74 75
76 77 78 79 80 81 82 83 84 85 86 87 88 89 90	76 77 78 79 80 81 82 83 84 85 86 87 88 89 90
91 92 93 94 95 96 97 98 99 100	91 92 93 94 95 96 97 98 99 100

Centre Pass

1	3
2	4

Name	1	2	3	4	Name	1	2	3	4
	GS					GS			
	GA					GA			
	WA					WA			
	C					C			
	WD					WD			
	GD					GD			
	GK					GK			
1					1				
2					2				
3					3				
4					4				
5					5				

Captain ______ Captain ______
Umpire ______ Umpire ______
Scorer/Timer ______ Scorer/Timer ______

Figure 8.21 Sample score card

STATISTICS SHEET (SAMPLE)

(to be completed by statisticians)

HOME TEAM

NO.	Player's name	Kicks	Handballs	Marks	Goals	Behinds
		1 2 3 4 5 6 7 8 9 10	1 2 3 4 5 6 7 8 9 10	1 2 3 4 5 6 7 8 9 10		
		1 2 3 4 5 6 7 8 9 10	1 2 3 4 5 6 7 8 9 10	1 2 3 4 5 6 7 8 9 10		
		1 2 3 4 5 6 7 8 9 10	1 2 3 4 5 6 7 8 9 10	1 2 3 4 5 6 7 8 9 10		
		1 2 3 4 5 6 7 8 9 10	1 2 3 4 5 6 7 8 9 10	1 2 3 4 5 6 7 8 9 10		
		1 2 3 4 5 6 7 8 9 10	1 2 3 4 5 6 7 8 9 10	1 2 3 4 5 6 7 8 9 10		
		1 2 3 4 5 6 7 8 9 10	1 2 3 4 5 6 7 8 9 10	1 2 3 4 5 6 7 8 9 10		
		1 2 3 4 5 6 7 8 9 10	1 2 3 4 5 6 7 8 9 10	1 2 3 4 5 6 7 8 9 10		
		1 2 3 4 5 6 7 8 9 10	1 2 3 4 5 6 7 8 9 10	1 2 3 4 5 6 7 8 9 10		
		1 2 3 4 5 6 7 8 9 10	1 2 3 4 5 6 7 8 9 10	1 2 3 4 5 6 7 8 9 10		

VISITING TEAM

NO.	Player's name	Kicks	Handballs	Marks	Goals	Behinds
		1 2 3 4 5 6 7 8 9 10	1 2 3 4 5 6 7 8 9 10	1 2 3 4 5 6 7 8 9 10		
		1 2 3 4 5 6 7 8 9 10	1 2 3 4 5 6 7 8 9 10	1 2 3 4 5 6 7 8 9 10		
		1 2 3 4 5 6 7 8 9 10	1 2 3 4 5 6 7 8 9 10	1 2 3 4 5 6 7 8 9 10		
		1 2 3 4 5 6 7 8 9 10	1 2 3 4 5 6 7 8 9 10	1 2 3 4 5 6 7 8 9 10		
		1 2 3 4 5 6 7 8 9 10	1 2 3 4 5 6 7 8 9 10	1 2 3 4 5 6 7 8 9 10		
		1 2 3 4 5 6 7 8 9 10	1 2 3 4 5 6 7 8 9 10	1 2 3 4 5 6 7 8 9 10		
		1 2 3 4 5 6 7 8 9 10	1 2 3 4 5 6 7 8 9 10	1 2 3 4 5 6 7 8 9 10		
		1 2 3 4 5 6 7 8 9 10	1 2 3 4 5 6 7 8 9 10	1 2 3 4 5 6 7 8 9 10		
		1 2 3 4 5 6 7 8 9 10	1 2 3 4 5 6 7 8 9 10	1 2 3 4 5 6 7 8 9 10		

Figure 8.22 Sample statistics card

help to define the sporting traditions in your school and can be used for comparison by future classes. Record keeping can be done manually or electronically. The following are examples of record-keeping forms for netball and Australian Rules Football.

CROWNING EVENT

Sport is all about finding out who is the best team or player for a particular season, and for others to mark their progress towards improvement. Crowning events mark the end point, or high point, of the season, and create an opportunity for festivities and celebration, which are important aspects of play and sport.

FESTIVITIES

Sporting competitions are occasions for celebration, from the major festivals associated with the Olympic Games, to Sunday football games and children's soccer matches. In your physical education class, teachers and students should work together to create a constant festival that celebrates improvement, trying hard and playing fairly. This can be achieved with posters, team colours, player introductions, award ceremonies, digital recording and so on.

There are many ways you can make your team more festive and a special part of your schooling.

Names

Teams should have names. As a team, choose a name to define your vision. Don't use names that project violence or sexism, or are insulting or inappropriate in any way. It's easy to come up with some fun and creative names for your team. One way is to think of a particular feature of the sport you are playing and add an adjective or place name in front of or after it. For example, if you were studying water polo you might name your team 'The Goal Seekers' or the 'The Flying Dolphins'. Ice hockey is a sport known for its heavy body contact and rough play. However, one of the most famous teams in that sport is 'The Mighty Ducks'.

Rituals and chants

Many teams also engage in team rituals, either to prepare them for a big match or to celebrate a successful game. You can increase the festivity of your team by having a team mascot, or a team song that you sing after every match. The 'chant' or song of many sporting teams date back to the club's beginnings. Australians have a proud tradition of creating songs and chants in support of their favourite teams, as do European and South American football (soccer) teams.

Alamy Stock Photo/Cultura Creative (RF)

Figure 8.23 Most sports fans have their own team rituals and chants.

Display the results

Each team can also have their own bulletin board or website to post their results, statistics or even photos of the team performance. You could even vote for a player of the week and post a profile of that player on your bulletin space.

Alamy Stock Photo/Mar ano Garcia

Alamy Stock Photo/Action Plus Sports Images

Figure 8.24 Sportspeople have their own celebrations and rituals that they perform when they win.

Team colours

Most importantly, most teams get to choose their team colours and wear clothing that helps create team identity and promote festivity. You can be involved in deciding how your team colours can be used in many ways:

- Choose the colour lanyard or vests your team wears.
- Choose a particular colour for headbands or wristbands.
- Create your own dyed or screen-printed shirts. (Hint: speak to the art department in your school to help you with this.)
- Design a logo that can be printed onto iron-on paper and attached to your sports uniform. (Check with your teacher and principal before you do this.)

Awards

Awards are a key part of keeping sport festive. It is important to recognise hard work, performance improvement, victories and fair play.

Here are some things to remember when giving awards to people in your team.

- Good performance doesn't just include players who perform well or teams that win. It can also include officials who referee tough games well, or duty teams that get fields or courts prepared successfully.

Shutterstock.com/nikkytok

Figure 8.25 Trophies celebrate individual or team achievement.

- It is very important to recognise and celebrate fair play. Think about recognising the fair play of an opposition player on your bulletin board or website.
- If awards are given, many people need to agree with the awards being given. Maybe set up an 'awards committee' at the beginning of each season, with members from each team agreeing who the big awards, such as 'Most Valuable Player' or 'Most Improved Referee', will be given to.
- Coaches and managers who fulfil their duties should be recognised with an award. It is difficult to coach and manage people your own age, and you should recognise students who perform in these roles especially well.

WELLBEING 

Identify

We all see different things as important. What about you? What do you value?

This activity has been developed in collaboration with Dr David Bakker of MoodMission

Understand

Research shows that people are happiest when they are doing things that they think are important. Sometimes these things are called values. Picking your most important values can help identify what you already are doing to live a good life, and can help clarify how to live a more valued life.

Practise

1 Below is a short list of personal qualities/characteristics that many people find important. Read through them all and circle your top five. These are the things that you believe are most important to living a good life:

a Having fun
b Staying active
c Being thankful
d Finding peace
e Understanding
f Asking for help
g Being creative
h Achieving
i Knowing things
j Being free
k Being kind
l Trusting others

8

2 What's something small that you could do that fits with one or more of these things? For example, if you picked Understanding, you could ask a teacher to explain something you don't know. Note down your idea.

Reflect

Does one of your top five jump out at you as being the most important? Why do you think it's important? Have you been taught this by your family or someone else? How are you already working towards your top value?

1 List the different types of surfaces that tennis is played on.
2 List the different types of surfaces that volleyball is played on.
3 In one sentence for each, describe how you score points in the following games or sports:
- soccer
- netball
- cricket
- tennis
- darts.

4 From what you now know about striking/fielding games, what tactical decisions will you be able to transfer from net/wall and target games the next time you play a striking/fielding game?
5 Some examples of popular crowning events include AFL/NRL/A-League grand finals, but it doesn't always have to be a grand final event. Make a list of crowning events you are familiar with, and describe the festivities and celebrations that occur at them.

1 Make a list of all the different sports you will be exploring during physical education this year.
   a Next to each of the sports, think of an animal, object or action that has an association with that sport. For example, in Sepak Takraw, elite players will frequently do a bicycle kick in the air to spike the ball into an opponent's court. It looks like a flip. So, an action you might associate with Sepak Takraw is 'flipping', and an appropriate team name could be the 'Bathurst Flippers'. In Australian Rules Football, many of the teams are named after birds or objects that fly because of the jumping action that players use to challenge for the ball or to take a mark. Team names include the Sydney Swans, Hawthorn Hawks, Adelaide Crows, West Coast Eagles and the Collingwood Magpies.
   b Using your list, come up with a new and creative team name you might use when you are participating in one of these sports.
2 Get some ideas for a team song by doing a web search of your favourite sporting teams and listening to their team song or chant.
   a Pick one of your newly created team names and search on the internet for song lyrics that contain one or more of the words that are in your team name.
   b See if you can change the words in the song to make them more appropriate to your team and sport.
3 Make a list of the awards that you think should be offered for the following achievements. Try to be as creative as possible with your award names.

Weblink
As an example to get you started, the NRL appropriated the words from the Hoodoo Gurus' classic song 'What's My Scene' to 'That's My Team'. Follow the link to watch the original then search online for the NRL version.

a Award names for filling various roles, including player, coach, manager, official, etc.

b Award names specific to the sports listed below (e.g. 'Golden Boot' for most goals scored in a soccer season):

- soccer
- volleyball
- archery
- softball
- basketball
- plus three more sports of your own choosing.

4 Imagine your teacher has decided that you will be assigned to one team for the whole school year. During the year you are going to participate in four different sporting seasons: soccer, baseball, volleyball and golf. Design an activity that your teacher could use to ensure that the players of each team represent a wide variety of skills, thus increasing the chance of healthy competition.

1 Design your own competition schedule. Based on your Health and Physical Education timetable over the coming term, allocate which lessons could be used for team practice and training and which could involve a competition match. Ensure that you have the opportunity to play each team, and that you leave enough time for the final event and festivities.

2 Sporting events can bring communities together, or sometimes tear them apart. There are many examples (both current and historical) where sport has brought out the best and worst in humanity.

a Research an instance when sport exemplified the best in humanity, and report back to your class. Be sure to identify particular details that separated the event from the thousands of others that occur around the community, country and world on a daily basis.

b Now think about when sport was used or appropriated for the wrong reasons. What other underlying social tensions must have existed for sport to be used in this way?

Quiz
What are the key features of a physical education program?

3 Reflect on the experiences you have had as a member of a team (it doesn't have to be a sporting team).

a Make a list of positive and negative experiences you had with that team.

b Looking at your negatives list, suggest what could have changed in order for that experience not to have been a negative one for you.

HOW DO YOU MODIFY GAMES APPROPRIATELY?

One of the more difficult tasks you will be asked to perform during your physical education lessons is to modify and create your own games. In order to have the skills to do this, you will need to recall two of the concepts explored earlier in this chapter:

- the fundamental rules of the game
- the games classification system.

MODIFYING GAMES USING THE FUNDAMENTAL RULES

By simply changing one or all four of the fundamental rules in any game, you will change the way the game is played. Remember, these are the fundamental rules:

- scoring
- players' rights
- freedom of action
- physical engagement.

Let's see what happens when each fundamental rule for a popular game is modified.

Modifying scoring

By simply changing the rule that governs who, when and how players can score, the way the game is played can change dramatically.

For example, what would happen in a game of soccer if you also got points for kicking the ball over the crossbar? What would happen in a game of Australian Rules Football if a behind was worth more points than a goal?

Modifying players' rights

Players' rights include equal chances to score points during a game. In some games, this refers to the number of turns each player is allowed to have, or how many balls or innings can be played. In other games, especially net/wall and territorial games, it refers to the right to start the game and to restart after a break down in play.

For example, what would happen in a game of cricket if a team only got to face 20 overs or as little as 100 balls, as they do in short-form versions of cricket, instead of the 50 overs normally faced in a one-day test or over two innings in a five-day test match? Why do the players play different shots in these shorter games? Why do the bowlers bowl differently?

What would happen in a game of Rugby League if you reduced the number of tackles each team is allowed? How would you play the game differently if you were given only three tackles instead of the normal six tackles?

Modifying freedom of action

Freedom of action rules govern the special actions that a player in possession of the ball can do that other players cannot. These rules give a game its unique character. For example, if you compare netball and basketball, they are similar in many ways. They both require teams to move a similarly sized ball up and down a court of similar size and put the ball through a similarly sized hoop to score points. But when watching a game of netball and basketball, you see how different they really are – those differences are in the freedom of action that the ball carrier has in each game.

In basketball, players can move with the ball as long as they bounce it with one hand while they are moving. Once they stop moving and hold the ball with two hands, they must pass or shoot the ball. In netball, the ball carrier is not allowed to move more than one step once they have planted a foot on the ground. They can only move the ball by passing to teammates. In netball, not all players can shoot. Only the goal attack and goal shooter are allowed to shoot goals; all the other players must pass.

In both games, ball carriers cannot be touched by defensive players, but extra freedom of action is given to a ball carrier in netball because defensive players are not allowed to come within one metre of them. In basketball, defensive players can stand as close as they like, without touching their opponent, and can even try to take the ball from their opponent's hands.

Alamy Stock Photo/ALAN EDWARDS

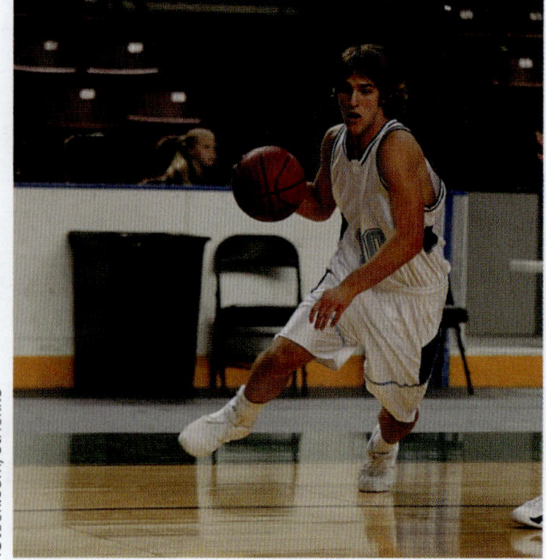

iStock.com/strickke

Figure 8.26 Netball and basketball look similar but have very different rules about the way players can move with the ball.

Modifying physical engagement

To ensure appropriate behaviour and that games foster fair play and sportsmanship, the rules of physical engagement are also important. These rules ensure that scoring, players' rights and freedom of action rules are respected and that cheating and poor sportsmanship are discouraged.

For example, most of the football codes involve some body contact. The physical engagement rules ensure that any contact between players also provides some protection to the player receiving the contact. There are restrictions on how players can be tackled, how many players can be involved and where on the body contact is allowed.

Other games also have physical engagement rules simply to ensure good sportsmanship. For example, according to the rules of tennis, players can be penalised for throwing or abusing their racquet during a match. Most sports restrict abuse towards officials and opponents.

MODIFYING GAMES USING THE GAMES CLASSIFICATION SYSTEM

When you want to modify a game or sport, you are not restricted to only changing the fundamental rules. You can also change aspects of each of the games based on the games classification system (page 320).

The games classification system is a useful tool for modifying different sports because many share similar features. You can change these features to make a game more or less difficult to play.

Modifying target games

Target games can be modified to suit the various learning needs and skill levels of different players.

Distance to the target

You can increase or decrease the distance to the target. When you increase the distance to the target, you also increase the level of skill and potential number of tactical decisions required in order to hit the target.

Figure 8.27 In archery, the distance, position and size of the target can be modified.

Size of target

You can increase or decrease the size of the target. Smaller targets are usually more difficult to hit and need greater skill and accuracy. By making the target bigger, you also make it easier to hit.

Position of target

Another way you can modify a target game is by moving the position of the target:

- You can place the target behind an obstacle, which is common in sports like snooker and lawn bowls. Placing the target behind an obstacle means greater skill and higher levels of problem-solving and decision-making are required in order to hit the target.
- Moving targets are usually more difficult to hit than stationary ones. Some target sports, such as trap shooting, have moving targets, but you can modify any number of target games by making the target a moving one.

Balls and scoring

Finally, you can also increase the difficulty and the variety of skill required in target games by changing the weight and/or size of the projectile or introducing a scaled and/or bonus scoring system.

UP AND MOVING

Buran or Yangamini

1 Go to the Australian Sports Commission website and search for the First Nations Australian game of Buran. Read the fact sheet about the game. Then, using each of the modifications you know you can make to target games, redesign the game to make it easier to play.

Illustration by Glenn Robey, reproduced by permission of the Australian Sports Commission.

Figure 8.28 Buran is a competition based on accurately throwing a boomerang.

OR

2 Also on the Australian Sports Commission website, you will find information about the First Nations Australian game of Yangamini. Using each of the modifications you know you can make to target games, redesign this game to make it easier to play.

Illustration by Glenn Robey, reproduced by permission of the Australian Sports Commission.

Figure 8.29 Yangamini is an object-throwing game (usually marbles or coins).

Modifying net/wall games

Net/wall games can also be modified to suit the various needs of players.

Modify the ball

The weight and/or size of the ball can be changed to make the game easier or more difficult to play. Think about playing handball at school. Instead of playing with a tennis ball, which is small and weighs a bit, the game would be easier if the ball was the size and weight of a volleyball. In the game of Sepak takraw (page 324), many beginners would not be able to play with the rattan ball to begin with. Why not play with a balloon and then gradually pick a heavier ball (like a volleyball), then a soccer ball, then eventually a size 3 soccer ball (which is about the size of a rattan ball). Some children in Thailand play Sepak Takraw with a tennis ball!

Shutterstock.com/Africa Studio

Figure 8.30 Comparative sizes of balls used for various sports

Tennis Australia runs a program for children called 'Hot Shots'. This program modifies the game of tennis to make it easier for children to learn. In Hot Shots, kids learning to play tennis use red, orange and green low-compression balls that don't bounce as high as the usual yellow tennis balls, making them easier for new players to hit.

Modify the racquet or bat

Another way of modifying net/wall games is to change the shape and/or size of the bat or racquet. Playing with an oversized racquet face can make the ball easier to hit. Many racquets and bats for net/wall games are now available with shorter shafts (handles) and smaller grips for younger, smaller players.

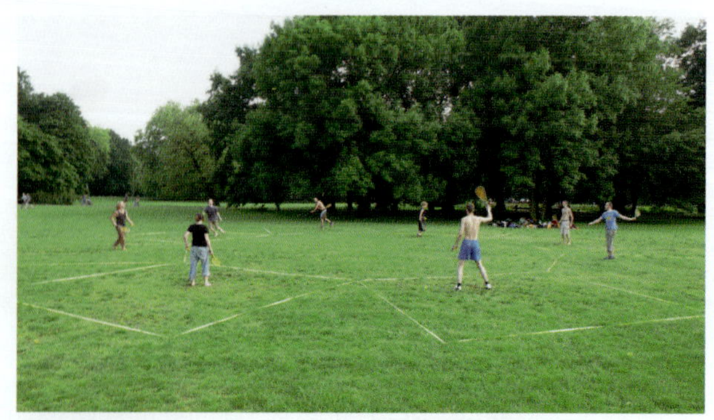

Speedminton Australia

Figure 8.31 Speedminton® is played without a net. The playing zones are usually marked by tape on the ground.

Modify the court

Many modified net/wall games will also change the size and dimensions of the court space to decrease or increase the difficulty of the game. In Hot Shots the courts are smaller, which makes them easier to play on because there is a smaller area to cover. This also helps new players develop good footwork patterns and encourages use of all parts of the court. A tennis court is not even necessary for Hot Shots – it can be played on any flat, hard surface.

Speedminton® is a new game that modifies the traditional net/wall game of badminton. Speedminton® has no net or adjoining court space. It is a classic example of modifying court space to increase or decrease difficulty of skill and tactical decisions in net/wall games.

Modify the net

A common modification to net/wall games is to adjust the net height. A very low net is easier to defend when players are close to the net; a high net is much more difficult. It is also harder to play offence shots like a smash on a high net.

Weblink
Do a web search for Speedminton. What equipment do you need to play this game?

Modify the rules

You may also choose to change the rules. For example, you may allow more bounces before the ball is returned, or a greater number of fault serves. Very few people would be able to play Sepak Takraw if the ball was not allowed to bounce. Try playing it in your class with a volleyball, and allow the ball to bounce once or twice before another person has to kick it. If you play one bounce/no fault handball at lunch and recess, why not allow a new player (or the teacher) to have two bounces in their square and one fault serve?

Modify the team

Finally, you can vary the number of teammates or opponents in net/wall games. Increasing the number of your opponents increases the difficulty of you being able to play the ball into open court space. It also increases the chances for the opposing team to return the ball. A common modification in tennis is American Doubles. In American Doubles, the better player plays on the singles court against two players on the doubles court.

Modifying striking/fielding games

Modify the equipment

The size of the bats and balls used to play a striking/fielding game can be modified. As with net/wall games, a larger bat or ball makes it easier to hit and catch.

Modify the field

The size and dimensions of the playing field can also be changed. This affects both the skills and tactical decisions needed to play. A larger field usually requires more skill to play a striking/fielding game because the fielding team must cover more open space and must pitch, bowl and throw further. For the batters, a larger field means they will have more open space to hit, but scoring points by hitting into designated scoring zones, such as into the boundary or over the fence, will be harder.

Modify hitting

Methods of hitting and delivery can be changed. Examples include having a bowler, allowing a bounce ball, hitting from a tee, allowing more time to hit (or allowing more strikes) and allowing more time to field the ball.

Modify the team

The number of players on each team can be changed. A fielding team with more people can cover more open space, and so improve its chances. In the same way, having more batting players gives a team more opportunities to score points.

Modify scoring

The scoring system in striking/fielding games can be controlled in many ways (e.g. bonus points for particular plays). To encourage players to attempt longer shots, the number of runs awarded for hitting the ball over the fence could be increased. To encourage fast running between the wickets in cricket or between the bases in baseball, each run between the wickets could be worth two runs, or a point could be given for every base a runner is able to occupy. Point incentives could be awarded to the fielding (defensive) team or runs could be taken away from the striking team if the fielders catch a player out. This will also affect the type of shot the striking team is likely to play.

Getty Images/Peter M. Fisher

Figure 8.32 Do you have your own rules for backyard cricket?

INVESTIGATION

HOW CAN I MODIFY TERRITORIAL GAMES?

Purpose

To modify the rules of a territorial game.

Method

1. Select a territorial game of your choosing.
2. Addressing each of the headings below, identify how you would make the full version of this game both more AND less complex.

Discussion

1. Modifying the equipment: The weight, size and shape of equipment can all be changed. This is common in modified territorial games (especially for juniors). It can be difficult to kick a full-sized football or throw a full-sized netball.
2. Modifying rules: Scoring targets, the way points are scored and other game rules (e.g. time allowed in possession, areas allowed in, types of passes and movements) can also be adjusted. If you change the point systems, this also changes the way the game is played. Why are **field goals** more likely to be

field goal a type of goal in sports like rugby or American Football, where a player tries to kick the ball between the two tall goalposts, and over the crossbar

attempted in a game of Rugby Union than Rugby League? Field goals are worth three points in Union and only one point in League, so players often shoot field goals in Union when their running attack loses momentum; in a League game they often only shoot field goals to break a deadlock.

3 Modifying the playing area: To work on a specific tactic in a territorial game (e.g. short passing), you can change the size of the playing areas.

4 Modifying teams: Another way to make territorial games harder or easier is to change the roles of players and/or the number of teammates or opponents in any game.

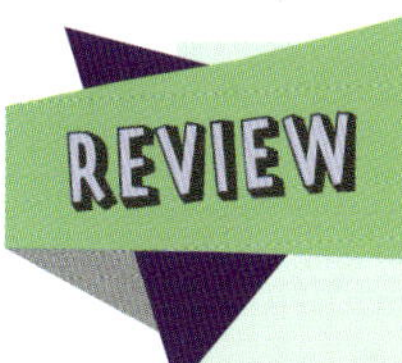

1 What is the difference in freedom of action rules for the following players?

a A soccer goalkeeper versus a soccer field player

b A T-ball batter versus a softball batter

c A tennis server versus a volleyball server

2 How do you think the games would be played if you changed the players' rights rules in the following sports?

a Volleyball, with the team that loses the point getting to serve the next time

b Baseball, if batters were only allowed one strike instead of three

1 How do you think the games would be played if you changed the scoring rules in either of the following sports?

a Rugby, by awarding a point for kicking the ball over the dead ball line

b Golf, by adding an extra stroke to your score if you hit the ball off the fairway

2 For each of the situations shown in the following diagrams, where would an attacking player be best positioned to create an overlap opportunity?

a Rugby

b Hockey

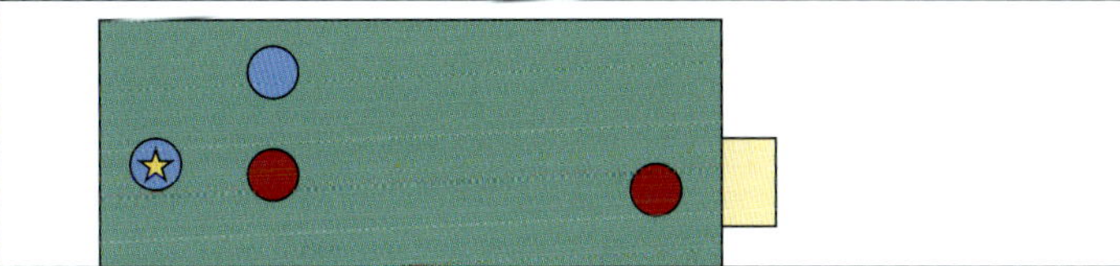

c Australian Rules

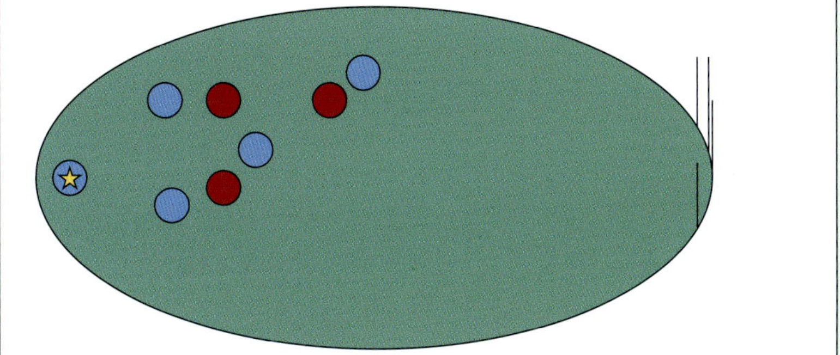

1 Territorial games are a culmination of tactical problems that people learn from playing target, net/wall and striking/fielding games.
   a As a group, select any territorial game and list the skills and tactics needed to be a competent and literate sportsperson.
   b Using that list, identify a game from each of the other games classifications that you could learn from and transfer into your participation in your chosen territorial game.

CREATING YOUR OWN GAMES

During this chapter, you have explored what it takes to become a competent, literate and enthusiastic sportsperson. You now know how games can be modified for different reasons and the importance of games and sports in some other cultures. Now you are ready to apply your knowledge of movement skills to solve strategic and tactical problems in many games and sports.

Worksheet 8.11

Children have been making up games to play with their friends since the dawn of time. Some of the best games you play are probably these informal schoolyard games that you created with your friends, or games passed down from generations of students before you. Now it's time for you to go and create your own game or sport legacy at your school!

Create a game

In this activity, you will create your own game with three or four classmates.

1 Select a game type from the games classification system (page 320).
2 List the scoring, players' rights, freedom of action and physical engagement rules for your game.
3 Collect the equipment needed to play your game and find a space to play. Remember, this is a schoolyard game, so it is better to have a smallish space and as little equipment as possible.
4 Instruct the rest of your class on how to play your game, then let them play it.
5 Watch how your classmates play your game, and change it as necessary to make it more or less difficult for different players. Remember, you can modify similar types of games using the games classification system.
6 Start to construct role descriptions for other participants of your game, such as umpire, commentator, coach, etc.
7 See if you can get other students at your school to play your new game during recess and lunch breaks. Then you will know how successful your new game is. Who knows, you may have even started a new sport that will be played in schoolyards for years to come! Congratulations!

Quiz
End of chapter

CHAPTER 8 REVIEW

1 Define what you understand a 'literate sportsperson' to be.
2 Summarise three qualities of a 'competent sportsperson'.
3 Write a definition of an 'enthusiastic sportsperson'.
4 Consider four roles you might be able to perform in your favourite sport.
5 What are the four 'classifications of games'?
6 Name at least two sports from each of the games classifications.
7 Recall the four fundamental rules of games.
8 Describe how two sports from the same games classification differ, based on your knowledge of fundamental rules (e.g. soccer and hockey, or cricket and baseball).
9 Propose why most sports are organised around 'seasons'.
10 What is a 'crowning event'?
11 Discuss the importance of crowning events in the culture of a sport.
12 Explain why maintenance of accurate records is so important when participating in a sport or game.
13 Describe a 'festivity' associated with your favourite sport or sporting event.
14 Based on your reading and additional research, what role does traditional sports and games play for First Nations Australians?

ENHANCING PERSONAL FITNESS THROUGH LIFELONG PHYSICAL ACTIVITY

9

IN THIS CHAPTER

You will discover that being active throughout your life has a major impact on your health and wellbeing.

iStock.com/pixdeluxe

By the end of this chapter, you should be able to:

- understand how lifelong physical activities enhance your health, fitness, and wellbeing
- determine which health-related and skill-related components of fitness are being developed in your life
- design, apply and evaluate a range of self-management strategies used to increase physical activity levels and reduce sedentary behaviour to enhance health and wellbeing outcomes for yourself and others
- explore and evaluate self management strategies to increase physical activity levels, health, and wellbeing outcomes
- apply a range of methods to measure your body's response to physical activity.

WHAT IS LIFELONG PHYSICAL ACTIVITY?

Quiz
Pre-chapter

Before you start, take the pre-chapter quiz to find out how much you already know.

One of the aims of health and physical education is to develop your fitness, health and wellness through participation in lifelong physical activities. Sometimes the terms 'health' and 'fitness' are used interchangeably, but they are two different things. The term 'health' is used a lot in this book, but the focus of this chapter is on lifelong physical activity and fitness.

iStock.com/pixdeluxe

Figure 9.1 Lifelong physical activities are activities you can do in your daily life, like skating at skate parks.

Lifelong physical activities

Improve health: physical wellness and absence of illness and disease. Includes social, emotional, spiritual, cognitive and physical health.

Improve wellness: how a person feels about life and how effectively they can function.

Improve fitness: the ability to perform moderate to vigorous levels of physical activity without fatigue, and the capability of maintaining such ability throughout life.

Figure 9.2 How lifelong physical activities link to health, wellness and fitness

LIFELONG PHYSICAL ACTIVITIES

Lifelong physical activities are also called 'lifestyle' or 'lifetime' physical activities, and should make up the majority of your physical activity on a daily or weekly basis. These are the types of activities that are part of your daily routine, such as walking, skateboarding, BMX riding, playing games like foursquare, playing games with friends, kicking the football, gardening, performing household chores, playing tennis, throwing a frisbee, swimming, dancing, playing golf and canoeing. All modern health and physical education programs include a range of lifelong physical activities, as well as team and individual sports.

Table 9.1 Lifelong physical activities

Lifelong physical activities
1 will improve your health if you perform them regularly
2 should be performed daily
3 can easily fit into your daily routine
4 may feel light and easy for you while you are young, but they will make you work harder when you get older

5 are easy to perform even with a low personal level of fitness or skill
6 include sports and recreational activities that are easy
7 use more energy than just sitting down
8 should be kept up throughout your whole life
9 require very little equipment
10 can be performed anywhere, anytime and even on holidays
11 can be performed by yourself or with one or two others
12 are more likely to become part of your daily routine if they don't make you 'huff and puff' really hard

WB Worksheet 9.1

FEELING FIT

People sometimes say they feel 'fighting fit', but what does this actually mean? To be 'fit' you need to think of your mind and body as if they are a high-performance race car. Like the engine in the racing car, your body needs high-quality fuel to perform. Your body is fuelled by food (as you read in Chapter 2); like an engine, it needs to be fine-tuned and maintained, and this is achieved through regular physical activity.

Dreamstime.com/Leigh Warner

Shutterstock.com/Pete Saloutos

Figure 9.3 Feeling fit requires fine-tuning, just like a high-performance race car.

tailored created to suit the individual needs, interests and context of a specific person or group

Fitness is specific to the daily needs of each person: their occupation, sports and daily roles and routines. Every person has a different level of fitness and has different needs. For example, a bricklayer who is physically active for most of their working day will have very different demands on their body compared to an office worker. Just as physical demands and fitness levels vary between people, so do the types of physical activities people participate in. There is no 'one size fits all' program to suit everyone. Training or physical activity programs need to be **tailored** specially to people's needs, interests, abilities and level of fitness.

Worksheet 9.2

Quiz
What is lifelong physical activity?

123RF/Daniel Kaesler

Shutterstock.com/imtmphoto

Figure 9.4 A person who does manual labour is far more active at work than an office worker.

1 Outline three characteristics of lifelong physical activities.

1 Refer to Table 9.1. Analyse several activities you engage in on a regular basis. Identify the essential elements of each and determine which of these activities meet the criteria to be considered lifelong physical activities. Then record each using a table and write true or false next to each statement in the table.

1 Analyse one of the main lifelong physical activities you engage in regularly, identify which essential elements of the activity are like the list for an activity to be considered a lifelong physical activity. Do some research to determine where you could engage in this lifelong physical activity within your local community, outdoor settings, and spaces.

Video
Fitness: What do you think of when you hear the word 'fitness'? Watch the video and join the discussion!

WHAT IS FITNESS?

Worksheet 9.3

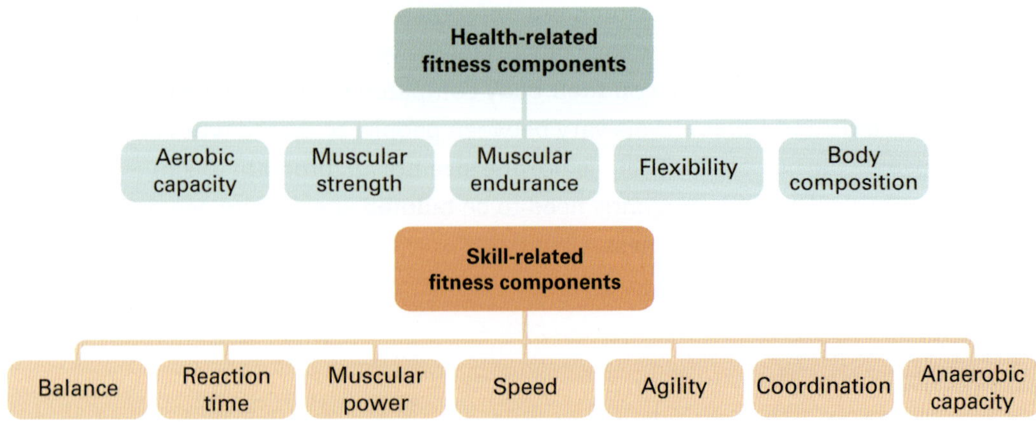

Figure 9.5 Health- and skill-related fitness components

9780170463096

UP AND MOVING Design an activity

1. Bring a range of items from home such as a broom handle, towel, empty plastic bottles, milk crates, tarps, buckets, boxes, long sports socks/stockings, pool noodles or inexpensive sporting equipment items (tennis balls, frisbees, skipping ropes). Use these items to create a physical activity you could do at home.
2. Later in the chapter, revisit the activity you designed to see if you can determine which health- and skill-related components of fitness it uses.
3. As a class, you could design a 'fitness circuit' consisting of several of the activities you have designed.

Worksheet 9.4

Worksheet 9.5

HEALTH-RELATED FITNESS COMPONENTS

Worksheet 9.6

Aerobic power

Aerobic power is also called aerobic fitness, cardiovascular fitness, cardiorespiratory fitness or aerobic capacity. Aerobic power is often considered the most important health-related fitness component. It is the ability of the heart, blood vessels and the **respiratory system** to supply fuel and oxygen to the working muscles, as well as the ability of the muscles to use the oxygen for sustained exercise. Aerobic power should be the foundation of any physical activity program because a high level of aerobic capacity helps you to run, walk, cycle or swim. It also helps you to recover after high-intensity exercise.

Excellent lifelong physical activities for developing aerobic power are brisk walking (power walking), jogging, running, cycling, swimming, participating in aerobics classes and dancing. The health benefits associated with aerobic physical activities help to prevent **chronic disease**.

aerobic power the body's ability to supply oxygen to muscles, and the ability of muscles to use that oxygen

respiratory system the system of organs involved with breathing, including the nasal passages, windpipe and lungs

chronic disease a long-lasting disease that is not usually cured completely

Muscular strength and endurance

Muscular strength is the capacity of the muscle or muscle group to exert force against a resistance. In contrast, muscular endurance is the capacity of the muscle or muscle group to exert a force repeatedly against a resistance. Table 9.2 outlines the difference between these fitness components.

Alamy Stock Photo/David R. Frazier Photolibrary, Inc.

Figure 9.6 Swimming is an excellent physical activity to increase aerobic power. What other components of fitness can be developed by swimming laps?

Alamy Stock Photo/Design Pics Inc

123RF.com/Jamie Roach

iStock.com/monkeybusinessimages

Getty Images/Dan Mullan

Figure 9.7 Muscular strength and muscular endurance are important for both sporting performance and everyday life.

Table 9.2 Muscular strength versus muscular endurance

Having good muscular strength	Having good muscular endurance
⇨ allows you to move objects or body parts with force (useful in sport and everyday life)	⇨ allows you to exert a force repeatedly, such as running up 12 flights of stairs
⇨ helps in weight loss and maintaining a healthy body weight, due to the increase in your **metabolism**	⇨ allows you to repeat movements for a long period without too much **fatigue**
⇨ allows you to control your own body weight when lifting or bracing against a force such as a football tackle	⇨ helps to avoid injury
⇨ is useful in sports such as weightlifting and in resistance training or lowering something heavy and pushing.	⇨ helps in activities such as rowing, cycling, swimming, push-ups and sit-ups.

metabolism the processes in the body that work together to use food for growth and energy

fatigue physical and/or mental exhaustion that can be triggered by stress, medication, overwork, mental and physical illness or disease

Keep going!

As a class, participate in an activity that requires muscular endurance, such as crunches, step-ups, half squats, push-ups, or walking or jogging upstairs.

Resistance training

Resistance training involves using free weights (dumbbells and barbells), bands, machines or your own body weight to place resistance on a muscle or a group of muscles. Resistance training is the ideal way to improve your muscular strength and muscular endurance. You will learn more about resistance training and circuit training in Years 9 and 10.

Getty Images/Cameron Spencer

Figure 9.8 Rowing requires muscular endurance of both the upper and lower body. Can you think of other physical activities that require muscular endurance?

FAST FACT

When you increase your muscular strength using resistance training, you actually increase the size of your muscles. The extra muscle, although it weighs more than fat, burns more energy, which aids fat loss. That is why people often put on weight or stay the same weight when they start doing resistance (weight) training, even though their clothes may become looser.

UP AND MOVING Push me

This activity shows the difference between muscular strength and muscular endurance. Find a classmate with a similar body size and strength to you, or watch a pair who are similar.

1. Push against each other (carefully) until one person takes two to three steps backwards.
2. Take each other's hands and cycle your hands in a circular motion carefully, as if you are winching up a sail on a yacht. Do this for two minutes.
3. Discuss which activity you think required muscular strength and which needed muscular endurance.

Worksheet 9.7

Flexibility

A flexible joint is one that can move through its full range of motion. Being flexible is important in your daily life because it allows you to do things such as throwing a ball for your dog at the park or bending over to tie your shoelaces. Joint flexibility is different for each joint. Just because someone has excellent flexibility in one joint doesn't mean they will have good flexibility in all joints. For example, you might have good shoulder flexibility but be unable to touch your toes. Being flexible allows you to move certain joints quickly and easily, and to avoid injury and chronic pain. The best way to improve your flexibility is to stretch regularly.

Shutterstock.com/Juriah Mosin

Figure 9.9 Stretching should be part of daily life.

Getty Images/Tom Werner

Alamy Stock Photo/WENN Rights Ltd

Figure 9.10 Flexibility is important to perform both sporting and everyday lifelong physical activities.

CASE STUDY  PILATES

Identify

Jacqueline Horne stumbled on to Pilates in her late 80s and now the Launceston resident said the exercise helps her with balance and strength.

Understand

Ms Horne took part in a study through the University of Tasmania eight years ago and after being shown a few different exercises, she was hooked.

'They showed us a few Pilates exercises which reminded me of my old ballet exercises,' Ms Horne said. 'I was a traditional ballet dancer as a child and then I took it up again when I had little daughters learning and I did ballet for a number of years.'

According to Ms Horne, the Pilates even contributed to her fast recovery from a couple of health issues she suffered in recent years.

'Partly it was my core strength and flexibility that had a lot to do with my very quick recovery from each of those dramas,' Ms Horne said.

The amount of control required for Pilates exercise is also something that appeals to Ms Horne, and she said over time this has made her feel more confident in her body.

'Being flexible and feeling strong and well balanced is a great thing,' Ms Horne. 'To be able to walk along and not worry that you're going to tumble over or reach for something or not being able to reach for something.'

Dr Marie-Louise Bird from the University of Tasmania who is studying the impact of exercise on elderly people said Pilates could be more than simply strengthening.

'One of the basic principles of Pilates is actually thinking about how you move and that movement awareness or mindfulness gives people a better understanding of where their body is in space,' Dr Marie-Louise said.

Discuss

1 Which fitness components did Jacqueline believe were being developed by her participation in Pilates?
2 Describe two other benefits of Jacqueline participating in Pilates.
3 Explain how Pilates helps people to understand where their body is in space.
4 Research where the nearest location and facility would be for you or one of your family members to go to participate in Pilates.

Body composition

Body composition is divided into two categories:

- fat-free mass, which is muscle, bone, water, connective tissue, organs and teeth
- fat mass, which is essential and non-essential fat stores.

FAST FACT

Health-related and skill-related fitness components contribute to improved health as well as fitness and lifespan. Both health and skill components are quite distinct though. Just because you are very flexible doesn't guarantee you would have good balance or coordination.

Essential fat

Essential fat is something that everybody needs. It is needed for normal body functions, including:

- regulating your body temperature
- absorbing shocks
- regulating nutrients.

Essential fat is found in the heart, lungs, liver, spleen, kidneys, intestines, muscles, central nervous system and bone marrow.

Non-essential fat

Non-essential fat is found in the **adipose tissue**, and it is known as body fat. While we all need non-essential fat, having too much adipose tissue can lead to health problems such as type 2 diabetes, cardiovascular disease and certain types of cancer.

adipose tissue stored fat that is used as a source of energy; it also cushions and protects internal organs

The safest way to improve your body composition is to do regular physical activity and eat a healthy balanced diet, including foods from all food groups (see Chapter 2).

SKILL-RELATED FITNESS COMPONENTS

Skill-related fitness components include:

- balance
- reaction time
- anaerobic capacity
- muscular power
- speed
- agility
- coordination.

Each skill-related fitness component is unrelated to the others. Just because someone is fast on the basketball court does not mean that they are also coordinated. Improvements to any (preferably all) skill-related fitness components are likely to increase performance in sport and other physical activities. When you watch elite athletes playing team sports, you see that they are often above average in most skill-related components of fitness.

Balance

Balance is one of the most important components of fitness for every activity you do, from your daily movements around the house, such as dressing, showering or going to the toilet, right through to specific sporting movements, such as throwing, jumping and landing.

Figure 9.11 shows that to maintain your balance (equilibrium), the external forces acting on your body – such as gravity, wind (to a very small extent), moving objects or people, and friction – must be constantly matched by internal forces, such as muscular contractions, otherwise you would fall over. Balance varies depending on whether you are stationary (still) or moving. Your muscles and joints make constant adjustments to maintain your balance every time you change your position or the environment you are in. For example, it is fairly easy to maintain your balance standing still on a flat stable surface, but if you were standing in waves where there was a strong undertow and water pushing against you, it would be very hard to stay on your feet.

Gravity (external force)

Forces applied by moving objects (external force)

Wind (external force)

Muscular contraction (internal force)

Friction (external force)

Alamy Stock Photo/Karl Phillipson/Optikal

Figure 9.11 To maintain balance, internal forces must equal external forces.

Improving your balance is very specific to the activity being performed. Generally, balance can be improved by lowering the centre of gravity and/or increasing the area of the base of support. For example, if you were canoeing or kayaking and you started to lose your balance, you could lower your hands and paddle to retain your balance.

iStock.com/Caziopeia

Figure 9.12 Equestrian events require excellent balance. Although horse riding is considered a lifelong physical activity, the unpredictable movements of a horse can lead to a loss of balance.

UP AND MOVING Balance

To do this activity, you will need to find a partner of similar build.

In a safe manner, person 1 tries to push person 2 off balance.

Now person 2 should attempt to become more stable. Person 1 again tries to push person 2 off balance. Next, swap the roles and do the activity again.

Discuss the changes person 2 made to increase their stability and stay balanced. Did they work? Why or why not?

Reaction time

external stimulus a factor that comes from outside your body

Reaction time is the time it takes your body to respond to an **external stimulus**. You use your reaction time every day. Every time you cross the road, you wait and check for traffic and use your reactions to decide when it is safe to cross. When driving a car, every time the traffic lights change colour or there is a change in the speed of the traffic, drivers use reaction time to remain safe. In sport, athletes need to assess external information quickly before responding. Think about a sprinter or swimmer reacting to the sound of the starting pistol to power quickly off their blocks. A netballer has to react quickly to a ball-up to ensure they have the first possession.

Decision-making

Reaction time is directly linked with your decision-making speed. The more options you have to choose from, the more information your brain and central nervous system have to process before you can react. For example, a netballer playing 'centre' looks around before making the next pass. They may have four different options. They will weigh up the position of their teammates and the opposition before making a split-second decision.

Distractions

Distractions can increase reaction time. The more information the brain is presented with, the more slowly it will respond. For example, many car accidents are caused when drivers or pedestrians are distracted by talking on the phone or looking at a text message. In Figure 9.13, the cyclist must process information collected by her senses to avoid a dog that runs out in front of her. The cyclist will need to use her balance and coordination to react safely and quickly to avoid the collision.

The best way to improve your reaction time is to practise in various challenging but safe situations.

Figure 9.13 A combination of information from the senses, coordination and balance will be used during reaction time.

UP AND MOVING: Drop it

Work in pairs or groups of three. Have one person hold out a ruler and drop it without warning. The person facing the ruler must try to catch it as soon as they can before it hits the ground. The distance between the release point and the catch point is an indication of your reaction time. A person who catches the ruler after it has fallen only a very short distance will have a faster reaction time than someone who catches it after it drops a longer distance, or who doesn't catch it at all. If you don't have a ruler you could use a tennis ball, a relay baton or even a smooth stick.

Anaerobic capacity

Anaerobic capacity refers to the total amount of work that can be done by the anaerobic systems, which have limited stored energy. To run a short distance at your fastest speed, your body has to release energy and produce energy very quickly, as there is not enough time for your body to produce this energy using oxygen.

Muscular power

Muscular power is the ability to exert a force rapidly and for a short time. Muscular power depends on both muscular strength and speed. Just because someone is strong doesn't guarantee they will also be powerful, although they are more likely to be than someone who is weaker. Muscular power is not necessarily used a lot every day.

Alamy Stock Photo/atsportphoto

Figure 9.14 Netball players perform many skills at a high intensity and short duration that require them to have muscular power.

An example of muscular power is a person picking up a toddler and lifting them above their head and then giving them a kiss. It is a fast movement, requiring powerful contractions of the muscles. If this same exercise was performed in slow motion, it would rely more on the person's strength. Other everyday activities using muscular power (strength plus speed) include running up a flight of stairs, lifting boxes, rising from a chair, walking, or playing with children.

In sport, muscular power is very important. It is the foundation of sprinting, jumping, lifting, swinging a golf club or bat, riding a bike uphill and throwing. Many of the physical activities you participate in during your practical physical education lessons will require muscular power.

The maximum power a muscle can generate occurs at a lower speed. During very high-speed movement there may not be enough time to develop maximum force, and therefore power can actually decrease.

The best way for non-athletes to improve muscular power is to train using circuit training or resistance training. When you want to improve your muscular power, it is important to use weights that are not too heavy, and to complete each repetition quickly. For example, you might do three to six repetitions of a bicep curl and complete this three times (three sets) using a medium weight, and allow two to three minutes' rest in between each set.

Figure 9.15 Bicep curls will help improve the power of the bicep muscles.

CASE STUDY

SPEED CLIMBING: MUSCULAR POWER IN ACTION

Identify

View the clip of Indonesian Aries Sasanti Rahayu, known as 'Spiderwoman', breaking the women's speed climbing world record by completing the 15-metre climbing course in under 7 seconds. Speed climbing, which requires incredible muscular power, will debut at the 2021 Tokyo Olympics.

Weblink
Aries Sasanti Rahayu speed climbing

Figure 9.16 Speed climbing requires muscular power.

Understand

Speed climbing is one of the three disciplines of the climbing event that will debut at the 2021 Tokyo Olympics. Think of it as a climbing triathlon: competitors will be graded on their combined results of speed climbing (a timed race between two athletes on a 15-metre wall), bouldering (a graded course with multiple routes of varying difficulty to be completed in four minutes) and lead climbing (how high a climber can go in six minutes).

Rahayu is a speed climbing specialist. Dubbed 'Spiderwoman' after a standout 2018 performance, the 24-year-old star was the lone Indonesian athlete to appear on Forbes Asia's '30 Under 30' list and will likely compete for gold in Tokyo …

To process how quickly Rahayu reached the top of the wall, she averaged a speed of roughly 7.71 km/h over 15 metre. That's the equivalent of running a 5K in 39 minutes – while doing pull-ups. Rahayu managed to keep that pace while running up a wall.

Source: 'Gone in seven seconds: "Spiderwoman" breaks women's climbing speed record', Gabriel Baumgaertner, 22 October 2019, *The Guardian.*

Discuss

1 Describe how Aries Sasanti Rahayu would use muscular power to climb the course so quickly.
2 Outline a physical activity you complete in a typical week that uses muscular power.

Speed

Speed relates to how fast you can move your body or part of your body from one point to another.

$$\text{Speed} = \frac{\text{Distance moved}}{\text{Time}}$$

If asked to imagine a really fast athlete, you might think of someone like Usain Bolt, the Jamaican sprinter who held the world records for both the 100-metre and 200-metre sprints for more than a decade. Speed is important for sprint swimming, cycling or running, and for all team sports.

The most effective way to improve speed is resistance training, improving your muscular power and interval training or circuit training. With interval training, you can work hard for longer because your body has time to recover between efforts.

Sprinting

When sprinting, you reach top speed in 20 to 30 metres, so to get to top speed quickly, you need to coordinate the following within a few seconds:

- Drive your lead arm.
- Drive out (if you were in a crouched position, e.g. starting blocks) at a 45-degree angle.
- Take a large first step and then extend.
- Drive the arms and hands down and back.
- Push the ground back and away.
- Ensure your feet hit the ground below or behind the hips.
- Keep your heels pretty close to the ground during the first six to eight steps.
- Let your upper body straighten naturally.

Alamy Stock Photo/PCN Photography

Figure 9.17 Usain Bolt set a world record of 9.58 seconds for the 100-metre sprint and 19.19 seconds for the 200-metre sprint in 2009.

UP AND MOVING — How fast is best?

Jump as high as you can in the air. Or, if you are outdoors, throw or hit a ball as hard as you can. Experiment with the speed at which you perform the activity. Try to do it fast, medium and slow.

1 If you can, measure the distance of your throw or hit, or the height of your jump.
2 Describe what happened as you changed the speed of the performance.
3 Was there an optimal speed of performance?
4 Discuss what happened when you performed the task too fast or too slow.
5 You could even graph your results using the three different conditions: fast, medium (normal speed), slow.

Agility

Agility is the ability of a person to quickly change the speed or direction of their body without losing control. Many team sports and racquet sports require players to accelerate (speed up), decelerate (slow down), quickly change direction to get away from an opponent or sprint to catch a ball or another player.

Alamy Stock Photo/Mariano Garcia

Figure 9.18 To score in basketball, players must exhibit exceptional agility to evade their opponents.

Coordination

Every sport and physical activity requires some coordination. Coordination is the ability to use the body's senses to perform **motor skills** smoothly and accurately. Coordination is sometimes referred to as either hand–eye, head–eye or foot–eye coordination, depending on which body parts are involved. High-performing athletes have good coordination, which allows them to perform skills accurately, proficiently and successfully.

motor skills a sequence of movements of the nerves and muscles that combine to perform a particular task

Coordination may involve:

- control of the body parts to perform a sequence of movements (e.g. a gymnastics routine)
- manipulation of equipment (e.g. a bat or club) and parts of your body (e.g. hitting a ball with a hockey stick).

FACE TO FACE Which type of coordination?

Write each of the following sporting activities on a piece of paper. In pairs or small groups, classify each activity in terms of the type of coordination used by placing a tick in the corresponding column (i.e. hand–eye, head–eye or foot–eye). Use your workbook or fill in the scaffold online.

Physical activity	Hand–eye	Head–eye	Foot–eye
Football (soccer) player heading the ball towards the goal			
Table tennis player returning a serve			
Baseball shortstop flicking a ball from their glove, using their throwing hand, to the second base player to turn a double play			
AFL player kicking for goal			

Scaffold
Coordination types

Practice makes perfect

When you watch an elite performer, their movements often appear effortless, smooth and as if they have all the time in the world. This, of course, is the result of years of experience, hard work and hours of practice during training, competition and recovery. You can constantly improve your coordination as you learn and practise skills over and over under varying conditions.

No matter how fit or strong someone is, if they are not coordinated they will never perform at a high level. Coordination is very specific. For example, being able to juggle a soccer ball doesn't mean you would be able to consistently make a set shot in basketball from the free-throw line.

Coordination is even specific to certain skills within the same sport or physical activity. For example, you might be terrific at driving a golf ball off the tee, but really poor at putting onto the green. Coordination can even vary enormously across variations of a skill. For example, in tennis a player may be excellent at performing a slice serve but unable to perform the kick serve. Coordination is influenced by the equipment used, fatigue, amount of warm-up, level of practice that week, mood, weather conditions and other factors.

Coordination improves dramatically with practice and experience. If you know how to ride a bike, think about when you were first learning to ride. You were probably very wobbly when you started out, and had to concentrate just to stay upright. Now that you are more experienced, you can coordinate the skill of riding a bike without thinking, leaving your mind free to concentrate on the external factors such as traffic, other cyclists, pedestrians and other possible dangers.

Quiz
What is fitness?

9780170463096

1 Identify which health-related and skill-related fitness components are being used in your life and specify in what context. Determine why agility is important for each of the following athletes:
- baseball catcher
- hockey player
- football (soccer) goalkeeper
- circus performer
- ballet dancer
- tennis player
- 100-metre hurdler.

2 Explain the difference between aerobic capacity and anaerobic capacity.

1 In small groups, discuss two sports you enjoy playing or watching, and talk about which parts of these games require players to be agile.

2 Go online to locate images that depict the concept of balance. They may be images of athletes playing sport or people in everyday lifelong physical activities. Add them to your online album or to the class collection.

1 As a class, go online and review action shots of the Perth Wildcats Basketball team. Examine which health-related and skill-related fitness components are important in basketball.

HOW DO I MEASURE MY BODY'S RESPONSE?

Most of the health-related fitness benefits of being physically active occur when you are working with at least moderate intensity. Moderate-intensity physical activity usually consists of sustained rhythmic movements. If you are working at a moderate intensity:

- you should be able to comfortably have a conversation
- you are working at 50–70 per cent of your maximum heart rate
- you are expending three to six times the energy that you would while at rest.

Moderate-intensity physical activities include brisk walking, bike riding, raking leaves, mopping the floor, sweeping, lifting weights, doing aerobics, golfing and paddling in a pool.

But how do you know if you are working at a moderate intensity? You could measure your heart rate, breathing rate or even your ability to talk while exercising.

MEASURING YOUR HEART RATE

You first need to know how to monitor your heart rate, also known as your 'pulse'. Every time your heart beats, it pumps blood into the arteries. If you hold a finger against an artery, the surge of blood you can feel is caused by the pulse. Table 9.3 describes the two easiest positions to locate your pulse.

Worksheet 9.8

Taking your carotid pulse is probably the easiest because the carotid artery is relatively close to the surface and the pulse is strong. (Don't use your thumb to count your pulse, because your thumb also has a pulse in it.) You need to stop exercising for a moment to take your pulse, or take it after you've finished exercising. To work out your pulse rate, count how many beats you feel in one minute. You can also estimate the beats per minute by counting for 30 seconds and doubling the count, or for 15 seconds and multiplying the count by four.

Table 9.3 Measuring your pulse

Location of pulse	How to count it
Radial pulse	Place your index and middle fingers on the inside of your wrist, just below the base of the thumb.
Carotid pulse	Place two fingers (index and middle fingers) on either side of the front of the neck, just below the jaw. Use a light pressure, but do not press too hard. If you measure your carotid pulse on the right side, use your right hand.

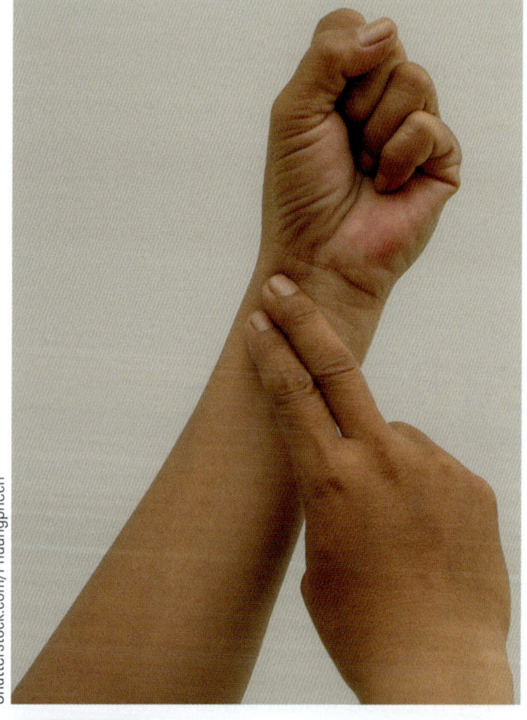

Shutterstock.com/Phuangphech

Figure 9.19 Radial pulse is taken on your wrist.

Shutterstock.com/New Africa

Figure 9.20 Carotid pulse is taken on your neck.

MEASURING YOUR BREATHING RATE

Another useful way to work out whether you are working at a moderate intensity is to measure your breathing rate. Sometimes your breathing rate is known as your 'respiratory rate'. It is measured in breaths per minute. You can measure your breathing rate by simply placing a hand on or under the lower rib cage area.

A normal healthy adult at rest will take between 12 and 20 breaths per minute. At rest, one of the best ways to measure your breathing rate is to ask a friend to count your breaths without you realising; your breathing rate can change if you try to count your own breaths. If you are exercising intensely, you can breathe up to 40 to 50 breaths per minute to supply your body and working muscles with all the oxygen required. Even when you stop exercising, your breathing rate can remain higher than when at rest for 20 to 40 minutes, to give the body a chance to recover. Table 9.4 shows the normal average resting respiratory rates for different age groups.

Table 9.4 Average resting respiratory rates by age

Age	Breaths per minute
Birth to 6 weeks	30–60
6 months	25–40
3 years	20–30
6 years	18–25
10 years	15–20
Adult	12–20

MEASURING YOUR ABILITY TO TALK (TALK TEST)

You can also work out whether you are exercising at a moderate intensity by using the talk test. If you can talk comfortably while being active, you are most probably exercising at a light to moderate intensity. If you are unable to talk comfortably, you are more likely to be working at a more vigorous intensity.

Worksheet 9.9

Shutterstock.com/giorgiomtb

Figure 9.21 Use the talk test to determine how hard you are exercising.

UP AND MOVING — Check your respiratory rate

Before standing up, take your resting respiratory rate and determine if it is within the average rate for your age group.

Next, complete a marching on the spot activity for 2 minutes. At the 90-second mark, take the talk text and estimate what intensity you are working at.

The final activity requires you to run on the spot for 1 minute as fast as you can. Towards the end of the minute, take the talk test again so you can estimate whether you are working at a moderate or vigorous intensity.

Quiz
How do I measure my body's response to physical activity?

FAST FACT

Your breathing rate during maximal exercise, such as running up a flight of stairs or a steep hill, can reach 40–60 breaths per minute.

1 Define moderate-intensity physical activity. Explain what intensity you would be working at if you were not able to hold a conversation comfortably during exercise.

1 Apply your knowledge relating to measuring your bodies response to physical activity by taking your resting respiratory rate (RR) breaths per minute and heart rate (HR) beats per minute. Now complete 1 minute of exercise such as jogging on the spot or sit ups. Calculate how much your RR and HR increased after exercise. Identify what procedures you used to measure RR and HR.

1 Create a model of the lungs using materials such as balloons, straws and pipe cleaners to depict what happens to the respiratory rate during exercise the respirator rate.

HOW CAN I DESIGN, APPLY AND EVALUATE A PERSONAL PHYSICAL ACTIVITY PLAN?

When creating a personal fitness plan within your overall individual activity plan, you need to be realistic. Start with a simple, personalised plan based on the activities you enjoy and that could be easily built into your everyday life. In Years 9 and 10 you will learn more about training principles, training methods and exercise plans. Table 9.5 shows an example of a personalised physical activity plan for a 14-year-old. The goal of this plan is to meet the Australian 24-hour Movement Guidelines for Children and Young People (5–17 years), and to improve aerobic power.

Table 9.5 Example of a weekly personalised physical activity plan for a 14-year-old

	6–9 a.m. (before school)	9–3 p.m. (school hours)	3–10 p.m. (after school)
Monday	Walk to school (10 min)		High Impact Training (HIT) class (60 min)
Tuesday		Physical education class (100 min)	
Wednesday	Walk to school (10 min)		After-school/interschool sport (volleyball; 90 min)

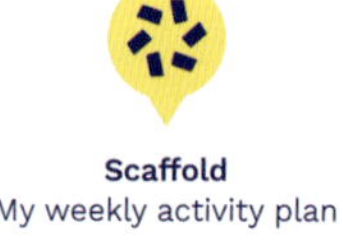

Scaffold
My weekly activity plan

9780170463096

	6–9 a.m. (before school)	9–3 p.m. (school hours)	3–10 p.m. (after school)
Thursday		Boxing stations (20 min)	Walk the dog (45 min)
Friday		Physical education class (50 min)	Home-based fitness circuit (45 min)
Saturday		30 push-ups Sit-ups (3 sets 25 reps)	Bike ride (90 min)
Sunday			Walk the dog (70 min)

INVESTIGATION: DESIGNING YOUR OWN ACTIVITY PLAN

Purpose

In this activity, you will design your own personalised physical activity plan.

Materials

Use the personalsied physical activity plan template provided, or a similar template to Table 9.5.

(Template) Personalised Physical Activity Plan

Method

Design your plan using these steps:

1. Determine how much physical activity is recommended for your age group based on the Australian 24-hour Movement Guidelines for Children and Young People (5–17 years).
2. Complete your table (like Table 9.5). Make sure you have included activities that allow you to address all aspects of the Australian 24-hour Movement Guidelines for Children and Young People (5–17 years).
3. Plan to use several of the **cognitive** and **behavioural** strategies that are discussed in the following section. Once you have read that section, explain which strategies you will use and how they will help you keep to your individual activity plan.

cognitive related to what you think or understand

behavioural related to behaviour; what you do

Discussion

1. How much physical activity is recommended for children and young people (5–17 years)?
2. Evaluate whether you are meeting the Australian 24-hour Movement Guidelines for Children and Young People (5–17 years). If not, plan to add some additional physical activity to your typical week and think about how you could reduce your sitting time and ensure you are getting adequate sleep.

3 Analyse whether your individual plan would enable you to meet the 24-hour movement guidelines if you completed it as intended.
4 Discuss which areas of your plan are most likely to be affected and not completed when you are tired or not motivated to be active.
5 Describe what cognitive and behavioural strategies (at least two of each) could be implemented to help you overcome barriers to being active, such as being tired or feeling unmotivated.

There is no such thing as 'one size fits all' when creating personal activity and fitness plans. Your strategies must be tailored to your own needs, interests and experiences. Use a range of cognitive strategies, including things that relate to what you think or understand and things that increase your knowledge or awareness. In addition, use a range of realistic behavioural strategies to increase or maintain your activity level. Behavioural strategies are the 'doing' things, like organising a friend to go for a walk with you. This is not something you just think about, you also do it.

COGNITIVE STRATEGIES IN AN INDIVIDUAL ACTIVITY PLAN

Cognitive strategies include:

- learning more about the importance of being active
- understanding the risks of being inactive and not doing some kind of activity every day
- caring about consequences (how your inactivity can affect yourself, your family and friends)
- understanding the benefits of being active and fit
- being more aware about opportunities to be active.

Although cognitive strategies are an essential first step, they are not enough to change your behaviour on their own. Behavioural strategies, the things that you actually do, are most important. A person must be regularly active to improve or maintain fitness.

WELLBEING CHECK IN

'THANKS, BRAIN'

Identify

Our brains are amazing things, but sometimes they can be pretty annoying.

This activity has been developed in collaboration with Dr David Bakker of MoodMission

Understand

We can have some really random thoughts sometimes. We didn't necessarily want to have the thought – our brain just made it up. Our brains can also get stuck on a loop of negative thoughts sometimes, especially if we get upset at ourselves for having a negative or upsetting thought. Accepting that our brains will have these kinds of thoughts can actually decrease how much we get caught in the loop, and can lead to having more positive thoughts and feelings.

Practise

1 The next time you have a negative or anxious thought, say to yourself 'Thanks, brain!' You're probably not actually thankful, but you can use sarcasm on your mind. For example:

- **a** *What I said was stupid.* 'Thanks, brain!'
- **b** *I'm an idiot.* 'Thanks, brain!'
- **c** *They're making fun of me.* 'Thanks, brain!'
- **d** *I'm feeling anxious.* 'Thanks, brain!'

Reflect

It's normal to have random upsetting thoughts. Some people picture themselves doing something out of character, like hitting someone, and then they feel bad and guilty because they would never do those things. Usually our brain has these thoughts to make sure we don't do something. Have you had some of these thoughts that you can say 'Thanks, brain!' to?

BEHAVIOURAL STRATEGIES IN AN INDIVIDUAL ACTIVITY PLAN

Behavioural strategies are the 'doing' things. Realistic behavioural strategies are essential to an individual activity and fitness plan. They include:

- substituting alternatives
- enlisting social support
- motivating and rewarding yourself
- committing yourself by setting realistic goals
- reminding yourself.

Substituting alternatives

A great place to start is to think about your current physical activity level. By completing a daily activity log, you will become more aware of your **sedentary behaviours**. Today, technology continually saves us from the need to move and expend energy. Many people sit in a vehicle to go to and from school or work. They then sit for most of the day, and then sit for many hours at home afterwards, watching television or using a computer, smartphone or tablet.

sedentary behaviour activities that involve sitting or lying down, not moving around, with little energy expenditure. Examples include watching TV, reading, or sitting in a car or on public transport

Have a look at the activities you do on a typical day and then identify up to 10 more active alternatives to some of your light intensity or sedentary activities. There are some examples on the worksheet called CLASS questionnaire. The whole idea of substituting alternatives is to replace a usual behaviour with a more active option, building this into your everyday life until it becomes a lifelong physical activity. Table 9.6 has some suggestions for substitute activities.

Worksheet 9.10

Table 9.6 Substituting more active alternatives in everyday life

Typical sedentary activities	More active alternatives
Park as close to the shops as possible	Park further away from the shops and walk
Watch television while sitting or lying down	Exercise while watching TV or exercise for a few minutes during the ads (sit-ups, push-ups, squats, lunges, stretching, bicep curls)

Typical sedentary activities	More active alternatives
Complete household chores slowly, such as picking up laundry or tidying your room	Dance or move vigorously to music while completing household chores
Drive in a car to watch the footy each week or watch sport on television	Play a recreational sport each week or take public transport to a watch a game
Drive to the local shops	Walk, cycle, scooter or skate to the local shops
Be driven to school	Walk, ride or take public transport to and from school at least once per week
Be driven to a friend's house	Walk, cycle, scooter or skate to a friend's house
Play a computer game with friends while sitting down	Play something active with your friends, or an active computer game such as Wii
Lie in bed before getting up or going to sleep	Stretch in bed before getting up or going to sleep
Sit while waiting for your toast to cook or the kettle to boil or while talking on the phone	Complete squats, lunges, leg raises or other exercise while waiting for your toast or the kettle or while on the phone
Sit at a desk while studying	Stand at a desk (place a box under your keyboard or buy a desk that changes height
Sit still in a car as a passenger	Perform some **isometric** exercises while riding in a car
Sit and talk during school recess and lunch break	Walk around while talking during recess and lunch break periods
Use a trolley to carry shopping bags to the car	Carry shopping bags to the car and do some bicep curls using the bags as your weights

isometric exercises that do not involve movement; only your muscles contract or change length. An example is to clasp your hands together and push them towards each other.

Enlisting social support

You are much more likely to exercise regularly if you have someone to do it with. It doesn't matter who that person is: it could be a family member, a friend or even the dog.

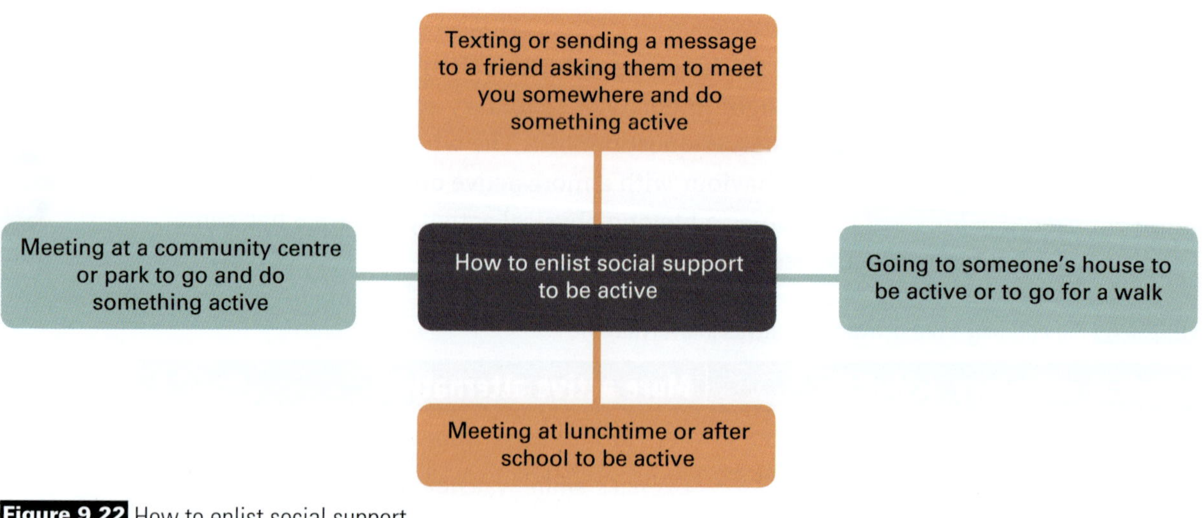

Figure 9.22 How to enlist social support

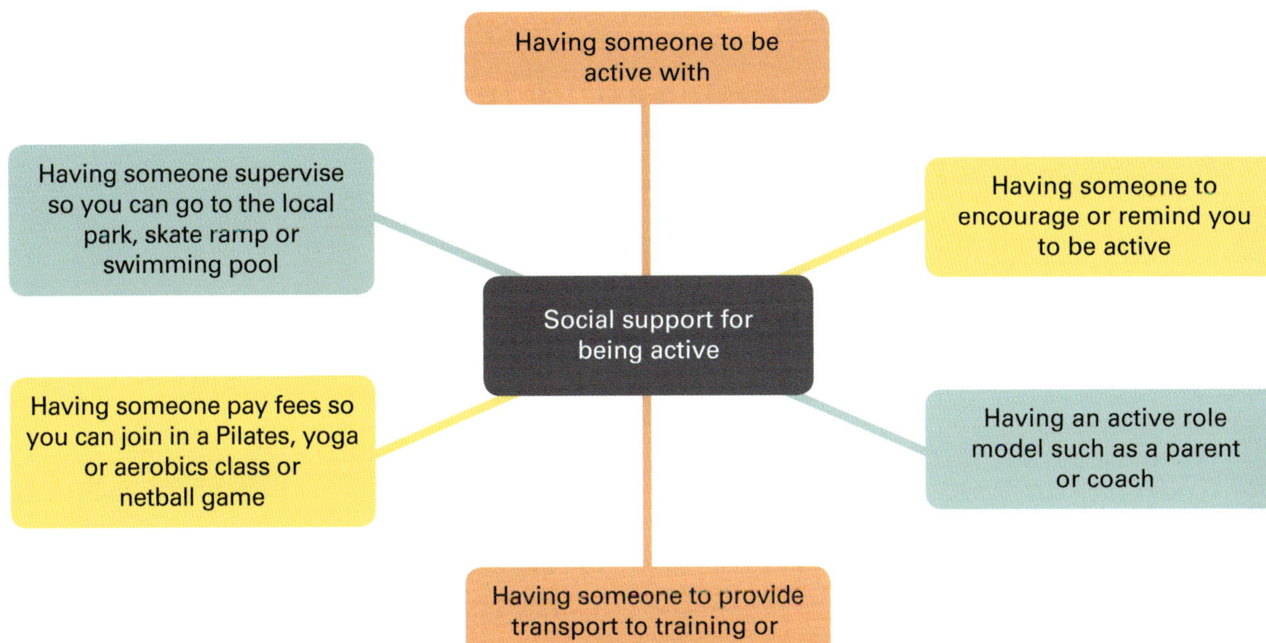

Figure 9.23 Social support for being active

Figure 9.24 Social support can take many forms: being active with someone else, providing encouragement, role modelling, providing transportation to an activity or providing supervision.

CASE STUDY

GETTING ACTIVE IN BUSSELTON

Identify

In this case study we will explore different physical activity opportunities for different age groups and consider who could provide you with social support to be active.

Figure 9.25 What are the physical activity opportunities at Busselton Jetty?

Understand

The Busselton Jetty which stretches nearly 2 km to an underwater observatory is an iconic tourist destination location on the southwest tip of Western Australia. Busselton is known for its sheltered beach and seasonal humpback whale populations and the train that runs along the jetty. The foreshore area has an incredible array of physical activity equipment, spaces and facilities for people of all ages. The stunning area has loads of cafes and the physical activity spaces include:

- A 2 km jetty to an underwater observatory
- A natural beachfront swimming pool area
- Walking and cycling paths and drink fountains and bike racks along the paths
- Bicycle hire
- Sporting grounds, basketball rings, sports pavilions, tennis courts
- Sea play by the bay adventure play space
- Jetty play toddlers' play space
- Skateboard Park
- Landscaped green spaces
- Accessible playgrounds and amenities for all abilities

Discuss

1 Examine the images of the Busselton foreshore. You could do additional research online to make a list of the types of physical activities available in this area. Identify who could participate in each using a coding system. (Use the following groups: older adults, adults, teenagers, children, toddlers.)

2 Notice the COVID-19 sign for the playground, if you were to play on an adventure playground like determine what strategies would you have to consider using during a pandemic.

3 Select three different physical activities you would like to do if you visited the Busselton foreshore and discuss what social support you might enlist as a self-management strategy to enable you to participate. (Consider parental supervision, transportation, cost, friends, or siblings, etc.)

4 For each of the physical activities you selected in question 3 outline several health and/or skill-related fitness components that could be developed by regular participation.

Weblink
Busselton Jetty

FAST FACT

If you want to talk to someone about any health concerns, there are a variety of people you could ask: your family, your friends, a coach, a teacher, your doctor, school nurse, physiotherapist, school counsellor, social worker, chemist or local council staff. The important thing is to communicate with people whenever you have any concerns about your health.

Motivating and rewarding yourself

When creating a personal physical activity plan, it is important to recognise both internal and external factors that motivate (reinforce) you to be active. Understanding what motivates you greatly increases your chances of success, and of keeping to a routine over time.

Table 9.7 Motivation factors

Internal motivation factors	External motivation factors
'I will feel healthier and have more energy'	'I will buy myself a treat after I complete a few sessions'
'I will feel happy when I reach my goal'	'My mum will let me buy a new song online if I walk home from school for the next couple of weeks'

Pedometers

One of the best devices you can use to assist you to both set goals and monitor whether you are achieving them is by wearing a pedometer. Most smart watches have an activity monitor with a pedometer function built in. Pedometers provide immediate feedback. They track your daily steps and can be very motivating. They also increase your awareness of how active (or sedentary) you are. It is recommended that adults walk at least 10 000 steps per day, and that young people walk at least 15 000 steps a day.

Step it up

To start, measure your daily steps for two days and then take an average of these. (To get your daily average, add the two daily totals and divide by two.) If you are under the recommended target, try increasing your daily step count by 10 per cent. For example, if your daily steps are 6000, then a 10 per cent increase will bring that to 6600 steps a day. After a few weeks, increase your daily steps again. Continue increasing them around every three weeks.

Alamy Stock Photo/Amy Cicconi

Figure 9.26 Wearing an activity tracker or a pedometer is a great way to become more aware of your activity level and to motivate you to reach your daily step goal.

Make a commitment

The key to success in creating your personal fitness plan or physical activity program is to make a commitment or set a goal. Being able to set goals is an essential self-management life skill. Goal setting is a motivational technique allowing you to set clear targets, priorities and expectations. Table 9.8 describes the guidelines for effective goal setting using the SMARTER approach. Your smart watch has a range of function to assist you to set goals, track your activity and remind you to be active. It can even give you daily goals calculated on the daily activity you accumulate.

Table 9.8 Developing effective goals using the SMARTER approach

SMARTER	Description	Example goal: cycling to and from school
Specific	Goals need to be specific and as clear as possible	To bike ride to and from school at least three days per week for a month
Measurable	You should be able to measure progress	To cycle 7.5 km to and from school at least three days per week
Accepted	Goals should be accepted by everyone involved	It is okay with your parent/guardian to cycle to school, and okay with your school to leave your bike there during the day
Realistic	Goals should extend you but be achievable	The path includes a hill; it is okay to get off the bike and walk up it
Time-based	Goals should include a specific end date	Until the end of Term 1
Exciting	You need to be challenged and inspired	Yes!
Recorded	Goals should be written down	Write what you did in your school diary each day

Remind yourself

Worksheet 9.11

Another essential component of creating a personal fitness plan is setting up an effective reminder system. Here are some simple tips you can use to remind yourself to be active and to complete your exercise plan:

- Keep physical activity clothing and shoes in your parent/caregiver's car.
- Set automated reminders in your phone or computer or write them in your calendar or diary.
- Place prompt signs around your house – on your fridge, on your desk or in your parents' car – saying things like 'Have you done your exercise today?'
- Email a reminder to yourself.
- Keep a spare pair of runners at the front door as a reminder to be active.

Quiz
How can I create a personal fitness plan?

1 Outline three cognitive strategies and three behavioural strategies you currently use or could use to help you be active.
2 Define each component of SMARTER goals.

1 To help you work out what motivates you to be active, examine five ideas or goals relating to your personal health, fitness and wellbeing that you would like to achieve. Determine the internal and external reinforcers you could use to help you achieve your goals. Fill in a table like this one, either online or in your workbook. Discuss the strengths and limitations of your ideas.

Scaffold
Achieving my goals

Task or goal	Internal reinforcer	External reinforcer
Example: I will walk to school three days this week	I will have more energy at school and won't feel as tired	My parents will let me buy a new song or game online if I reach my goal

1 Evaluate how effectively you utilise social support strategies to encourage you to be regularly active. Examine at the examples provided in the chapter and determine which of these strategies you already use regularly and make a judgment about the strengths and limitations of each. For example, I regularly ask my mum to take me to and supervise me at the local swimming pool, where I swim and hang out with my friends in summer.

HOW CAN I USE MY LOCAL ENVIRONMENT TO IMPROVE MY HEALTH?

In Chapter 3 you learnt how your environment can influence your physical activity. People who live in more supportive environments (e.g. near parks, swimming pools or bike paths) are more likely to meet the Australian 24-hour Movement Guidelines for Children and Young People (5–17 years). Here are some strategies for exploring your local school and community.

UP AND MOVING

Get up and moving

1 Find a map of your local community and calculate the distance you would need to travel to school. Is it a walkable distance, or could you ride your bike? Outline the factors you would need to consider if you were to use active commuting (walk or ride) to and from school. On your map, identify all the physical activity facilities and places you could be active within a short distance of school and home. Look for things such as local parks, bike paths, walking trails, swimming pools, sporting facilities such as a recreation centre, tennis courts and so on.

2 Go for a bike ride around your local community or on a local bike trail and take photos of buildings and the natural environment that either enable or are barriers to physical activity. Describe ways to increase physical activity opportunities. Place these photos in an electronic portfolio and describe whether you have used any of these facilities, outdoor setting or spaces in your local area to be active.

3 Design your own 'ultimate playground' concept for your school. Send it to your school principal or school council, or to your local council, either as an idea for a new community park or to improve an existing park.

School audit

As a class, your teacher may take you outside to audit your school grounds and facilities. Examine which features of your school (grounds and buildings) encourage you to be active. Identify things that make it difficult for you to be active (e.g. areas where ball games are not allowed). What features within your school encourage you to be active?

FINDING INFORMATION AND SERVICES THAT SUPPORT YOUTH

There is a lot of health information and many services available to help young people stay healthy and active as they grow older. You can obtain information about your health, wellbeing and ways to be active from your school, your GP, the local community centre, recreation centres and your local council.

FACE TO FACE My community centre

Working in small groups, research your local community centre and your local recreation centre. Investigate the facilities that are accessible and what programs are offered to different age groups. Use the 'My recreation centre' scaffold table provided to enter the information you find.

For programs aimed at 12- to 14-year olds, collect information relating to their cost, timetables and how to get there (walking distance, or need to take the bus or ride a bike).

Scaffold
My recreation centre

iStock.com/FatCamera

Figure 9.27 There are many opportunities in your local community to participate in physical activity.

INVESTIGATION

EXPLORING MY LOCAL COMMUNITY RECREATION AND LEISURE CENTRE

Purpose

The purpose of this investigation is to explore and evaluate all the physical activity opportunities at your local community recreation and leisure centre for various age groups.

Method

Go to your local recreation centre, either as a class excursion or for homework. Draw or obtain a map of the facilities on offer, participate in a fitness class or activity, and collect information about all the programs they offer for students your age. Use the table provided online in Nelson MindTap (like Table 9.9) as a starting point. Ensure you design your results table to include the different programs offered, the various age groups each program is for, the various facilities and any other important data you can include in your report, such as price.

Discussion

1 Complete a data analysis of your findings. Examine the data collected and determine whether or not you think there is a good variety of physical activity opportunities for people of different age groups and abilities to participate in.

2 Describe three active programs you would be interested in participating in.

3 Explain what the strengths and limitations would be in relation to participating in these activities within each setting during the COVID-19 pandemic where you live. (e.g. consider ventilation, space, rules, etc).

4 Evaluate whether or not your local opportunities would cater for most age groups and discuss the strengths and limitations of the opportunities available for people of various age groups.

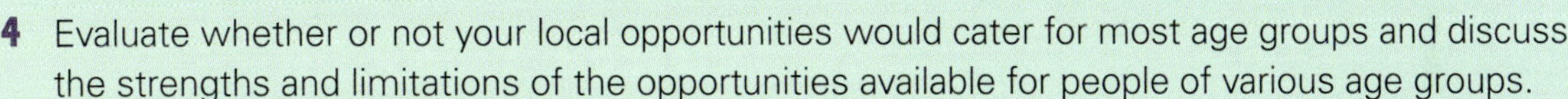

5 Recommend two additional programs that could be implemented within your local recreation and leisure centre that would be of interest to you. Also include what facilities would be used or would need to be built for these programs to be implemented.

Scaffold
My recreation centre

Table 9.9 Programs and facilities in recreation centres

Program	Facility	Outcomes		Age group the program is designed primarily for				
		Fitness	Social	Young children	Children	Adolescents	Adults	Older adults
Group fitness classes (e.g. cycle/spin, body balance, body attack, pump, Pilates, yoga)	Fitness studios	✓	✓			✓	✓	
School holiday programs	Pool and multipurpose room	✓	✓		✓	✓		
Personal training	All areas	✓	✓				✓	✓
Learn to swim	25-metre pool & wave pool	✓		✓	✓	✓	✓	
Crèche	Crèche			✓				
Aqua play	Toddler pool	✓	✓	✓				
Birthday parties	Pool and multipurpose room		✓	✓	✓	✓		
Swim squad	25-metre pool	✓			✓	✓	✓	
Resistance training for older people	Gym	✓	✓					✓
Aqua aerobics	25-metre pool	✓	✓			✓	✓	
Gentle exercises to music	Rehab pool	✓	✓					✓

iStock.com/Bojan89

Newspix/Mark Stewart

Figure 9.28 Recreation centres provide activities for people of all ages.

WELLBEING CHECK IN

SHOWER WARRIOR

This activity has been developed in collaboration with Dr David Bakker of MoodMission

Identify

How good does it feel to jump in the pool when it's a hot day? Maybe you can get that feeling without needing a pool.

Understand

Cold water can activate the part of our body's nervous system that makes us feel alert and awake. It can also calm us down by activating something called the dive reflex. The dive reflex helps mammals when they dive into water. It lowers the heart rate and brings calm so the body doesn't use its supply of oxygen as quickly. If we put something cold on our face and hold our breath for 20 seconds, we can activate the dive reflex and start to feel calmer.

Practise

1. The next time you have a shower, end it by turning the water fully cold.
2. Try to stay under the cold water for at least 5 seconds.
3. To get the full effect, try to keep your face under the water and hold your breath for 20 seconds.
4. Shut the water off, get out and towel off to warm back up.
5. On a scale from 0–10, how awake do you now feel?

Reflect

Aside from a really hot day, when might this be good to try? Some people do it if they're tired or stressed. Do you think it could help when you feel those things too?

PROMOTING HEALTH AND WELLBEING IN SCHOOL COMMUNITIES

Your school has an important role in promoting health and wellbeing among its students. In order to provide the best programs, schools need to continually assess their programs to make sure they are achieving the right results. To assist schools in this assessment, in 2012, the Centers for Disease Control and Prevention (CDC) in the USA published a 'School Health Index' (SHI). The SHI is designed as a way for schools to assess their own 'health', based on the eight interactive parts shown in Figure 9.29. The SHI helps schools to meet the following goals:

- Identify the strengths and weaknesses of their programs and policies on health and safety promotion.
- Develop an action plan to improve student health and safety.
- Involve the school community in regularly improving school policies, programs and services.

Worksheet 9.12

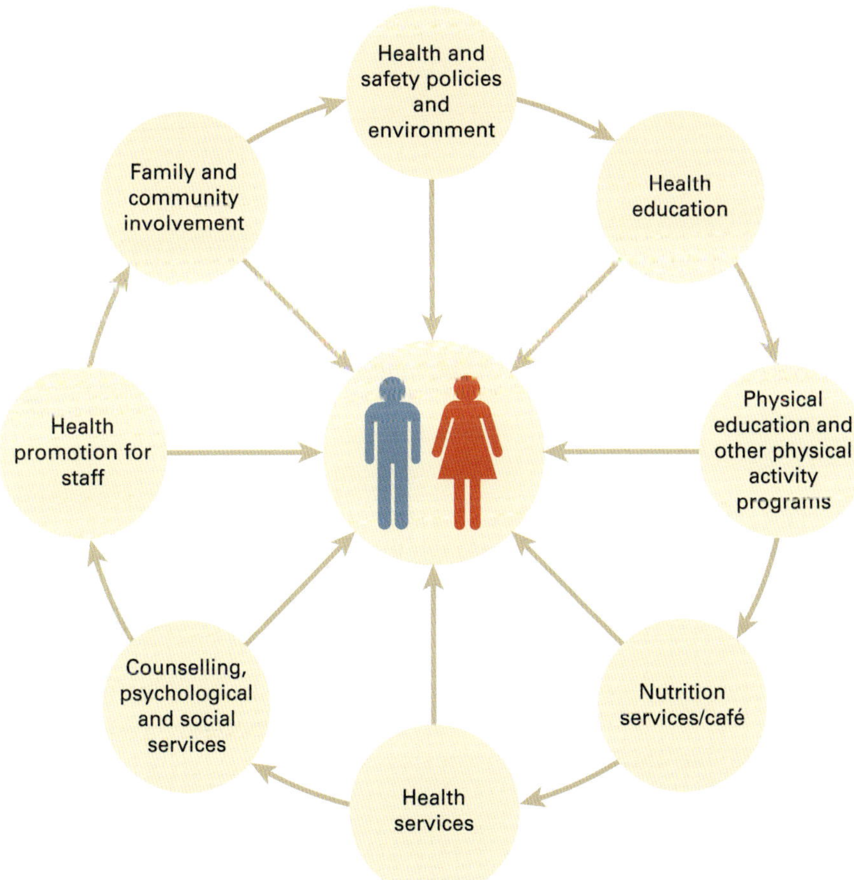

Figure 9.29 Interactive parts of the School Health Index (SHI)

FACE TO FACE School Health Index

Weblink
School Health Index

Worksheet 9.15

Quiz
How can I use my local environment to improve my health?

Search for 'School Health Index' (SHI) on the internet or use the weblink provided.

Have a look at the School Health Index for Middle and High Schools developed by the Centers for Disease Control and Prevention, in the United States.

The School Health Index recommends schools bring together a team of people to design, implement and evaluate policies and programs to promote health and safety. The team could be made up of people within your school, such as:

- principal or assistant principal
- health and physical education teachers
- guardians or caregivers
- students and parents or other family members
- health and physical education coordinator and sport coordinator, coaches
- canteen manager, cleaner and maintenance manager
- school counsellor, psychologist, student welfare coordinator and school nurse
- representatives from health organisations and local council.

Once the SHI team is assembled, they meet and complete the self-assessment checklist. They then fill in the overall score card to identify strengths, weaknesses and areas for improvement. At the next few meetings, the team creates a school improvement plan to prioritise the areas needing improvement while retaining the strengths.

1 Make a table that lists each of the health-related and skill-related fitness components and their corresponding definitions. Print out a hard copy, cut the terms and definitions up and mix them up. With a partner or in a small group, match each of the health-related and skill-related components of fitness with the correct definition.

1 Search for three services in your local community that provide physical activity opportunities for youth, but that aren't arranged through the local recreation centre.

1 Propose that you are the owner of a local recreation centre. Several key buildings make up the area around the recreation centre (noted below). Create a weekly timetable of classes to run at your centre that provide everyone nearby with the opportunity to engage with lifelong physical activities.

Close to the recreation centre are:

- an aged care facility
- a senior high school
- a parents' group
- a professional swim team.

CHAPTER 9 REVIEW

1 Outline 10 characteristics of lifelong physical activities.

2 Identify each health-related and skill-related fitness component.

3 Explain the difference between muscular strength and muscular endurance.

4 Describe why you need essential fat in your body.

5 What can occur as a result of having excessive adipose tissue?

6 Give an example of when having a fast reaction time would be advantageous for a sport you enjoy participating in.

7 Outline two examples of when an athlete uses hand–eye coordination.

8 Describe how you would measure your carotid pulse.

9 Identify the average resting respiratory rate for a 14-year-old.

10 Explain what is meant by the term 'cognitive', and provide five examples of cognitive strategies that could be used in an individual activity and fitness plan.

11 Describe three behavioural strategies you could use to create and sustain an individual activity and fitness plan.

12 Outline five sedentary behaviours and five more active alternatives.

13 Apply the SMARTER approach to draft a goal for yourself in relation to being active.

14 Discuss which two reminder systems would suit you best for your personal fitness plan.

15 Explain what the School Health Index can be used for.

JUST DANCE!

IN THIS CHAPTER

You will learn about the important role of dance in various cultures and how to work in a group to create your own dance.

By the end of this chapter, you should be able to:

- examine your own motivation to dance
- examine the impact that other people have on your dance choices
- investigate reasons why people dance
- examine the role of dance in the culture of First Nations Peoples
- investigate the role of dance in cultural identity
- explore the cultural contribution of dance to the mind-body-spirit connection to health and wellbeing
- explore the use of the elements of dance
- compose and perform a sport dance routine
- develop a safe and effective dance routine, applying selected criteria
- take on a role within a group to assist with the choreographic process
- propose and use feedback based on selected criteria to improve your dance and performance level
- develop leadership and collaborative skills
- investigate opportunities to participate in dance in the wider community.

Shutterstock.com/Nicetoseeya

WHY DO PEOPLE DANCE?

Quiz
Pre-chapter

Before you start, take the pre-chapter quiz to reflect on your experiences, thoughts and feelings around dance.

An important feature of human beings is the ability to communicate, not only through speech and writing but also through physical expression. While there are many ways for people to communicate, one of the most popular methods through the ages has been dance.

The ability to speak allows people to communicate effectively, but can all of your ideas, emotions and feelings be expressed through speech alone? The physical and visual portrayal of ideas, stories and emotions through movement can convey so much more than just words. For many people, dancing is about self-expression – how you feel usually influences how you move.

Since ancient times, dance has been used to convey how people think and feel. It can help to tell a story, to celebrate or to express cultural, spiritual or emotional issues. Dancing can also provide many social opportunities.

Importantly for many people, dance is now becoming a very popular form of exercise. Dancing just to move your body can also benefit your health.

Shutterstock.com/chaoss

Figure 10.1 Dance is used to communicate and express yourself.

FAST FACT
Dance is one of the most popular physical activities for Australians aged 9–14.

FACE TO FACE Do you dance?

Discuss in small groups whether each of you dance. If you do, why? If not, why not? The answer to these questions might not be as straightforward as you think.

WELLBEING CHECK IN

HAVE A DANCE

Identify

Ever wanted to dance like no one's watching? Now's your chance!

MoodMission

This activity has been developed in collaboration with Dr David Bakker of MoodMission

Understand

Dancing can be a great way to feel good. Dance allows opportunities for individuals to express how movement and music makes them feel. Dancing doesn't have to be fancy or rehearsed to get those benefits, and you don't have to consider yourself a dancer. If you've tapped your foot along to the beat before, that's a type of dancing. Now it's time to let the music move the rest of your body too.

Practise

1 Consider a place where you can turn up the volume without disturbing others. Maybe you can do it in your bedroom.
2 Find a song that makes you feel good and want to dance. Dance allows opportunities for individuals to express how movement and music makes them feel.
3 Now close the door and dance! Let yourself get caught up in the music and let your body move however it wants.

Reflect

Summarise how you feel after having a dance. Is it the same as the feeling you have after exercise, or different in some way?

REASONS TO DANCE

People dance for many reasons. These range from the desire to continue cultural traditions, to participating in dances that are part of traditional ceremonies and events, to opportunities to socialise and exercise.

- Dance is often part of the activities of ethnic or migrant groups because it is a way for people of the same ethnic background to stay together and keep their traditions alive. You can learn a lot about another culture by participating in or studying their dance activities. Many community ethnic groups have regular dance nights or educational programs that are designed to teach people about their cultural traditions.

CASE STUDY ⇨ CONNECTION TO LAND: INDIGENOUS AUSTRALIANS BRINGING BACK LIFE TO THE DARLING RIVER

Identify

Red dust rises around Jason Dixon's feet as he pounds the earth in rhythm with clap sticks. He has danced and played on this earth since he was a young child.

Figure 10.2 Dancing on the banks of the Darling River

Weblink
View the video clip at the link (Darling River dance) and read the dialogue below to explore the following questions.

Understand

When he dances, Mr Dixon says he is thinking about a childhood spent fishing, camping, learning and playing along the Darling River …

Mr Dixon is preparing to do a special dance, one he hopes will be part of growing efforts by Indigenous people along the Darling River to honour and ultimately rescue a waterway their cultures have revolved around for thousands of years.

'When we're out there dancing, we're dancing on Mother Earth, letting her know that we are still here,' he said …

Mr Dixon, a Ngamba and Wangkumara man from Bourke in western New South Wales, is dancing in the Yaama Ngunna Baaka Festival, a gathering of dancers from three states. They are performing in five locations along the banks of the Darling River. Mass fish kills in the river in 2018 and 2019 made international news, sparking protests, inquiries and documentaries. Mr Dixon said dancing was a way of bringing attention to the Darling and the cultures connected to it, but was also an expression of mourning, anger and hope.

'When we do this stamping, we want you to understand we're not just hurting our feet, it's telling you a story,' he said. 'We're crying out, we say, "We're mourning with you, we want you to be strong with us, we want you to hold on in the hope they will realise one day that enough is enough".' …

Three hundred kilometres downstream from Bourke, the Barkindji dancers are practising on the Wilcannia football oval and there is a buzz building in the town … Wilcannia has been one of the communities hardest hit by a lack of water in the river. It is also the heartland of Barkindji culture, which revolves around the Darling or Barka River.

'Our storyline lies within the river system and without the river there's no us, there's no story, there are no Barkindji people,' [Barkindji man and dance leader] Mr Whyman said … With no sign of water coming down the river, Mr Whyman said a gathering to dance and sing in language was just what they needed.

'Dancing on country is a mighty feeling; every time you dance you feel it here in your heart,' he said.

Discuss

1 Consider why these Indigenous Australian dance groups performing are these dances. What issue/issues are they trying to draw attention to?
2 Explain why connection to land, through dance, is so important to their cultural identity.
3 Interpret the symbolism of feet pounding/feeling the ground during ceremonial dance for Indigenous Australians.
4 'Our storyline lies within the river system and without the river there's no us, there's no story, there are no Barkindji people,' Mr Whyman said.
Use the quote above to explore how these dance performances impact on the mind/body/spirit connection for Indigenous Australians.

Newspix/Renee Nowytarger

Figure 10.3 Cultural dance performance

Dancing is also considered exercise and can be a great way to improve fitness and coordination. Some styles of dancing are especially good for this purpose because they require a lot of energy! Salsa and Bollywood dancing classes have become very popular, and there is a strong 1950s rock'n'roll dance culture in many areas. Dancing can also exercise your brain: it can take a lot of concentration and practice to perform some of the more difficult dances, and this can be especially beneficial for older dancers.

9780170463096

Alamy Stock Photo/National Geographic Image Collection

Figure 10.4 Partner dancing

FAST FACT

As well as improving physical health, dance has been found to reduce stress, anxiety and depression!

Weblink
Find out more about how dance improves brain health with 'How dancing improves brain health' article.

Dancing is also an important form of social interaction in many societies. In the past, these dances were often one of the only ways that single young adults could mingle and meet, even though their interactions were heavily supervised. In rural Australia, church dances or barn/bush dances were important events.

Figure 10.5 Bush dancing

Dances are often important parts of **rituals**, ceremonies and celebrations. Many cultures have specific dances associated with important milestones in peoples' lives, such as coming-of-age ceremonies in First Nations Peoples' cultures, or debutante balls in some Western cultures. Have you been to a wedding and seen the bride and groom take part in their first dance? This is another type of ritual that is part of a celebration. Some cultures incorporate dancing into funeral processions as a way to celebrate a person's life.

ritual a ceremony that consists of certain actions carried out in a certain order

Alamy Stock Photo/Ninette Maumus

Figure 10.6 'Jazz funerals' in New Orleans, USA, are known for their long processions, jazz music and dancing.

Shutterstock.com/Robert Przybysz

Figure 10.7 Self-expression is an important reason why many people dance.

Many people dance because they find it the best way to express themselves. In the same way that painters or writers express themselves through artworks and writing, dancers convey emotions such as joy, sorrow and fear through movement. Often, the expression of these emotions through movement alone has a greater impact on an audience than more traditional forms of expression, such as speech.

CASE STUDY — WHEELCHAIR DANCING

Identify

Strictly Wheelchair are a wheelchair dancing group in Queensland, bringing joy to participants and challenging perceptions about disability.

Keeara and Reece Photography

Figure 10.8 Strictly Wheelchair performing

Understand

In the 73-year history of this eisteddfod [Welsh festival], it's the first time a group of wheelchair dancers has competed ... The group, Strictly Wheelchair, dances to the song This is Me, made popular by the Hugh Jackman film, *The Greatest Showman* ... It's taken more than a year of work to prepare for this moment – performing among thousands of people at the Mackay Eisteddfod in north Queensland ...

The chance to dance and perform has made a big impact on 19-year-old Lauren Carey's life. Her carer Kay Marchant said the rehearsals were something they both looked forward to.

'Lauren's life is music and dancing and we just love it,' she said. 'It's been a wonderful thing for people to realise that disabilities don't have barriers anymore.' ...

As well as being fun, and providing social connection, the dancing has helped participants improve their mobility.

'They really bring their all to this and they really extend themselves,' [choreographer] Ms McMurtrie said. 'Recently I've seen people do movements they haven't done before. Things like stretching, bringing their arms together in a way I haven't seen them do before.'

Peter Sumpter from Sporting Wheelies said the improvement had not happened by chance.

'This year we've quite deliberately brought in a stretch session at the start,' he said. 'Their range has improved, and their fine motor skills as well.' ...

Last year Strictly Wheelchair was invited to perform a showcase, or a 'curtain raiser', for this section of the eisteddfod. This year they are competing …

As well as giving the dancers a chance to try something new and build new friendships, the group is helping to challenge ideas about disability. 'It helps show they are indeed valuable functioning citizens and they all have something to offer,' Mr Sumpter said … For Ms McMurtrie, the group is also about bringing joy to lives, including her own.

Discuss

1 In the context of this video and the article, what does the term 'adapt' mean?
2 Explain the physical, social and emotional benefits the participants have experienced from being a part of these sessions.
3 Identify the possible benefits to the local community of the Strictly Wheelchair program.

Weblink
Watch the video about Strictly Wheelchair in action before answering the following questions.

FACE TO FACE Why do people dance?

1 Write down your reflection on why you dance.
2 Then, with a partner, examine the following questions and give three responses to each.
   a Why do people dance?
   b Who and what influences your dance choices?
3 As a class, tally your results to create a class graph of the most frequent responses.
4 With your partner, investigate places in the local community offering dance classes that you might be interested in. List three places, including where they are situated and what genre of dance they offer.
5 As a class, create a local Dance Directory of places that offer dance classes that can be distributed to your class or to others in your school or community.

Worksheet 10.1

REVIEW

1 What are the most common reasons why people dance?

REFLECT

2 How can dance impact on the physical, mental, emotional and spiritual wellbeing of people?

EXTEND

3 Reflect on the term 'inclusion' and propose how dance opportunities for young people could be made more inclusive.

DOES CULTURE INFLUENCE DANCE?

First Nations Australians have cultures that are many thousands of years old. A large part of this culture is based on recognising and acknowledging the relationship with Country. Through their history and spiritual connection to their environment, First Nations Peoples have become unique and impressive storytellers.

Figure 10.9 Aboriginal people participating in a ceremonial dance at Groote Eylandt, Northern Territory

Getty Images/Penny Tweedie

One of the ways that First Nations Peoples tell stories is through dance. Over thousands of years, they have used dance as a way to creatively express their lives, communities, beliefs and history. Dance is a traditional and essential part of life within Indigenous communities and an important way of passing on culture and identity from one generation to the next.

EXPLORING ONE OF AUSTRALIA'S LEADING INDIGENOUS DANCE COMPANIES

The history and traditions of Australian Indigenous dance are now being widely recognised, both nationally and internationally, through the Bangarra Dance Theatre.

Bangarra Dance Theatre is recognised around the world for the high standard of its **choreography** and performance. It celebrates Indigenous culture and presents it through contemporary dance works.

choreography the process of designing a dance

Fairfax Syndication/Karleen Minney

Figure 10.10 The Bangarra Dance Theatre productions are world-class representations of Indigenous culture. This image is from their recent *Dark Emu* performance, based on the book by Bruce Pascoe.

FACE TO FACE Bangarra Dance Theatre

The Bangarra Dance Theatre is one of Australia's leading Indigenous performing arts organisations.

Take a few minutes to explore the Bangarra Dance Theatre website to develop an understanding of what the company is and what it does. As a class, discuss the following idea.

Do you feel that a company such as Bangarra Dance Theatre is an important part of Australian culture?

CASE STUDY

REKINDLING YOUTH PROGRAM

Identify

Watch the mini documentary 'Rekindling', about Bangarra's youth program, then reflect on the questions in your workbook.

Weblink 'Rekindling' documentary

Understand

Rekindling is a youth program launched by Bangarra in 2013. The aim of the initiative is to inspire young First Nations students with in-school residencies from Bangarra artists. Alongside local community elders, students research stories from their community and use these to create their own dance performances and events. The initiative is run by one of Bangarra's most acclaimed dancers, Sidney Saltner.

Discuss

1 Consider what the Bangarra Dance Company is trying to achieve through its Rekindling Youth Program.
2 Explain what is meant by the term 'custodians of culture'.
3 Consider which of the three dances participants usually select and propose a reason for this.

4 What is the story behind 'Lost City'?
5 Explain the significance of visiting 'Lost City'.
6 Investigate if 'Lost City' is more than just a dance. Summarise your findings.
7 Many dance movements are used in everyday life and in many sports. In the vision of the dance movements we see participants performing a jeté, or leap. In what other sports is a leap performed?
8 Define what is meant by the term 'choreography'.
9 How were the participants feeling about their performance?
10 Suggest what you believe the participants have gained by being a part of the Rekindling Youth Program.

Worksheet 10.2

WELLBEING CHECK IN ⇨ YOGA MOVEMENTS

Identify

This activity has been developed in collaboration with Dr David Bakker of MoodMission

Yoga is a cultural practice originally established in Ancient India. Western countries focus more on the physical exercise properties of yoga, but within India, yoga focuses more on spiritual health and meditation, with strong links to the Hindu faith. Yoga involves movements (poses) and meditative states to improve mental wellbeing.

Understand

Yoga can make us feel good in different ways. Firstly, it's a type of exercise, and we know that exercise can release feel-good chemicals throughout the body. Secondly, it can be quite relaxing because it's slow and quiet. And thirdly, it's a way of being present and mindful. When doing a yoga pose, it's important to pay attention to how you feel in your body. Yoga helps us feel our bodies non-judgementally.

Practise

1 Kneel on your hands and knees, so your hands are in line with your shoulders and your knees are in line with your hips. Lift your head so your neck and spine are straight.
2 Move your hands forward so they are slightly in front of your head and bring your big toes together. Press into your palms while moving your knees to the edge of your mat, one leg at a time (if you're not using a mat, your knees should be slightly wider than your hips).
3 Inhale, look ahead of you, and begin to exhale as you bend slowly to sit on your heels. Keep the side of your torso long and your arms stretched out straight in front of you with your palms flat on the mat.
4 Inhale, reach forward and exhale as you sink back onto your heels, bringing your head and chest to the mat with relaxed chest, shoulders and elbows.

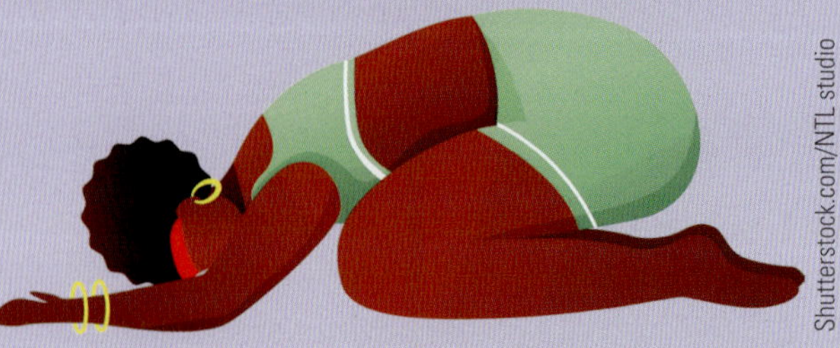

Shutterstock.com/NTL studio

5 Hold this position for 30 to 60 seconds. You can slowly rock your hips from side to side if you wish.
6 Repeat the above steps to move in and out of the pose one more time.

Reflect

This was just one yoga pose. How might you feel if you did a whole class? There are some classes on YouTube that you could check out.

INVESTIGATION  HOW DO OTHER CULTURES EXPRESS THEIR IDENTITY THROUGH DANCE AND MOVEMENT?

Purpose

Now that you have explored Australian Indigenous dance, let's investigate how other cultures express their identity through dance and movement. By researching and investigating the cultural significance of dance within different communities from around the world, we can learn a great deal about their cultural identity.

Method

1 View the video showcasing traditional Georgian dance.
2 If you thought that was impressive you might like to investigate what inspires this culture to perform these dances and movements. Alternatively, you might like to choose a different culture to research and explore.

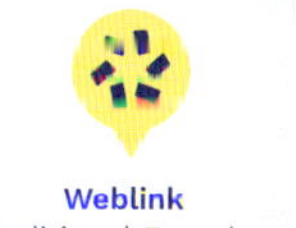
Weblink
Traditional Georgian dance

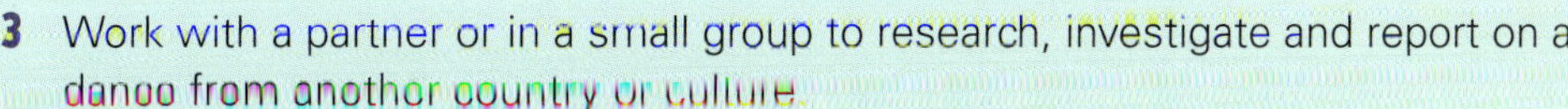
3 Work with a partner or in a small group to research, investigate and report on a dance from another country or culture.

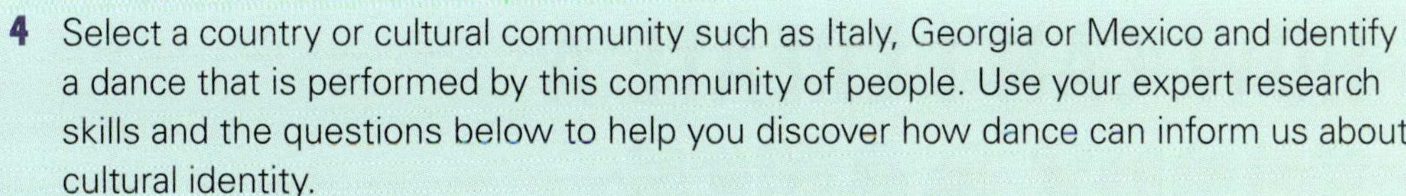
4 Select a country or cultural community such as Italy, Georgia or Mexico and identify a dance that is performed by this community of people. Use your expert research skills and the questions below to help you discover how dance can inform us about cultural identity.
5 Once you have all your information, create an expert report. This could be in the form of a PowerPoint, Keynote or film documentary in response to the discussion questions below. Your teacher will guide you on the format the report should be presented in.

Discussion

1 What is the name of the dance?
2 From which country or cultural community does the dance originate?
3 Does the dance tell a story? If so, what is it?
4 Why is the dance performed?
5 Is the dance part of a ritual, celebration or ceremony?
6 Is there a traditional aspect to the dance? If so, explain what this is.
7 Is the dance still performed in our modern culture?
8 Where is the dance performed?

9 Who can participate in the dance and why?
10 Is there special clothing or equipment required to perform the dance?
11 Is the dance performed individually or does it provide an opportunity for social connection?
12 Does the dance provide physical fitness benefits to the participants? If so, explain how.
13 Is there emotional or spiritual significance to the dance? If so, investigate how participating in the dance could impact the emotional or spiritual wellbeing of the participant.
14 How might you describe the overall cultural significance of this dance to the identity, beliefs and values of the people who created it?

Worksheet 10.3

1 Discuss how dance traditions and practices influence cultural identity.
2 Explain what Bangarra is.

1 Explore and discuss the term 'cultural dance'.
2 Explore and analyse how cultural dance can reflect cultural identities.

1 As a class, investigate, learn and perform a cultural dance. Pay particular attention and respect to the cultural traditions of the people who created it.

HOW CAN I CREATE A SPORT DANCE?

Some of the best dances, both to watch and perform, are those that tell a story. By telling stories, dancers can create meaningful experiences, both for themselves and for the audience.

Telling a story through a sport dance can be as simple as performing known movement patterns relating to a particular sport or sharing movements that express feelings about your own sporting experiences.

When creating a dance, a choreographer will choose moves that represent and signify an action, idea, thought or feeling. Most importantly, the movements have to reflect what the choreographer is trying to show.

Most sports involve a series of movements that are performed continuously in order to achieve a certain goal. For example, a serve in tennis allows the game to begin. If the serve is not performed, the game cannot begin. It is then up to the receiver to play the next shot to continue the game, and so on. It is a bit like having a conversation with the other player, but using movements instead of speech. The next time you watch your favourite sport, follow the movement conversations that the players have and note how this can almost flow into a dance within itself. After all, dancing is just about expressing yourself through movements and movement patterns. Sport is very similar, except that in a game situation, you usually choose your next move as a result of what your opponent has just done.

Shutterstock.com/dotshock

Figure 10.11 A movement conversation within a game of volleyball

CHOOSE A SPORT

Throughout the rest of this chapter you will work in groups of three or four to create a sport dance based on your favourite sport or activity. Your dance can be about one sport, or it could combine movements from a range of different sporting activities. For example, your group might decide to focus the dance just on tennis; another group might want to include a move from each group member's favourite sport, such as football, basketball, volleyball and cricket. It is entirely up to your group to decide.

Worksheet 10.4

Start by thinking about the sports and activities you participate in and then about which moves and movement patterns can be developed into a super sport dance move! Remember, everyone has different interests and talents. You might be great at surfing and have some awesome surfing moves that could be used in your group's dance.

The important thing to remember is that your dance will be a **mime** of sport movements or movement patterns and therefore no equipment will be used. There is no need to bring in your favourite surfboard or cricket bat for this activity!

mime to express actions or emotions without using words

It is also important to remember that if your dance reflects a contact sport, for the safety of others, the dance won't include any body contact such as tackling or taking marks over another student! If you want to include these movements in your dance, they must be performed safely as an individual mime movement only.

MUSIC

You will need to choose some music to use as part of your sport dance. The music will need to be of a **consistent tempo** so each group can practise at the same time.

consistent tempo a regular speed or rhythm

FACE TO FACE Creating a playlist

Create a class playlist that includes each group's songs using Spotify, Apple Music, YouTube or iTunes.

By using a playlist, each group can practise their dance to each song, and then choose one to use for the final sport dance performance.

When choosing music, it is important to ensure that the music is suitable for listening to at school. Your teacher will provide a playlist that is sequenced and at a continuous beat for the entire album.

GETTING STARTED

Now that you have started thinking about your own favourite sport or activity and some movements to include, take some time to check out examples of sport dances to give you some extra ideas.

Video
Basketball sport dance

UP AND MOVING Basketball sport dance

Check out this video in Nelson MindTap and have a go at the basketball sport dance as a class.

The basketball sport dance is an example of a dance that has been choreographed specifically to a complete song. You will not be required to create a dance as long as this, but after you complete this unit, your teacher may give you another opportunity to extend your dances further in a follow-up unit.

SAFE DANCE PRACTICES

proactive to be in control of a situation rather than reacting to it

Participating in dance requires you to think and be **proactive** about safety in the same way as you would with any other sporting activity. It is always essential to consider safe dance practices before undertaking any dance-related activity.

Consider and apply the following when dancing and creating your sport dance:

- Have a safe dance space, including your own personal space and a safe surface to dance on.
- Consider the personal space of others when working in a group.
- Wear appropriate footwear (runners) and clothing.
- Have your asthma puffer with you if necessary.
- Participate in a warm-up, cool down and stretch in each session.
- Choose appropriate and suitable movements: everyone should be able to perform all your moves. There should be no equipment, no body contact and no gymnastic-type movements.
- Drink plenty of water.

Shutterstock.com/Stokkete

Figure 10.12 Dance moves, just like any other sporting moves, need to be performed safely and require specific warm-up activities.

No breakdancing or gymnastics moves

Breakdancing and acrobatic/gymnastic styles of movement require specialised instruction and matting. Therefore, the following movements are **not** to be performed or included in your dance:

torso the mid-section of the body, not including the head and limbs

- rotation of the **torso** over the back, head and neck
- body support using the back, head or neck
- gymnastics rolls, cartwheels and handstands, etc.

PLANNING YOUR SPORT DANCE

Before you can start planning your sport dance you need to know a little bit more about what you are required to do.

Basic structure of the dance

With your group you will create a sport dance that:

- has at least four steps (moves)
- uses a **16-count phrasing** for each step (move)
- has at least two **formations**
- begins and ends with a sporting pose.

You may not be familiar with some of these terms yet, but they will all be discussed as you progressively plan and develop your sport dance. But at least now you know what you are going to do, and it should be lots of fun!

16-count phrasing equivalent to16 beats

formation a shape created by a dancer or group of dancers

What is choreography?

Choreography is the process of designing and composing a dance, and involves making decisions about what goes into a dance. It focuses on elements of dance such as the story or idea, the movements that will be included, the timing and length of a dance and its patterns and formations.

The term choreography can also mean dance notation or dance writing, which includes recording the dance in a written format.

Choreographing or creating a dance involves bringing together a number of different choreographic areas. In designing your sport dance, you and your group will focus on the following areas:

- idea or storyline
- steps or moves to be included
- timing and phrasing of the dance
- formations within the dance
- presentation of the dance.

As you learn about each area, you and your group will make decisions about how to develop each of these choreographic areas to create a fun and effective sport dance.

Alamy Stock Photo/Dmitriy Shironosov

Figure 10.13 A choreographer at work

Working as a group

To create a dance, your group will need to work together cooperatively and develop some effective decision-making skills. When creating a dance as a group, it is important that everyone feels that they have had some input into the creation. Having input and assisting in the development of the dance will give each group member some ownership of it. It is no fun when one person or a couple of people make all the decisions because it becomes their creation and does not represent all the group members.

It is also important to remember that you are not going to end up with a dance that will have everything you want in it. It will be a combination of ideas that have been chosen by the group, using a decision-making process.

DECISION-MAKING PROCESS

Throughout the dance creation process, your group will need to make many decisions. One way to help make the best decisions about your sport dance is to use the democratic decision-making guidelines (DMG).

The democratic decision-making process is based on the principle that every group member is entitled to provide input and ideas throughout the creation of the dance. It is then up to the group to discuss and try each idea and then vote to make a final decision.

FACE TO FACE Decision-making guidelines (DMG)

To assist in making decisions, your group can use the following steps.

- Step 1: Brainstorm all ideas – ensure each group member has had an opportunity to contribute.
- Step 2: Discuss and try each idea in your dance.
- Step 3: Vote.

If you have a situation where there is a tie in votes, the area coordinator will make the final decision. When voting, remember that you are trying to create a fabulous dance and not just to include your own personal ideas!

The following example illustrates the three steps of the democratic decision-making guidelines. Your group might be trying to decide on the sport pose to include at the beginning of the dance. As a group, you would brainstorm each group member's ideas, ensuring that every person has the opportunity to contribute (Step 1).

Getty Images/RCWW, Inc.

Figure 10.14 The democratic decision-making guidelines will help your group make decisions.

You would then discuss and try each suggestion, looking at how each pose contributes to the overall presentation of the dance (Step 2).

As a group, you would then vote on the most appropriate pose for inclusion into your dance (Step 3).

Now that you have some ideas about how to make decisions in your group, it is time to get started!

Your teacher will help you and your class form groups of three or four. It is important to only have three or four students in your group, as you will each take on a specific role to assist the choreographic process. If you only have three students in your group then one student will have to take on two roles.

Once you have a group, you can begin the planning process for developing your sport dance. This process is made up of five parts:

1 Create a story
2 Steps and moves
3 Timing/phrasing
4 Formations
5 Presenting the sport dance.

1. CREATE A STORY

The first thing you need to decide on is your idea or story. As a group, brainstorm and discuss the sport or sports to be included in your sport dance. Some points to discuss include:

- Is the dance going to be about just one sport, or a combination of sports and activities that group members participate in?
- If it is a combination of sports, how many sports will be included?
- Will the dance have a particular storyline or will it just be a presentation of sport movements?

Remember, this is a story about your experiences, so there are no right or wrong answers, just decisions to be made. Use the democratic decision-making guidelines (DMG) to assist your group. Because you don't have a group coordinator for this first activity, if you can't agree within your group, you may need to ask your teacher to help with the final decision.

FACE TO FACE Start planning

Get together with your group to begin planning. By the end of this activity each group will have chosen a sport or a group of sports to focus its dance on.

Sometimes it helps to have music on in the background to provide some inspiration!

Take on a coordinating role

As well as being a part of a group, each group member will need to take on a coordinating role. Table 10.1 shows the four choreographic areas and accompanying roles each person is responsible for. If there are four people in your group, each person will become the coordinator for one choreographic area. If you have three people, one person will take on two roles.

Table 10.1 Choreographic areas and roles

Choreographic area	Roles	Responsibilities
Steps (moves) to be included	Steps coordinator	Leads the process that ensures the dance has at least four steps or moves that are repeated twice.
Timing or phrasing of the dance	Timings coordinator	Leads the process that ensures the dance has a 16-count phrasing for each step or move.
Formations within the dance	Formations coordinator	Leads the process that ensures the dance has at least two formations.
Presentation of the dance	Presentations coordinator	Leads the process that ensures the dance begins and ends with a sporting pose.

Being a coordinator means that when it's time to make decisions about this part of your dance, you will each have an opportunity to develop your leadership skills by ensuring that all group members are involved in the dance creation process. Coordinators are responsible for leading discussions and ensuring the democratic decision-making process is followed.

For example, the steps coordinator's responsibility is to:

- lead the brainstorming of ideas, ensuring each group member has the opportunity to contribute
- lead discussion and try each idea in the dance
- organise the vote. Importantly, if the group does not reach a decision from the vote, the coordinator will make the final decision.

A good leader will keep the group focused on the task, and provide motivation and encouragement for all group members and extra ideas for their specific choreographic area.

When your leadership opportunity arrives, make sure you give it your best shot!

FACE TO FACE Assign roles

Now it is time to get into your group and decide who will take on each coordinating role.

plyometrics a type of exercise that involves fast movements

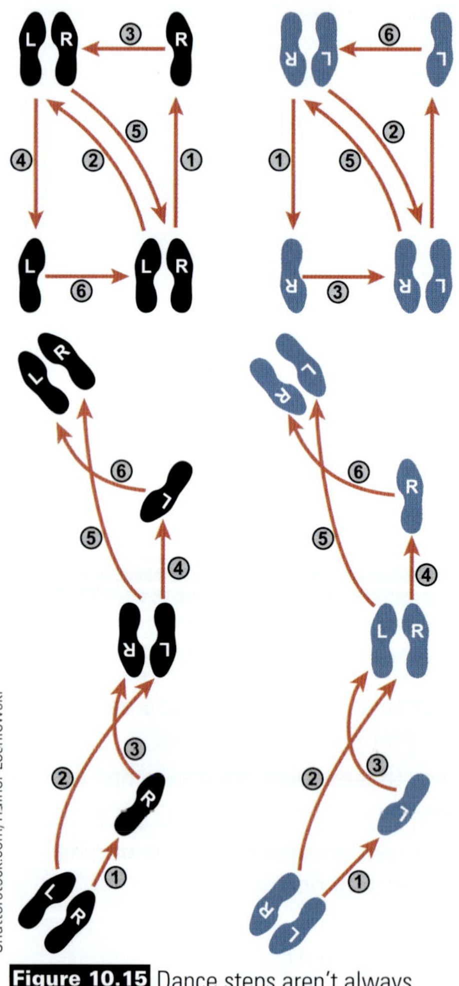

Figure 10.15 Dance steps aren't always complicated.

2. STEPS AND MOVES

Now it is time to begin adding the steps or moves to your dance. The terms 'step' or 'move' can be used interchangeably to mean the same thing. The term 'step' in dance can sometimes seem related to really complicated dance movements, but dance isn't just about intricate steps and patterns. A dance step is just a movement or a set of movements that create a movement pattern. Sometimes dance steps are performed without you even knowing it – every time you perform a sport move, you are performing a movement or movement pattern that could easily become a dance step.

For example, the 'Grapevine' is a dance step, but how often do soccer players cross their feet over when moving sideways in a game? The 'Carioca' is another exercise used as a **plyometrics** or agility warm-up for many sports! You can search the internet for videos on how to do the Carioca. The 'Running man', sometimes known as the 'Shuffle', is another example of a sport-based dance move that can be used as a dance step.

Another option for creating a sport dance move is to link a dance step with a sport move, such as performing a step-touch with your feet and performing a handball movement with your arms. Or walking forward for three counts and pretending to kick a ball on the fourth count.

There are many options for you and your group to explore. When you are creating your dance you need to think creatively and outside of the box! Just make sure that your sport movements are safe.

If you are having a bit of difficulty getting started, the best way to get everyone involved in the dance and the group creation process is to have each group member contribute a movement for consideration. Once you get started, the ideas will begin to flow.

FACE TO FACE Select the moves

As a group you need to select the steps or moves to include in your dance and the order that these will be performed in.

Remember, by the end of this activity your sport dance should have at least four steps or moves.

Steps coordinators, it is now time to facilitate your leadership! Off you go!

3. TIMING/PHRASING

Music is a great motivator for getting people up and dancing. Just think of the last time you chose to dance. Was it because you heard a favourite song with a catchy tune and a great beat?

It is generally the beat of the music that initially gets you tapping your feet and wanting to get up and dance. Good dance music is designed to get you moving through the constant beat and rhythms that underpin the musical melody. The beat and rhythm of the music provide the timing for your movements to flow with the music, and allow movements to be performed together when a group is performing a dance.

Alamy Stock Photo/Peter D Noyce

Figure 10.16 Drums are often used to drive the beat of the music.

Finding the beat

The beat of a piece of music is like its pulse or heart rate – it is the underlying and constant sound that continues from the start to the end of the song.

Weblink Carioca video

In your group, choose one of your favourite songs and tap out the beat together. You can tap the beat with your hands or feet or take it in turns to lead the tap using different moves.

When counting the music, one musical beat is equal to one count. Once you've found the beat, you can then break it into useable sections or **phrases** so that you can create timing blocks to fit the steps into. A musical phrase can be any length of beats, but you will use a 16-count music phrase for this group dance. The benefit of using a set phrase throughout the dance is that it allows you to know when to start and finish a step or move.

phrase a series of musical notes

For example, if you are going to include the Grapevine into your dance, you would have 16 counts in which to perform it.

In your group and using your chosen song, tap out 16-count phrasings. You can tap the beat with your hands or feet while counting out 16 beats. Continue counting a 16-beat phrasing until you and your group are confident at counting in this way.

UP AND MOVING Join your phrase to your steps

It is now time for your group to use a 16-count phrase for each of the four steps or moves you have chosen for your dance. You can do this by counting how many beats/counts there are in a step.

For example, one Grapevine is equal to four counts, so four repeats of the Grapevine can fit into a 16-count phrase.

Having a set phrasing helps all members of the group to know how many of each step or move to do, and when each step or move begins and ends.

As a group you have 16 counts for each of the steps. You need to count out how many repetitions of each step you can perform within the 16 counts.

Remember, by the end of this activity your sport dance should have a 16-count phrasing for each step or move.

Timing coordinators, it is now time to facilitate your leadership! Off you go!

4. FORMATIONS

Now that your group has the basis of a sport dance with steps and timing organised, the next step is to look at ways to make the dance more interesting. Space can be used as an element to create formations and patterns in your dances.

Shutterstock.com/Master1305

Figure 10.17 The placement of dancers on the stage creates the formations and patterns.

When creating a dance for a performance it is important to think about the audience and how the dance can be made more interesting from its point of view. One way to do this is to incorporate patterns and formations using the dance space available. A dance that incorporates the use of space with different patterns and formations will be more interesting to an audience than one that is performed in one straight line for the whole performance. The use of space as an element of dance can create greater visual interest for the audience. In simple terms, a formation is generally a shape created by a dancer or group of dancers and a pattern is the movement across the floor or space.

For example, a group of four dancers who begin in a straight line might use a ball bounce step with two dancers moving forwards and two dancers moving backwards to create a square formation. The movement to the formation creates the pattern.

There need to be at least two different formations in your group's dance.

FACE TO FACE Develop your formations

It is now up to your group to use the formation ideas shown in Figure 10.18 to help you develop two or more formations to include into your dance.

Remember, by the end of this activity your sport dance should have at least two formations within the dance.

Formations coordinators, it is now time to facilitate your leadership! Off you go!

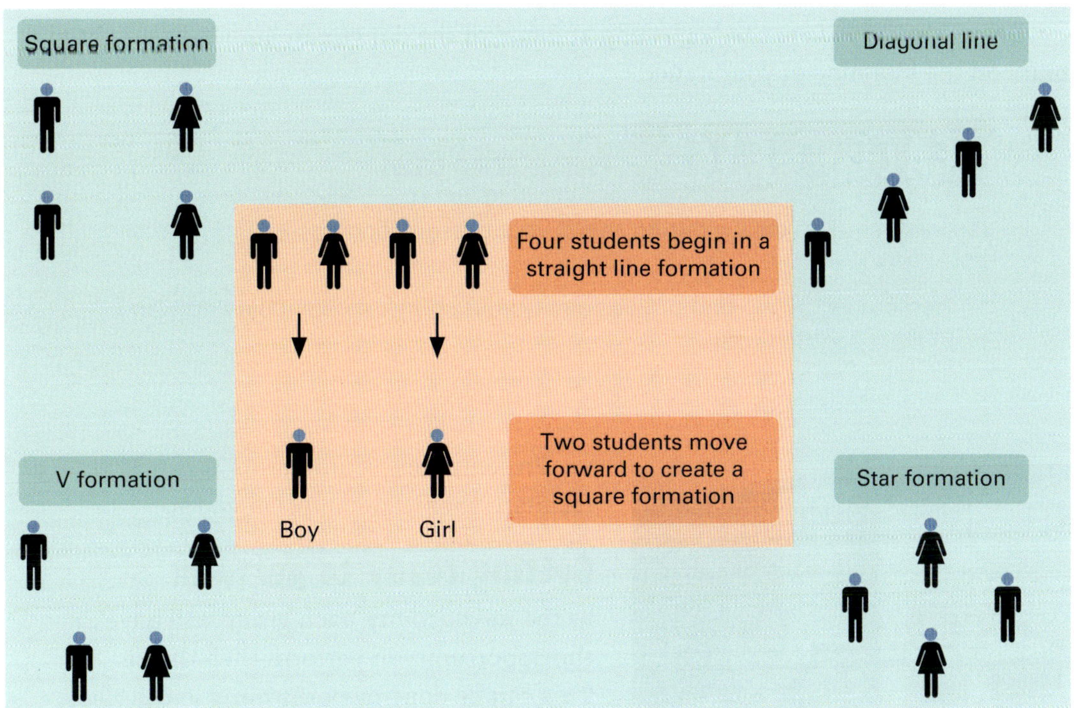

Figure 10.18 How to move into a different formation and formation ideas

5. PRESENTING THE SPORT DANCE

By now your sport dance will be looking really good. It is time to put the finishing touches to your creation to make the end product as exciting and interesting as possible to a prospective audience.

The presentation of your dance is just like putting the icing on a cake. There are many ways to improve a dance through presentation:

- costumes
- make-up
- props
- scenery.

prop short for 'property', props are items (not including scenery) used by actors during a performance

These elements all help to tell the story or idea of the dance and paint a picture for the audience.

Getty Images/Hill Street Studios

Shutterstock.com/Gromovataya

Figure 10.19 Make-up artists help tell the story of a dance for the audience. What story does this make-up tell?

Now that you have created a sport dance, one way to add to your presentation is to begin and end the dance in a creative sport pose that captures the audience's attention.

Using a sport pose at the beginning and end of the dance can help to highlight and build on the storyline of your dance.

FACE TO FACE **Presentation poses**

It is now time for your group to select your presentation poses. Ensure that the poses are safe!

Remember, by the end of this activity your sport dance should begin and end with a sporting pose.

Alamy Stock Photo/Marmaduke St. John/

Figure 10.20 Filming allows you to watch your performance and self-assess.

Presentations coordinators, it is now time to facilitate your leadership! Off you go!

Getting ready to perform

In the next activity, each group will have the opportunity to perform their dance. This can be done by performing to the class, to another group or just to your teacher. Importantly, all members of your group need to feel comfortable about the performance, as dancing should be fun! If groups don't wish to perform their dances to a live audience, a video recording of the dance could be taken for your teacher to assess.

Filming

Filming your dance can also help your group to self-assess and make any necessary changes to the dance to meet the set criteria. As this is a school activity, only school-based equipment can be used for filming this task.

Live performance

For those groups who do wish to perform their sport dances, keep in mind that it takes a lot of courage to get up in front of a group of people and perform, so it is important to support and encourage everyone's efforts.

1 Define the four areas of choreography explored in the making of your sport dance.
2 Explain one decision-making process.

1 Reflect on how well you performed and managed your leadership opportunity. Summarise the top two highlights and challenges.

9780170463096

1 Use your sport dance as a base for a longer dance, or combine different group dances together to create a class dance.

HOW CAN I SUPPORT OTHERS AND PROVIDE EFFECTIVE FEEDBACK?

Dancing is about self-expression. Although a group might be performing the same steps, the way that each individual student interprets and delivers these steps can be unique to each person.

FEEDBACK CRITERIA

If your class chooses to provide feedback to other groups, it must be based only on the four set criteria for the dance.

Feedback can only be provided on whether the sport dance meets the following goals:

- Has at least four steps (moves)
- Uses a 16-count phrasing for each step (move)
- Has at least two formations
- Begins and ends with a sporting pose

Alamy Stock Photo/H. Mark Weidman Photography

Figure 10.21 Applause is a form of feedback for performers.

HOW TO PROVIDE FEEDBACK

It is always nice to receive encouragement, so try to use the following tips if you are asked to provide feedback:

- Begin with a positive feedback statement about how the dance met the four set criteria.
- Provide a positive feedback statement about a highlight of the dance.
- Provide a constructive comment, only if necessary, on how the dance could be improved to meet the set criteria.

Providing feedback relating to the set criteria ensures that the dances are assessed fairly, without taking personal dance styles into consideration.

Worksheet 10.5

HOW CAN I DEVELOP A PERFORMANCE?

One of the many reasons people dance is to celebrate. The performance of your sport dance should be a celebratory session where your group get to display your hard work and acknowledge your achievements.

Your performance can be to the whole class, to just another group or you can have your dance filmed for your teacher to view. Whichever way you choose to perform your sport dance, it should be a fun and enjoyable culmination of the group effort.

CASE STUDY

YIRRAMBOI FESTIVAL

Identify

Under incoming creative director Caroline Martin, a future-focused First Nations festival Yirramboi burst through Melbourne's streets and arts institutions earlier this month [May 2019], redefining the way we view the city. In its second year, the biennial festival presented 100 events across 25 venues and public spaces, guided by an Elders Council who provided cultural knowledge.

Understand

Many of the performances looked towards future generations, starting with questions about who we are today and imagining the steps required to address issues of land rights, climate catastrophe and queer identity …

At Flinders Street station, blak dance by Tagalaka and Kurtijar woman and choreographer Jasmin Sheppard interrogated the impact of colonisation on women's bodies.

Beneath the streets, in the Dirty Dozen Gallery, Alan Stewart's exhibition Maaman traversed culture, time and space, from his ancestral roots in Taungurung country to inner city Collingwood, where he currently lives …

The cultural takeover transformed what we expect to see in city streets, reinserting blak people into the landscape and reminding those who may have forgotten that Melbourne was and continues to be Birrarung-ga, its traditional Woiwurrung name.

At the city's spine, Swanston Street, the State Library Victoria was transformed by the Dhungala Children's Choir, led by Yorta Yorta soprano and composer Deborah Cheetham. The children's voices soothed the colonial architecture, softening the authoritarian statues of English explorers that framed the choir on the grey stone steps …

The variety of live performances in unlikely city spaces – from shop fronts to trams – helped to draw audience members towards the Birrarung Marr, where a blak takeover was about to occur.

Source: 'Yirramboi Festival uses First Nations arts and culture to reimagine and disrupt Melbourne', Timmah Ball, 15 May 2019, *ABC News*. Reproduced with permission by Timmah Ball.

AAP Image/DAVID CROSLING

Discuss

1 Describe two different performance spaces used during Yirramboi festival.
2 Yirramboi festival transformed Melbourne street corners into performance spaces. What effect did this have?
3 How can you transform outdoor spaces into a performance space? Travel around your school – Investigate potential areas to perform in and suggest ways to redesign areas for safe performance. What message would it give to perform there?

COSTUMES

For your final performance you might want to select the one song your group enjoyed practising to the most. You may also consider wearing costumes.

For example, your group could wear matching singlets or shorts if the dance is about basketball, or you could all be dressed in individual sport outfits, depending on the theme of your dance. Importantly, any costume should add to the overall presentation of the dance and should not hamper your ability to move. Make sure your costumes are safe to dance in.

AAP Image/DAVID CROSLING

Shutterstock.com/Rus S

Figure 10.22 Simple costumes can enhance the presentation of your dance.

Quiz
End of chapter

UP AND MOVING Go for it!

Now it is time to perform your sport dance! Remember to have fun and enjoy all your hard work. Go for it!

REVIEW

1 Why was it important to only provide feedback on the criteria supplied?

REFLECT

1 Why is performing the dance important?
2 Examine how you felt about being involved in the sport dance process.

EXTEND

1 Explore opportunities for your class to perform your dances, such as a flash mob opportunity.

CHAPTER 10 REVIEW

1 Outline three reasons people dance.
2 List as many reasons as you can for how dance might be beneficial to health and wellbeing.
3 What is Bangarra Dance Theatre?
4 Consider how the Bangarra Dance Theatre is attempting to educate people about the significance of dance for First Nations Peoples.
5 Explain how two different cultures experience dance.
6 Discuss why storytelling is important in cultural dance.
7 Explain how sport moves can also be used as dance moves.
8 Describe the four areas of choreography used in the sport dance process.
9 Reflect on your participation in dance and list ways you demonstrated leadership.
10 Reflect on how your group made decisions in regard to your sport dance.

INDEX

D

E

F

G

H

I

9780170463096

O

P

T